NEURAL AND ENDOCRINE ASPECTS OF BEHAVIOUR IN BIRDS

NEURAL AND ENDOCRINE ASPECTS OF BEHAVIOUR IN BIRDS

Edited by

PETER WRIGHT M.A., D.Phil.

PETER G. CARYL M.A., Ph.D.

and

DAVID M. VOWLES M.A., B.Sc., D.Phil.

Department of Psychology
University of Edinburgh
Edinburgh, Great Britain

ELSEVIER SCIENTIFIC PUBLISHING COMPANY
AMSTERDAM / OXFORD / NEW YORK   1975

ELSEVIER SCIENTIFIC PUBLISHING COMPANY
335 Jan van Galenstraat
P.O. Box 211, Amsterdam, The Netherlands

AMERICAN ELSEVIER PUBLISHING COMPANY, INC.
52 Vanderbilt Avenue
New York, New York 10017

ISBN: 0-444-41356-1

Printed in The Netherlands

LIST OF CONTRIBUTORS

R.J. ANDREW
Ethology and Neurophysiology Group, School of Biological Sciences, University of Sussex, Falmer, Brighton BN1 9QY, Great Britain

L.D. BEAZLEY
Department of Psychology, University of Edinburgh, 60 Pleasance, Edinburgh EH8 9TJ, Great Britain

P.G. CARYL
Department of Psychology, University of Edinburgh, 60 Pleasance, Edinburgh EH8 9TJ, Great Britain

C.J. ERICKSON
Psychology Department, Duke University, Durham, N.C. 27706, U.S.A.

M.B. FRIEDMAN
Department of Psychology, Carnegie-Mellon University, Pittsburgh, Pa., U.S.A.

M.J. GENTLE
Agricultural Research Council Poultry Research Centre, King's Buildings, West Mains Road, Edinburgh EH9 3JS, Great Britain

D. HARWOOD
Department of Zoology, University of Manchester, Manchester M13 9PL, Great Britain

J.B. HUTCHISON
M.R.C. Unit on the Development and Integration of Behaviour, Sub-Department of Animal Behaviour, University of Cambridge, Madingley, Cambridge CB3 8AA, Great Britain

E.M. MACPHAIL
Laboratory of Experimental Psychology, School of Biological Sciences, University of Sussex, Falmer, Brighton BN1 9QY, Great Britain

M.C. MARTINEZ-VARGAS
Laboratories for Reproductive Biology, University of North Carolina, Chapel Hill, N.C., U.S.A.

R.K. MURTON
*Institute of Terrestrial Ecology, N.E.R.C., Monks Wood Experimental Station, Abbots
Ripton, Huntingdon PE17 2LS, Great Britain*

D.M. PARKER
*Department of Psychology, University of Aberdeen, Taylor Building, King's College,
Old Aberdeen AB9 2UB, Great Britain*

F.W. PEEK
Department of Biology, Nazareth College, Rochester, N.Y., U.S.A.

R.E. PHILLIPS
Department of Animal Science, University of Minnesota, St. Paul, Minn. 55101, U.S.A.

E.A. SALZEN
*Department of Psychology, University of Aberdeen, Taylor Building, King's College, Old
Aberdeen AB9 2UB, Great Britain*

P.J.B. SLATER
*Ethology and Neurophysiology Group, School of Biological Sciences, University of
Sussex, Falmer, Brighton BN1 9QY, Great Britain*

N.J. WESTWOOD
*Institute of Terrestrial Ecology, N.E.R.C., Monks Wood Experimental Station, Abbots
Ripton, Huntingdon PE17 2LS, Great Britain*

D.G.M. WOOD-GUSH
*Agricultural Research Council Poultry Research Centre, King's Buildings, West Mains
Road, Edinburgh EH9 3JS, Great Britain*

D.M. VOWLES
*Department of Psychology, University of Edinburgh, 60 Pleasance, Edinburgh EH8 9TJ,
Great Britain*

P. WRIGHT
*Department of Psychology, University of Edinburgh, 60 Pleasance, Edinburgh EH8 9TJ,
Great Britain*

H. ZEIER
*Swiss Federal Institute of Technology, Department of Behavioural Sciences,
Turnerstrasse 1, CH-8006 Zurich, Switzerland*

R.E. ZIGMOND
*M.R.C. Neurochemical Pharmacology Unit, Department of Pharmacology, Hills Road,
Cambridge, Great Britain*

PREFACE

In memory of Danny Lehrman

The papers on which this symposium is based sprang from a conference with primarily European participants, which was held in the Department of Psychology in Edinburgh during July, 1974. The main chores of organisation, administration and editing fell entirely to Peter Wright and Peter Caryl, and they would like to thank all the contributors for producing their manuscripts so promptly.

The 1974 conference was the second we had held — the first being in 1970, also in Edinburgh. It had then seemed timely to Danny Lehrman, Robert Hinde and myself to get together a small group of researchers who were actively working in the field relating a bird's behaviour to hormones and brain mechanisms. This small group, relieved from the density of a large conference and the pressure of publication, could present informal papers about work in progress and in prospect, with ample time allowed for discussion. In the event about 30 people attended and the meeting was rated such a success that the host was pressed to repeat the occasion.

It is notable that the majority of participants in the first conference were drawn from three main groups — Danny Lehrman and his colleagues from Newark, Robert Hinde's group from Madingley, and an Oxford–Edinburgh group (the Oxford work had just undergone fission after a period of rapid growth) leavened by a few independent workers. Many of the British contributors and nearly all the Americans had worked with Danny, and the majority of papers were devoted to his North Atlantic species, the Barbary or blonde ring dove. This was entirely appropriate since Danny was one of the main bridge builders between European ethology and American psychology. As a fine ornithologist he could meet most ethologists on their own grounds, and by choosing the reproductive behaviour of the ring dove as a test bed for their ideas on the endogenous basis of drive, he revitalised the study of the interaction between hormones and behaviour. In particular he pinpointed the role of prolactin and progesterone, and demonstrated not only that hormones affect behaviour, but also that behaviour affects hormones. This has clear implications for the role of the brain in integrating these relationships.

By the time of the second Edinburgh conference many new techniques, such as autoradiography and immunoassay, were available to develop and test ideas in this field. A body of well established avian work had been accumulating and was now worth publishing in review papers, for fruitful comparison with mammals. Although it was a sadness to us all that Danny Lehrman had died unexpectedly a year before this conference, the continued vigour of the field is a tribute to his classic work. At the meeting we still spoke of his opinions in our discussions, with happiness and respect, as if he were present. For he was a scientist whose ideas stimulated experiment, a critic who illuminated discussion without provoking enemies, a listener whose enthusiasm restored one's belief in oneself, and a friend who shared our affections. We hope this book may be a tribute to him.

David Vowles
Edinburgh, 1974.

CONTENTS

ETHOLOGICAL AND PSYCHOLOGICAL APPROACHES
TO THE STUDY OF BRAIN—BEHAVIOUR RELATIONSHIPS

PETER CARYL and PETER WRIGHT

Department of Psychology, University of Edinburgh (Great Britain)

It is now almost a quarter of a century since the publication of Tinbergen's book 'The Study of Instinct' (Tinbergen, 1951), which brought ethological ideas to the attention of English-speaking workers in related disciplines, such as psychology and neurophysiology. These disciplines traditionally concentrated on mammalian behaviour; indeed, at that time 'comparative' psychology had for years been largely preoccupied with the behaviour of a single mammalian species, the laboratory rat (Beach, 1950). In contrast, ethological ideas were derived from work on birds and a variety of lower animals, in addition to mammals.

To understand the differences of approach and emphasis between the literature on avian brain mechanisms and the much larger mammalian literature, it is necessary to appreciate the enormous influence that ethological concepts have had in the analysis of avian behaviour. A further point of importance is that avian social behaviour depends primarily on visual and auditory signals which can be easily perceived by a human observer; in contrast, signals which humans cannot readily detect, (such as scent, and also ultrasonics in the common laboratory rodents), play a major role in the communication of many mammalian species. As a consequence, birds have a clear advantage as subjects for the study of social behaviour, and both Lorenz and Tinbergen have noted that it was for this reason that the ethological analysis of avian social behaviour made rapid progress. This literature has provided a stimulus for much of the subsequent analysis of avian brain mechanisms, whereas in the mammalian field analyses of motivational systems have instead been concentrated to a large extent on feeding and drinking behaviour.

At first, there was very little neurophysiological work with birds to support or refute the ideas which arose from these behavioural studies. Von Holst and St. Paul (1963) pioneered this field in a paper in which they attempted to analyse in ethological terms the behaviour produced by electrical stimulation of the chicken brain, but it was not until several years later that the results of more detailed research on avian brain mechanisms became available. An undoubted stimulus for such studies has been the extensive neuroanatomical work of Karten (Karten, 1969; Zeier and Karten, 1971; Cohen and Karten, 1974), and in particular the publication of a detailed stereotaxic atlas of the pigeon brain (Karten and Hodos, 1967). Work on the Barbary dove or blonde ring dove, *Streptopelia risoria* (more correctly called *Streptopelia roseogrisea*, as Murton and Westwood point out, p. 55), which Lehrman chose for his studies of the interaction between hormones and behaviour (for review see Lehrman, 1965), has been similarly aided by the unpublished atlas of Vowles *et al.*, which now appears as an appendix to this volume. The recent stereotaxic atlas of the canary brain (Stokes *et al.*, 1974) may provide a similar opportunity for neurological studies to complement the existing behavioural and endocrinological work (*e.g.* Hinde and Steel, 1966) on this small songbird. The stage in the analysis of avian brain mechanisms is now approaching at which comparisons and contrasts with the much larger volume of information on mammalian brain mechanisms may be fruitfully made, and there are a number of examples in later chapters of this book which show the value of this approach.

This research on avian brain mechanisms also provides evidence on the basis of which the status of established concepts, stemming from purely behavioural analyses, can be assessed. A reconsideration of such concepts is particularly appropriate where one discipline is beginning to lay emphasis on concepts similar to those already established in another discipline. For instance, one of the most characteristic features of ethological work has always been the emphasis on describing behaviour in terms of motor patterns, sometimes species-specific; recently, the role of such motor patterns in behaviour has also been emphasised in theoretical papers in psychology (Glickman and Schiff, 1967; Valenstein, 1970). The interest of ethologists in motor patterns stems from early work which showed that such patterns might be units, the organisation of which was independent of the patterning of external stimuli (fixed-action patterns; Hinde, 1970). In many instances, such motor patterns were species-specific, and relatively unaffected in form by variation in early experience. However, more recent ethological work has cast doubts on the status of such patterns as units of behaviour, since many show considerable variability in form and may even be separable into elements which vary independently (Stamps and Barlow, 1973), although in other cases the behaviour patterns may indeed be highly stereotyped (Wiley, 1973). This situation clearly raises questions about the way in which the behaviour is organised in the central nervous system. Tinbergen recognised the possible value of brain stimulation in this context, and in 'The Study of Instinct' he laid considerable emphasis on Hess' observations in the cat that complex natural units of behaviour could be elicited by simple electrical stimulation of the brain (ESB). The question of the neural basis of behaviour that would be described by ethologists as fixed-action patterns is discussed by Phillips and

Youngren (1971) who emphasise in their studies that it is only fragments of more complex behaviour which are usually elicited by electrical brain stimulation. They suggest that the more complex sequences of behaviour depend on the organisation of such elementary units by the stimulation, in conjunction with the appropriate environmental stimuli, rather than on the triggering of some locus in which they are organised as a whole.

This brief example illustrates the way in which the more complex evidence which is now available can lead to a reappraisal of basis concepts; other examples will be apparent in later chapters.

Since discussions of the context and implications of the authors' own work are included in later chapters of the present book, it is unnecessary in this introductory chapter to attempt to set them in a broader context; much additional background material can be found in Pearson (1972). We would, however, like to draw attention to the ethological influences, noted earlier, which can be seen in many chapters, although some fall within the main psychological tradition. It might also be appropriate to note the influence, in many of the chapters on reproductive behaviour, of the work of Lehrman (1965). In this area, the availability and use of techniques for assaying circulating hormones has allowed a new approach to the problems he tried to solve. There are still many problems in interpreting these measurements, particularly in the light of short-term variations in hormone output, and the topic of the circadian release of hormones is of considerable importance. However, more detailed comment would be inappropriate here in the light of the observations on this topic in Chapter 4 of this volume, by Murton and Westwood.

Rather than discuss the material covered in later chapters, it seems appropriate to turn briefly to a more specific discussion of the use of ESB in the analysis of behaviour, since this topic receives little attention in subsequent chapters, despite its considerable importance as a technique. While the problems of interpreting lesion data have always been recognised (Gregory, 1961), ESB has seemed a more direct and less problematical technique for analysing brain function. However, Valenstein and his collaborators have recently questioned many of the assumptions about the neural basis of electrically stimulated behaviour and it is now clear that there are considerable problems of interpretation in brain stimulation studies. A comparison of the approaches to these problems adopted by workers on the avian and mammalian brain suggests that ethological techniques might prove valuable in resolving some of the current problems in the mammalian literature.

ESB has proved, in mammals, to be a powerful tool for the investigation of functional neuroanatomy. The electrically elicited behaviour is often 'stimulus-bound' in the sense that the behaviour does not outlast the duration of the stimulating current by more than a few seconds, and does not appear in control sessions without ESB. The classic examples of such behaviour come from hypothalamic sites in mammals, and, in particular, from studies of electrically elicited eating in the rat (Sheer, 1961; Steinbaum and Miller, 1965). In birds, stimulus-bound effects are more often examples of agonistic, courtship or grooming behaviour (Pearson, 1972; Vowles, 1975), rather than of appetitive behaviour directed to food and water. Indeed, as discussed in a later chapter, electrically elicited eating is a singularly difficult pattern to obtain reliably from brain stimulation in birds.

In mammals, the stimulus-bound behaviour which could be obtained with ESB had seemed to allow a simple interpretation of the mechanisms that were being tapped, in terms of specific neural circuits. However, Valenstein (1970) and his collaborators (Valenstein *et al.*, 1970) have argued that because stimulus-bound behaviour in the rat can be switched from one food object (*e.g.* food) to another (*e.g.* water), then the assumption that the behaviour reflects the activity of specific neural circuits underlying separate motivational systems is incorrect.

Valenstein's argument for hypothalamic plasticity has not gone uncriticised, but the alternative explanations for such effects have been in terms of altered electrical thresholds (Wise, 1968) or have involved models of central inhibition similar to those which account for localisation and fine discrimination in sensory systems (Teitelbaum, 1973). In addition to these neural explanations, there are also behavioural considerations which raise problems of interpretation, as can be seen from the following summary of Valenstein's procedure.

In these experiments of Valenstein *et al.* (1968), the rats were initially screened for the presence of stimulus-bound behaviour, and if none was apparent they were stimulated throughout the night in their home cages. Such 'night schedules' involve 30 sec of ESB every 5 min throughout a 12-h period, and this procedure may be employed on two or three occasions before abandoning the attempt to produce stimulus-bound behaviour in a particular rat. With this procedure, 25% of the animals eventually exhibit stimulus-bound eating, drinking, or gnawing of pine-wood blocks. Valenstein *et al.* therefore started with a mixed population of animals, some of which presumably exhibited stimulus-bound effects early in the testing, and others which came to *acquire* the response through repeated stimulation during the night schedules.

A high proportion of these animals can then be switched from one form of stimulus-bound behaviour to another by subjecting them to a similar procedure of night stimulation in the *absence* of the preferred goal object.

Essentially, this experiment may merely demonstrate that behaviour which is acquired as a result of repeated stimulation can then be modified by preventing the performance of the acquired response. There is little evidence to suggest what mechanism is involved, and in this connection it is particularly unfortunate that although Valenstein uses ethological terms in his discussions of the possible mechanisms involved, he has not adopted ethological techniques in the analysis of these mechanisms. Direct observation of the animals during the night schedule could provide evidence which would be highly relevant in our consideration of the mechanisms underlying the change in responsiveness. Repetitive stimulation in the absence of the goal object, as in the night schedules, is likely to produce both frustration and more general changes in arousal. Consequently, the appearance of displacement and redirected activities might be expected. There is a considerable ethological literature on the changes in responsiveness to external stimuli which occur in situations of frustation or conflict, and which lead to the appearance of displacement activities (for review see Hinde, 1970), and the behaviour that has been studied by ethologists in these situations shows a parallel with phenomena of more direct psychological interest, *e.g.* adjunctive behaviour (Falk, 1971), and 'superstitious behaviour' (Staddon and Simmelhag, 1971). While such work suggests a number of mechanisms which might lead to the changes that Valenstein observes, questions as to the mechanism can best be answered in the light of evidence from direct observation.

There is also the question of the degree of motivational involvement in the responses in the test periods. Although ESB often produces stimulus-bound effects in the rat, even within the lateral hypothalamic area there is evidence for a dissociation of motor and motivational components of the elicited behaviour. Morgane has reported (Morgane, 1961; Morgane and Jacobs, 1969), that although sites within the far-lateral and mid-lateral hypothalamus both support stimulus-bound eating, there is a clear separation of these two areas in terms of whether the same sites will support instrumental responses for food reward. Whereas the majority of far-lateral sites can also elicit performance of operant responses for food reward, only two of the 12 mid-lateral sites can do so. Such demonstrations of operant performance provide a more definite demonstration of motivational involvement in the behaviour elicited by ESB, and are available for feeding (Coons *et al.*, 1965), drinking (Andersson and Wyrwicka, 1957), and attack (Roberts and Kiess, 1964). In the absence of such information for stimulus-bound effects in Valenstein's studies, the relation of his results to hypotheses about the neural basis of *motivation* remains questionable.

A particularly elegant analysis of ESB in birds, which draws heavily on ethological concepts, is the work of Vowles on agonistic behaviour in the Barbary dove. Stimulation at hypothalamic and forebrain sites was shown to elicit either aggressive, defensive, escape or fleeing behaviour, and it was not possible to predict which of these types of behaviour would appear on any one occasion (Harwood and Vowles, 1967). The behaviour was stimulus-bound in the sense that, provided the stimulating current was sufficiently high, one of these types of behaviour would appear, but it was not possible to relate a particular behaviour to a specific anatomical locus. It is precisely this problem which Valenstein (1969) has tried to answer in terms of his prepotency hypothesis. Whereas Valenstein sought to dissociate motivational influences from the observed behaviour, Harwood and Vowles demonstrated that despite the variability in the elicited behaviour, analysis of the after-effects of stimulation did reveal reliable changes in the electrical threshold for eliciting avoidance behaviour, and these were clearly related to the nature of the display elicited during the preceding strong stimulation. If the dove responded to strong ESB with either aggressive or defensive behaviour, then this was followed by a period in which the electrical threshold for eliciting avoidance was raised. Conversely, if the dove responded by escape or fleeing, then the avoidance threshold was lowered. The same relationship between a high-intensity display and the sign of the after-effect also existed when the eliciting stimulus was behavioural rather than ESB. A similar after-effect has been described in juvenile gulls by Delius (1973).

The work by Harwood and Vowles (1967) clearly indicates the way in which it is possible to integrate an ethological analysis at the behavioural level with the use of ESB in the study of motivation, and it would be unfortunate if Valenstein's arguments were to decrease the use of ESB as a tool in such studies. Work at the purely behavioural level is equally open to interpretation in terms of contradictory hypotheses, and the ethological literature provides examples of the value of ESB in resolving such disputes. A particularly clear example concerns the causation of displays, which act as signals, and which have received considerable attention in birds because they are conspicuous features of social behaviour. An influential hypothesis about their causation has been the suggestion that they depend on motivational conflicts; thus threat displays are postulated to depend on a conflict between tendencies to attack and flee from the opponent. However, despite the usefulness of the hypothesis in birds and lower vertebrates (see Cullen, 1972), it has been found less fruitful when applied to mammals, (Andrew, 1972b), and there are problems in interpreting the behavioural evidence. While experimentally produced conflicts *do* cause birds to give displays (Blurton-Jones, 1968), it has not been established that they are *necessarily* present when displays occur. Indeed, Blurton-Jones demonstrated that thwarting a bird's attack would also produce some form of threat. Brain stimulation might be expected to throw more direct light on this question, which is difficult to resolve at the purely behavioural level, and has produced little supporting evidence for the conflict hypothesis. Many of the behavioural observations on which this hypothesis was based has come from gulls, but Delius (1973) using brain stimulation in juvenile gulls found that threat appeared to be related to attacking alone. Similarly, Phillips and Youngren (1971) found an almost total separation between the system controlling *attack and threat*, and that controlling fleeing, in the chicken. Earlier, Brown and Hunsperger (1963) had pointed out that the evidence available for cats did not accord with the conflict hypothesis, and the more recent avian work using ESB has left little doubt that the hypothesis does not adequately describe the causation of the displays at the neural level, even though it may be intuitively helpful in making sense of their form and the situations in which they are given.

The implication is clear. ESB is an important tool for the understanding of behaviour. Valenstein's (1970) hypothesis of hypothalamic plasticity raises the spectre of Lashley, and it would be unfortunate if its impact on the physiological psychologist was to decrease the use of ESB in the investigation of brain function. The alternative is to take up the challenge which is presented by such studies, and which was so clearly outlined in the pioneering work of von Holst and St. Paul (1963).

In this chapter, we have found it necessary to emphasise some of the differences in approach between ethologists and psychologists. It must be stressed that while there are still some differences in emphasis between these disciplines, there has been a growing rapprochement between them over the past quarter of a century, exemplified by the synthesis presented by Hinde (1970). Some of the techniques and experimental situations used initially by ethologists have subsequently proved ideal for the analysis of topics of interest to psychologists; for example, the imprinting situation has proved useful for the study of problems concerning the biochemical changes associated with learning, circumventing some of the problems of interpretation in more conventional learning situations (*e.g.* Bateson, 1970 and 1974). Conversely, ethologists have adopted techniques developed by psychologists; for instance, studies of the reinforcing value of materials used in resting behaviour have appeared in the context of ethological and endocrinological work, (*e.g.* Hinde and Steel, 1972; *cf.* Hinde and Steel, 1966). The interaction of the two disciplines is now so well developed that in many cases it is hard to determine whether the work should be regarded as psychological or ethological in character, and the terms have clearly lost their distinct meaning. This difficulty is apparent, for example, in studies of the reinforcing value of naturally significant stimuli (*e.g.* Bateson and Reese, 1969; Stevenson, 1969), in the studies of the Sussex group of hormonal effects on behaviour (Archer, 1974) and persistence (Andrew, 1972a), and in many of McFarland's studies of dove behaviour (for review see McFarland, 1971). This integration between the disciplines, in which Lehrman played such an important role, is particularly apparent in the field of avian behaviour, and the remaining chapters of this book illustrate some of the fruits of the interaction of these disciplines.

REFERENCES

Andersson, B. and Wyrwicka, W. (1957) The elicitation of a drinking motor conditioned reaction by electrical stimulation of the hypothalamic drinking area in the goat. *Acta. physiol. scand.*, 41, 194—198.
Andrew, R.J. (1972a) Recognition processes and behaviour, with special reference to

effects of testosterone on persistence. In D.S. Lehrman, R.A. Hinde and E. Shaw (Eds.), *Advances in the Study of Behaviour, Vol. 4*. Academic Press, New York, pp. 175—208.

Andrew, R.J. (1972b) The information potentially available in mammalian displays. In R.A. Hinde (Ed.), *Non-Verbal Communication*. Cambridge University Press, London, pp. 179—206.

Archer, J. (1974) Testosterone and behaviour during extinction in chicks. *Anim. Behav.*, 22, 650—655.

Bateson, P.P.G. (1970) Are they really the products of learning? In G. Horn and R.A. Hinde (Eds.), *Short Term Changes in Neural Activity and Behaviour*. Cambridge University Press, London, pp. 553—564.

Bateson, P.P.G. (1974) Specific and non-specific brain events associated with learning. *Biochem. Soc. Trans.*, 2, 189—193.

Bateson, P.P.G. and Reese, E.P. (1969) The reinforcing properties of conspicuous stimuli in the imprinting situation. *Anim. Behav.*, 17, 692—699.

Beach, F.A. (1950) The snark was a boojum. *Amer. Psychol.*, 5, 115—124.

Blurton-Jones, N.G. (1968) Observations and experiments on causation of threat displays of the great tit *(Parus major)*. *Anim. Behav. Monog.*, 1, 75—158.

Brown, J.R. and Hunsperger, R.W. (1963) Neuroethology and the motivation of agonistic behaviour. *Anim. Behav.*, 11, 439—448.

Cohen, D.H. and Karten, H.J. (1974) The structural organisation of the avian brain: an overview. In I.J. Goodman and M.W. Schein (Eds.), *Birds, Brain and Behaviour*. Academic Press, New York, pp. 29—73.

Coons, E.E., Levak, M. and Miller, N.E. (1965) Lateral hypothalamus: learning of feed-seeking response motivated by electrical stimulation. *Science*, 150, 1320—1321.

Cullen, J.M. (1972) Principles of animal communication. In R.A. Hinde (Ed.), *Non-Verbal Communication*. Cambridge University Press, London, pp. 101—125.

Delius, J.D. (1973) Agonistic behaviour of juvenile gulls: a neuroethological study. *Anim. Behav.*, 21, 236—246.

Falk, J.L. (1971) The nature and determinants of adjunctive behaviour. *Physiol. Behav.*, 6, 577—588.

Glickman, S.E. and Schiff, B.B. (1967) A biological theory of reinforcement. *Psychol. Rev.*, 74, 81—109

Gregory, R.L. (1961) The brain as an engineering problem. In W.H. Thorpe and O.L. Zangwill (Eds.), *Current Problems in Animal Behaviour*. Cambridge University Press, London, pp. 307—330.

Harwood, D. and Vowles, D.M. (1967) Defensive behaviour and the after effects of brain stimulation in the ring dove *(Streptopelia risoria)*. *Neuropsychologia*, 5, 345—366.

Hinde, R.A. (1970) *Animal Behaviour: A Synthesis of Ethology and Comparative Psychology* (2nd Ed.). McGraw-Hill, New York.

Hinde, R.A. and Steel, E.A. (1966) Integration of the reproductive behaviour of female canaries. *Symp. Soc. exp. Biol.*, 20, 401—426

Hinde, R.A. and Steel, E.A. (1972) Reinforcing events in the integration of canary nest-building. *Anim. Behav.*, 20, 514—525.

Holst, E. von and St. Paul, U von (1963) On the functional organisation of drives. *Anim. Behav.*, 11, 1—20.

Karten, H.J. (1969) The organisation of the avian telencephalon and some speculations on the phylogeny of the amniote telencephalon. *Ann. N.Y. Acad. Sci.*, 167, 164—179.

Karten, H.J. and Hodos, W. (1967) *A Stereotaxic Atlas of the Brain of the Pigeon (Columba livia)*. The John Hopkins Press, Baltimore, Md.

Lehrman, D.S. (1965) Interaction between internal and external environments in the regulation of the reproductive cycle of the ring dove. In F.A. Beach (Ed.), *Sex and Behaviour*. Wiley, New York, pp. 355—380.

McFarland, D.J. (1971) *Feedback Mechanisms in Animal Behaviour*. Academic Press, New York.

Morgane, P.J. (1961) Distinct 'feeding' and 'hunger-motivating' systems in the lateral hypothalamus of the rat. *Science*, 133, 887—888.

Morgane, P.J. and Jacobs, H.L. (1969) Hunger and satiety. *Wld. Rev. Nutr. Diet.*, 10, 100—213.

Pearson, R. (1972) *The Avian Brain*. Academic Press, New York.

Phillips, R.E. and Youngren, O.M. (1971) Brain stimulation and species-typical behaviour: activities evoked by electrical stimulation of the brains of chickens (*Gallus gallus*). *Anim. Behav.*, 19, 757—779.

Roberts, W.W. and Kiess, H.O. (1964) Motivational properties of hypothalamic aggression in cats. *J. comp. physiol. Psychol.*, 58, 187—193.

Sheer, D.E. (Ed.) (1961) *Electrical Stimulation of the Brain*. University of Texas Press, Austin, Tex.

Staddon, J.E.R. and Simmelhag, V.L. (1971) The 'superstitious' experiment: a re-examination of its implications for the principles of adaptive behaviour. *Psychol. Rev.*, 78, 3—43.

Stamps, J.A. and Barlow, G.W. (1973) Variation and sterotypy in the displays of *Anolis aeneus* (Sauria: Iguanidae). *Behaviour*, 47, 67—94.

Steinbaum, E.A. and Miller, N.E. (1965) Obesity from eating elicited by daily stimulation of the hypothalamus. *Amer. J. Physiol.*, 208, 1—5.

Stevenson, J.G. (1969) Song as a reinforcer, In R.A. Hinde (Ed.), *Bird Vocalisations*. Cambridge University Press, London, pp. 49—60.

Stokes, T.M., Leonard, C.M. and Nottebohm, F. (1974) The telencephalon, diencephalon, and mesencephalon of the canary, *Serinus canaria*, in stereotaxic co-ordinates. *J. comp. Neurol.*, 156, 337—374.

Teitelbaum, P. (1973) On the use of electrical stimulation to study hypothalamic structure and function. In A.N. Epstein, H.R. Kissileff and E. Stellat (Eds.), *The Neuropsychology of Thirst*. Winston and Sons, Washington, D.C., pp. 143—154.

Tinbergen, N. (1951) *The Study of Instinct*. Clarendon Press, Oxford.

Valenstein, E.S. (1969) Behaviour elicited by hypothalamic stimulation: a prepotency hypothesis. *Brain Behav. Evol.*, 2, 295—316.

Valenstein, E.S. (1970) Stability and plasticity of motivation systems. In F.O. Schmitt (Ed.), *The Neurosciences: Second Study Program*. Rockefeller University Press, New York, pp. 207—217.

Valenstein, E.S., Cox, V.C. and Kakolewski, J.W. (1968) Modification of motivated behaviour elicited by electrical stimulation of the hypothalamus. *Science*, 159, 1119—1121.

Valenstein, E.S., Cox, V.C. and Kakolewski, J.W. (1970) Re-examination of the role of the hypothalamus in motivation. *Psychol. Rev.*, 77, 16—31.

Vowles, D.M. (1975) An introduction to neuroethology. In P.B. Bradley (Ed.), *Methods in Brain Research*. Wiley, New York. In press.

Wiley, R.H. (1973) The strut display of male sage grouse: a 'fixed' action pattern. *Behaviour*, 47, 129—152.

Wise, R.A. (1968) Hypothalamic motivational systems: fixed or plastic neural circuits? *Science*, 162, 377—379.

Zeier, H. and Karten, H.J. (1971) The archistriatum of the pigeon: organisation of afferent and efferent connections. *Brain Res.*, 31, 313—326.

2

TEMPORAL PATTERNING AND THE CAUSATION OF BIRD BEHAVIOUR

P.J.B. SLATER

Ethology and Neurophysiology Group, School of Biological Sciences, University of Sussex (Great Britain)

INTRODUCTION

This chapter is concerned with the temporal organisation of behaviour and with quantitative methods for describing its fine structure. It will concentrate in particular on two concepts of behavioural organisation which are often mentioned, but rarely analysed in any detail: the idea that many behaviours occur in bouts and the idea that animals show short-term cycles of behaviour. Discussion of these will be illustrated largely by results obtained from work on adult male zebra finches which have been observed in isolation both from other birds and, as far as possible, from changing environmental stimuli (apart from a standard 12-h light—12-h dark lighting regime). This approach was adopted in the hope that the description of behaviour obtained would provide a base-line against which to compare behaviour in less constant conditions and a source of hypotheses about the mechanisms underlying the organisation of behaviour.

At first sight this subject may seem a curious one to find in a book of this sort, as neither neural nor endocrine mechanisms will be referred to directly. There are two reasons, however, why studies of temporal patterning are relevant to work at a more physiological level. The most obvious of these is that a detailed knowledge of the fine structure of behaviour poses questions about the way in which its organisation is achieved, the answers to many of which will ultimately lie in the study of nervous and hormonal mechanisms. A good understanding of the temporal pattern of behaviour thus helps to define more clearly the phenomena which need to be explained. The second contribution which this type of work can make is a methodological one. As more precise and quantitative methods for describing behaviour are developed, it will become possible to detect increasingly subtle departures from normality in animals subjected to neural and endocrine manipulations. Much of current experimental work on behaviour is concerned with detecting changes in the frequency of particular acts following treatments. Closer examination might reveal that an increased frequency of a behaviour resulted from the occurrence of more bouts of the normal length, or from the occurrence of the same number of bouts of a greater average length. In either case the likelihood of detecting such a change will be raised if the bout, as defined by the observer, is closely related to the actual switching on and off of the behaviour by the animal; how this may be achieved will be considered below. If a behaviour recurs cyclically, a treatment may also cause a change in its frequency by leading to an alteration in cycle length. The causal factors underlying cycles, or less regular fluctuations in the probability of a given behaviour pattern, may be specific or non-specific (Fentress, 1973). At one extreme they may affect that behaviour only, while at the other they may involve an arousal-like variable which affects the probability of many, or all, behaviours. Cycles approximating to both of these types seem to influence the behaviour of isolated zebra finches, and a preliminary examination of them will be presented later in this article. An understanding of features of behavioural organisation such as these should help in attempts to discover how experimental manipulations exert their effects.

To many people the most obvious and useful method for analysing the temporal pattern of behaviour is sequence analysis. This approach has been applied to the behaviour of zebra finches (Slater and Ollason, 1972), and its usefulness and limitations have been considered in detail elsewhere (Slater, 1973), so only a few words of caution will be mentioned here. A major difficulty involved in the analysis of sequences is that of non-stationarity: the tendency for the probability of occurrence of a behaviour to vary with time rather than to be simply dependent on the sequence of acts which preceded it. In behavioural terms, a major source of non-stationarity arises when the motivation of the animal shifts because internal and external causal factors change during the course of time. Non-stationarity, the bug-bear of the sequence analyst, is thus the bread and butter of anyone interested in motivation. As stationary sequences of behaviour in vertebrates are unusual it can be said, without much exaggeration, that behavioural problems become interesting when sequence analysis becomes impracticable. It is an aim of this article to explore ways of looking at behaviour where motivation is known to change, first in relation to individual behaviours and their distribution in time, and later in relation to the way in which behaviours are associated with each other.

THE BOUT ORGANISATION OF INDIVIDUAL BEHAVIOURS

Methods of bout definition

Many behaviour patterns are almost identical on separate occurrences and between individuals and, as they are of short duration and vary little in intensity, their distribution in time can be studied as if they were point events. For most behaviours an analysis of this sort reveals that events are not randomly distributed in time but tend to cluster or occur in bouts. Thus, if an animal has just scratched its head, taken a peck at food or flown from one perch to another, the most likely thing for it to do next is a repetition of the same behaviour. Reasons why this should be the case are not hard to find. In general, unless a behaviour is very effective at removing the causal factors which led to it, these are likely to persist and so cause further events. In the case of feeding, for example, a single peck at food will do little to remove a deficit and its influence will take time to operate. Furthermore, in this case, as in others, part of the bout organisation may stem from persistence of stimuli in the outside world. The bird which has just taken a peck at seed is likely to be still looking at the food dish, the appropriate stimulus for this, rather than any other, behaviour.

It is thus easy to see why behaviours should often take place in bouts. But how should one determine when one bout ends and a new one begins? Given that gaps between pecks at food are of variable duration and may or may not contain a variety of other behaviours, how should one decide which ones are within a bout of feeding and which are between bouts? An obvious, but unsatisfactory, solution to this problem is to say that a bout ends when the animal changes to doing something else. One reason why this gives a poor bout definition is that it depends to a large extent on what other categories of behaviour have been scored: if mandibulation of seed was one, then all successful pecks at food would occur in separate bouts. On the other hand, if only feeding and drinking were being studied (as by McFarland, 1970), some drinking would have to occur between them before two meals were regarded as separate. Another problem, which is apparent from the sequence analysis carried out on zebra finches (and from the work of Fentress (1972) on grooming in mice), is that a change in behaviour is often followed by a reversion to that shown previously, so that this cannot be regarded as having been totally switched off during the interruption. This is not just true amongst closely associated behaviours which may stimulate each other and so give rise to alternation. In the zebra finch, two behaviours which very seldom occur together in sequence are locomotion (Loco) and preening (Pr). In a sequence of nearly 9000 events of different types it was calculated from the frequencies of Loco—Pr and Pr—Loco that the triplet sequences Loco—Pr—Loco and Pr—Loco—Pr should have occurred six times and once respectively: the observed figures were 19 and 5. This surplus of observed over expected, which was also true of most of the other A—B—A type triplets studied, suggests that bouts which are defined as delimited by changes from and to other behaviour patterns are themselves clustered in time and not statistically independent of each other. For these reasons this is a poor way of defining bouts when studying the temporal pattern of individual behaviours.

14

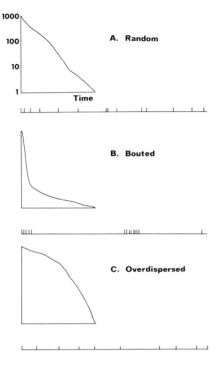

Fig. 1. The use of the log survivor function for plotting the distribution of intervals between events. In this type of plot, the log of the number of intervals greater than a particular length (ordinate) is plotted against that length (abscissa). This yields an approximately straight line if the events are randomly distributed (A); a concave plot if the events occur in bouts or clusters (B); and a convex plot if the events are overdispersed (C).

A more satisfactory way is to choose a time interval between events and define gaps shorter than this as within bouts and longer ones as between them. A useful way of choosing an interval appropriate to the way in which the behaviour itself is organised is derived from plotting the intervals between successive events of the same type in the form of a log survivor function (Fig. 1). If acts are randomly distributed in time the distribution of the intervals between them should follow a negative exponential. Plotted as a log survivor function this appears as a straight line, the slope of which is proportional to the probability of occurrence of an event with the passage of time since the last event. Where events are clustered, the slope is high to begin with and later drops, often to a fairly constant level, indicating that beyond that gap length the probability of a new event is low and relatively uninfluenced by the timing of the last. The point at which the probability becomes constant can be chosen with reasonable confidence as a suitable interval for bout definition. While this involves some degree of judgement, it is less arbitrary than other methods in common use. It may be made more accurate in the future by the application of line-fitting procedures which are currently being developed (Machlis, personal communication).

As can be seen from Fig. 2, the log survivor functions for most behaviours are concave and in some cases (*e.g.* preening, feeding and stretching in this bird) they appear appropriate for analysis in terms of bouts and gaps because the switch to a low probability is fairly abrupt, suggesting that this point is a significant one for the way in which the behaviour of this bird is organised. Other behaviours are less appropriate for analysis of this sort as the log survivor function tends to be smoother. In some of these cases the most economical description may be in terms of a renewal process (Heiligenberg, 1973); for song this is suggested by the fact that successive gaps between songs are, if anything, negatively correlated with each other. This is not compatible with a bout model which suggests an alternation between a high-probability state and a low-probability state each of which persists for some time. In other cases it has been suggested (Slater, 1974a) that more than one level of bout may exist, so that tightly-knit groups of acts may occur and these are assembled into more loosely-knit bouts; this too would lead to a rather smooth log survivor function.

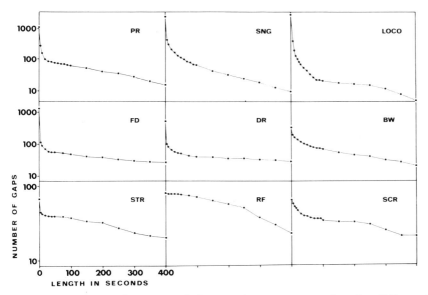

Fig. 2. Log survivor functions of the gaps between events for nine different behaviour patterns shown by one bird during 8 h of observation. All plots are up to a maximum of 400 sec, each point indicating the number of gaps recorded which were longer than the time shown on the abscissa. The abbreviations used are as follows: PR, preening; SNG, song; LOCO, locomotion; FD, feeding; DR, drinking; BW, bill-wiping; STR, stretching; RF, ruffling and SCR, scratching. Reprinted by permission of Slater (1974a).

The distribution of acts in time may also depart from randomness because they are overdispersed, the probability of a new event increasing with the passage of time since the last (Fig. 1). Ruffling in most birds shows this pattern (see Fig. 2). It is also a pattern which might be expected to occur in the distribution of intervals between bouts in those behaviours showing bouts. This is the equivalent of saying that one might expect the probability of a new bout beginning to increase with the passage of time since the last. This will occur where causal factors rise during non-performance and lead to the log survivor function curving downwards at long interval lengths.

Detailed analysis of feeding and preening

Once a behaviour has been split into bouts and the gaps between them, a number of ways of analysing the data further become possible. Detailed analysis has so far been restricted to feeding (Slater, 1974b) and preening (Slater, 1974a). The results obtained will be briefly summarised here to allow discussion of some interesting differences and similarities between these behaviour patterns, and of hypotheses about the organisation of behaviour which have been derived from the two studies.

Zebra finches eat in tight meals so that a bout criterion is not hard to derive from the log survivor function of intervals between pecks, though those chosen varied between 20 and 80 sec for the nine birds studied. Finer details of the temporal pattern of feeding show strong differences between birds, which are to some extent constant over time. This is shown in Table I where some of the results obtained by Slater (1974b) on five birds are compared with those of Holden (unpublished) who studied the same birds two years later without knowing which was which. Not only were bout

TABLE I

Bout criterion chosen and mean bout and gap lengths found for five birds, the feeding of which was recorded by two different observers two years apart

Bird number	Slater's results (July—November 1971)			Holden's results (October—November 1973)		
	Criterion (sec)	Mean gap (20-sec units)	Mean bout (pecks)	Criterion (sec)	Mean gap (20-sec units)	Mean bout (pecks)
9	20	24.3	19.5	20	33.3	26.1
31	20	54.6	43.9	40	57.3	47.2
14	80	35.7	32.8	60	60.6	35.0
30	20	38.1	11.8	20	39.9	21.2
6	80	56.1	41.6	60	75.5	50.1

lengths and gap lengths remarkably similar between the two studies, but it is gratifying to find that the two independent observers also chose rather similar bout criteria for each bird.

Despite the strong differences between individuals, Slater (1974b) found that his birds fell into two groups, within each of which feeding patterns were similar. In the smaller group the probability of a new meal starting did not increase with time since the last meal, nor did the probability of a meal ending change as it proceeded. There was no evidence for cycles in the feeding of these birds, and the number of pecks in a meal tended to correlate strongly with the length of the gap before it rather than with that after it. The simplest hypothesis to account for these results was that these birds fed sufficiently to make up their deficit whenever they came across food. By contrast, the majority of the birds conformed to another pattern. The longer it was since they last fed, the more likely were they to start feeding, and in most cases the probability of stopping feeding increased as a meal became longer. These birds thus tended to show meals of a rather constant length. In four cases feeding was also periodic, with a cycle length of 24—30 min. Meal length tended to correlate with that of the gap following rather than that before. Most of the results for these birds could be explained on the hypothesis that they went to the food hopper when they had accumulated a threshold deficit, rather than feeding when they happened to come across it during their movements round the cage.

Individual differences in the pattern of food intake are obvious to the observer watching isolated zebra finches and are also apparent in the distribution of gaps between pecks plotted as log survivor functions. It was clear from these that each bird should be assigned a bout criterion based on its own data rather than a general one being chosen for all birds. However, this leaves the possibility that some of the other differences found may stem from the fact that different criteria were used for different individuals. Slater (1974b) could find no evidence that the two different patterns of feeding mentioned above were related to the bout criteria which the birds showing them had been given. More recently the same data have been partially analysed using five different bout criteria for each bird: 20, 40, 60, 80 and 100 sec. The results of this did not suggest that the feeding pattern of any bird would have appeared to be markedly different from that originally found had it been assigned another criterion.

A more interesting result came when correlations were carried out between bout length and the gaps before and after it for each of the five different bout criteria. These were examined to see which bout criterion gave the most significant correlation. There were 18 cases (gap before and gap after for each of nine birds) and in nine of these the most significant result was at the criterion originally chosen. By chance this should have been true in less than four cases ($\chi^2 = 10.4, P < 0.01$). This highlights the point that, if the criterion is carefully chosen so that it is closely related to the temporal structure of the behaviour, the chance of detecting organisation during subsequent analysis is maximised. With arbitrarily chosen criteria, or a common criterion for all birds, fewer of these correlations would have been detected, not because they were not there, but because the criterion used was inappropriate to the animal in question.

In contrast to the feeding behaviour of most birds, the analysis of preening showed few individual differences and a surprising lack of temporal organisation (Slater, 1974a). As with feeding, each bird was given a different bout criterion depending on the form of its log survivor function. These criteria varied from 25 to 70 sec amongst the ten birds studied. In no bird was the probability of preening starting found to change with the passage of time since the last bout, and only a negligible number of correlations between bout lengths and those of the gaps surrounding them were found to be significant. It is thus very hard to predict from its previous behaviour when a bird will start preening. The organisation of bout lengths was more striking, with a surplus of very short and very long bouts compared with the numbers predicted by simple mathematical models. The simplest such model, the geometric distribution, which assumes that the probability of a further event is independent of the number of previous events within a bout, gave a very poor fit to the data. The logarithmic distribution, in which the probability of continuing increases as the bout becomes longer, gave a better fit but still consistently underestimated the number of bouts of one and two events compared with that observed.

Perhaps the simplest hypothesis to account for this finding is the suggestion that when preening bouts start they are very likely to stop within one or two events but, if they get beyond this stage, their probability of continuing increases in a rather complicated manner. According to this hypothesis, long and short bouts would be similar in causation, so that it is reasonable to mass them for analysis. Short bouts are simply potential long bouts which, for some reason, stop early. An alternative hypothesis, which deserves further study, is that long and short bouts differ from each other in causation and that the apparently complex bout organisation results from massing together two phenomena which should be kept separate. There is some evidence in favour of this latter explanation: short bouts are significantly more likely to be preceded by locomotion than are long ones and the movements comprising them also give the impression of being more hurried than are those at the beginning of long bouts. These differences in form and context suggest that a causal difference may exist between long and short bouts so that, while each may have a comparatively simple bout length distribution, when they are massed the resultant distribution is a complicated one. The curious finding that the probability of preening does not increase with the passage of time since it last occurred, and that bout and gap lengths are not correlated with each other, could also be accounted for more easily if two disparate phenomena had been lumped together under the single heading of 'preening'. The possibility of such a causal difference between long and short bouts of preening will be discussed further below.

Discussion

It is hoped that the above section will have illustrated the usefulness of analysing the temporal pattern of individual behaviours, particularly in terms of bouts and the gaps between them in cases where this is appropriate. Many of the methods used could be easily, and usefully, applied to detecting differences between experimental and control groups of animals. An advantage which such studies would have over those discussed here is that looking for differences between two groups of animals is much more simply carried out than is testing a single group against random models of one sort or another. Now, however, we will go a little further towards considering the hypotheses for causation which can be derived from studies of the sort described above.

From the detailed analyses of feeding and preening, and work on other behaviours which has been referred to more briefly, it is clear that behaviour patterns differ from each other in some of the basic features of their temporal organisation. Why should this be the case? One consideration which is likely to be of major importance is the extent to which internal and external causal factors affect them and how these act. The term 'causal factor' is an unsatisfactory one to use in any explanatory sense, for it can embrace such a multitude of different things and is usually just a cloak for ignorance. But, by subdividing the blanket term, one can see that different behaviours are affected by different general types of causal factor. For example, solitary song in birds, of the sort shown by isolated zebra finches, is spontaneous in the sense that it does not require any specific external stimulus for its elicitation. While its frequency may vary with external factors such as levels of light and noise, it is primarily internally motivated; hormones clearly play a part here, but its short-term temporal pattern is probably largely a neural phenomenon (Lemon and Chatfield, 1973). The temporal pattern of feeding is also obviously influenced strongly by internal factors, but it differs from song in that it requires an appropriate external stimulus. This is an added complication, for an animal whose deficit is not sufficient to make it seek out food may, nevertheless, feed when it comes across it. A further situation is illustrated by grooming movements for not only does external stimulation appear to have an important part to play in eliciting these (Andrew, 1956), but the animal also carries the appropriate stimuli around with it.

The extent to which different behaviours are affected by internal and external causal factors, and the mode of operation of these, may thus account for the differences which are found in their temporal patterns. It has also been suggested here, however, that the same behaviour may have different causation in one situation from that in another. The two different patterns of feeding found were most easily explained on the hypothesis that one group of birds were induced to feed largely by the external situation (*i.e.* when they came across food), whereas the other fed in a more regular manner, suggesting that they were induced to feed largely by their internal state (*i.e.* when they had built up a certain deficit). Some birds of the latter group, as well as taking regular long meals, sometimes took a few brief pecks at food when they passed the food hopper during their movements round the cage. It seems likely that these meals were caused largely by the sudden appearance of the appropriate external stimulus, as has been suggested for the majority of meals in birds of the first group. The possibility that short meals are differently motivated from long ones has also been raised by Duncan *et al.* (1970) in their study of the feeding of domestic hens in Skinner boxes. During part of their analysis they ignored meals consisting of a single reinforcement on the grounds that these were likely to be 'motivated by factors other than hunger'.

In the case of preening, the surplus of very short and very long bouts compared with the numbers predicted by simple mathematical models suggested that long and short bouts might differ from each other in causation. The hurriedness of short bouts and their association with locomotion, the random distribution of gap lengths and the lack of correlation between bout and gap lengths all fitted in with this hypothesis. Slater (1974a) suggested that such a difference might occur if short bouts dealt with specific sources of irritation which might arise at any time, whereas long bouts resulted either from generalised irritation, perhaps building up slowly between them, or from a central programme. Short bouts in this case would be expected to deal more often with those areas of the body most prone to being dishevelled by other activities, while long bouts would cover all areas. According to this hypothesis short bouts occur when causal factors for preening suddenly rise so that the behaviour takes place briefly, the source of irritation is removed and the previous behaviour can continue. This hypothesis, therefore, proposes that ongoing behaviour is temporarily inhibited and then disinhibited by preening. It is, therefore, the opposite of the disinhibition hypothesis usually proposed to account for the occurrence of displacement activities (Andrew, 1956; van Iersel and Bol, 1958; McFarland, 1966). It is worth considering this alternative possibility because these short bouts do have certain characteristics in common with behaviours normally classed as displacement activities. The two features of displacement behaviour which survive from those given by Tinbergen (1952) are its hurriedness and its appearance out of context; if these are used to define it, then the short bouts of preening described here clearly fall under the displacement umbrella.

The disinhibition hypothesis, as set out by McFarland (1966), suggests that the conflict or thwarting, thought to underlie displacement, leads to a switch in attention to peripheral stimuli so that subsidiary behaviour is disinhibited. More recently he has suggested that disinhibition may be an important mechanism of switching between acts in normal sequences of behaviour (McFarland, 1969). In zebra finches, isolated from all other individuals, it seems unlikely that short preening bouts are associated with conflict or thwarting, though it is not in general beyond the ingenuity of ethologists to postulate a conflict of one sort or another in any situation where one seems to be required! Nevertheless, it could be that causal factors for the ongoing behaviour are lowered for a time for some other reason, so allowing preening as the second in priority act to appear briefly. The dominant behaviour would thus disinhibit and then reinhibit preening according to the general mechanism proposed by McFarland (1969). There are two objections to this. First, it is difficult to accommodate the hurriedness of the disinhibited behaviour into such a model (Duncan and Wood-Gush, 1972). Second, preening and locomotion normally occur in very different behavioural contexts (hence the idea that preening during locomotion is out of context) so that preening would be unlikely to be the second in priority act during periods of locomotion.

While a decision between these two models must await experimental evidence, it is important to bear in mind that disinhibition is not the only mechanism whereby short bouts of behaviour may appear out of context. In the present situation the inhibition hypothesis appears more likely and it is probable that it is appropriate in some cases of displacement also. In the displacement preening of terns, studied by van Iersel and Bol (1958), for example, the main behaviour pattern shown was head-shaking and, as this frequently followed displays involving pronounced head movements, it would seem more parsimonious to suggest that it was induced by these (its causal factors rising to inhibit them) than to postulate an internal conflict.

A further feature of the proposals being made here for the causation of short bouts of preening and that of the two feeding patterns found is the idea that the same behaviour may, on different occasions, be differently motivated. This idea has not of recent years been a popular one and, once again, views of the causation of displacement activities provide an interesting example. The conventional view has, in this case, moved from one extreme to another. Tinbergen (1951) saw displacement behaviour and its normal example as totally different in causation. Later, it was discovered that external stimuli appropriate to the normal behaviour also affected displacement (van Iersel and Bol, 1958; Sevenster, 1961). The disinhibition hypothesis was born and with it the idea that displacement activities were the same as normal behaviour but switched on in a different way. While the latter is clearly a more realistic view, it may be too simple. The proposal being made here is that if the occurrence of a behaviour depends to some extent on both internal and external factors, or both central and peripheral ones, it may appear on some occasions when one group is high and the other low and on others in the reverse situation. It is misleading in cases like this to suggest that the causation is always the same. The neatest demonstration of a system like this in operation is that by Baerends *et al.* (1955) on courtship in the guppy. They found that three different displays would be shown by males of high internal motivation (as assessed by the extent of their colour patterns) to small females, but an increase in the size of the female was necessary to elicit them from males of lower motivation. Thus a variety of mixes of internal and external factors could lead to the behaviour being shown.

To discuss ideas such as these in relation to purely descriptive work on the behaviour of zebra finches may seem unduly speculative, as all of them will need experimental work before they can be tested. However, one of the main aims of the descriptive project was to provide hypotheses about the causation of the organisation which was discovered; it is hoped that this discussion has illustrated how this has been realised.

SHORT-TERM CYCLES OF BEHAVIOUR

The phenomenon

It will be clear from the results presented so far that the probabilities of different behaviour patterns change with time, and that in some cases these changes are ordered enough to allow some prediction as to when the behaviour will next occur. But analysis in terms of bouts and gaps is not easily applied to all behaviour patterns and, even where it is appropriate, it does not provide a complete picture of the organisation in time of the behaviour being studied. One feature which may be missed is any tendency for the probability of a behaviour to change in a cyclical manner, although this can be detected from bout and gap analysis if there is usually only a single bout per cycle. Nevertheless, methods of looking for cycles provide a useful additional tool in the search for regularities in the organisation of behaviour.

That there are cycles in animal behaviour has been recognised for a long time and there are, of course, extensive studies on daily cycles, breeding cycles and annual cycles (Sollberger, 1965). The existence of cycles of less than a day has sometimes been referred to in passing (*e.g.* Crook, 1961; Rowell, 1961; in birds), but there have been few attempts at detailed study beyond those of Richter (1927) and Wells (1951). One problem in making such an attempt is that, while the rotation of the earth provides both an entraining stimulus and a selective advantage for cycles of behaviour 24 h long, short-term cycles would not in general be subject to such external constraints. Variability, both between animals and with time, is therefore to be expected.

The idea that cycles might occur in the behaviour of zebra finches came originally from the results of sequence analysis (Slater and Ollason, 1972). These showed that the animals fluctuate between periods of high activity, during which ingestion takes place, and low activity, which is associated with preening and drowsiness. Song tends to occur on the borderline between the two. In order to test whether such fluctuations were regular enough to be termed cycles, it was necessary to obtain long records and these were only practicable with two behaviours which could be recorded automatically: feeding and locomotion. As has already been mentioned, feeding in some zebra finches was found to be periodic, tending to occur every 24—30 min. The results of the locomotion analysis will be presented below, as well as those of some preliminary experiments aimed at finding out the extent to which other behaviours show cycles and, if they do so, how their cycles fit in with one another. From the theoretical point of view this last aspect is an important one. At one extreme all behaviours might show the same cycle length, so that the animal went through a set sequence of activities once in each cycle, suggesting that they were influenced by a single non-specific variable. At the other, each behaviour might have a cycle length of its own, suggesting that specific factors were more important, although there would obviously be some interactions between behaviours because many of them cannot occur at the same time as each other. If cycles are found in a number of different behaviour patterns, the extent to which they are of similar length and can be teased apart from each other experimentally will provide a useful means of assessing the importance of non-specific variables in the causation of behaviour. Such variables have often been postulated, under such names as arousal, activation and responsiveness (Andrew, 1974). While their reality is not easy to either confirm or disprove, the effort is worthwhile, for the existence of any such variable would obviously have an important influence both on the organisation in time of individual behaviours and on their associations with one another.

Cycles of locomotion

The data used to look for cycles of locomotion took the form of counts of the number of perch changes (gauged by microswitches under the perches) during each 3-min unit of the 12-h day. The data obtained were analysed using autocorrelation. With this technique the same data are placed in two columns and correlated with each other with progressively larger lags introduced between the figures in one column and those in the other. Before a lag is introduced the correlation coefficient obtained is +1.0 because the two sets of data are identical and each figure sits opposite itself. As progressively larger lags are used the value of the correlation coefficient falls but, if a cycle is present, it rises to peak again at a lag equivalent to the cycle length. This is because at this lag the peaks present in one set of data have been moved forward by one cycle so that they lie opposite the previous peak in the other set. While this technique is a useful way of detecting cycles, it involves some difficulties. First, the lag used should not exceed 10% of the observation period (Weiss *et al.*, 1966), which limits the length of the cycle which can be detected and often precludes the discovery of a second peak at twice the cycle length, which is otherwise a useful confirmation that the cycle found is a real one. Second, because the successive figures which are correlated are not independent of each other, there are difficulties in statistical interpretation (Sollberger, 1965). Third, it is affected by long-term trends in the data which tend to lead to high correlations at short lags and low ones at long lags regardless of whether a cycle is present or not. In the case of locomotion this last point is a difficult one because the behaviour shows a strong daily cycle, with a great deal of activity in the morning and little in the afternoon (Ollason and Slater, 1973). In the results to be presented, the effects of this were minimised by correlating not the raw data, but the difference between each figure and the mean of the 21 consecutive figures of which it was the eleventh.

A pilot study was first carried out on nine birds, with correlations being calculated for all lags between 3 and 45 min. As expected the results showed considerable individual variation, with some birds providing no evidence for cycles, while in others the main peak in the autocorrelation plot varied from 12 to 45 min. In those birds showing a peak at short lags subsidiary peaks were found at multiples of that lag. As several of the birds in this pilot study showed pronounced peaks close to the maximum lag which was used, more systematic data were later collected from 14 individuals using lags of up to 1 h. The results are summarised in Fig. 3, and suggest a tendency amongst these birds to show cycles of locomotion approximately 45 min in duration.

These results should be regarded as preliminary, and a more detailed study is planned to see whether cycle length in the same individual varies with time of day and over the course of several days and to see also how it relates to the overall activity level of the individual. Nevertheless, at this stage one can conclude that zebra finches show short-term fluctuations in locomotion and that in some birds these changes are regular enough to be termed cycles. While individual variation is great, the cycles found are most commonly around 45 min long.

The relationship of other behaviours to changes in locomotion

Slater (1974b) was unable to find any consistent relationship between locomotion and feeding by cross-correlating between them. This is in line with the fact that the commonest cycle length found for locomotion, 45 min, is longer than those found for feeding. These two behaviours, therefore, appear relatively independent of each other in their temporal patterns.

The main problem in detecting how other behaviour patterns fit in with the fluctuations in locomotion is that these fluctuations are so different between birds. This prompted some pilot experiments in which attempts were made to induce different animals to show cycles of the same length by subjecting them to a cyclical input from the outside world. In the first such attempt, a tape-recording was played to isolated birds on two consecutive

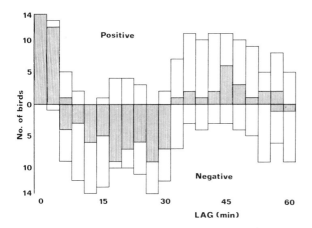

Fig. 3. Summary of the autocorrelation results for locomotion recorded from 14 birds for one day each. The open columns indicate the number of birds showing positive and negative correlations at each lag. The shaded columns indicate the number in which the correlation found was significant ($P < 0.05$, using the parametric method described by Weiss *et al.*, 1966).

days for 6 h. In each 19.5-min period there were 9.5 min of silence followed by a period of 10 min during which a brief tone was played at a rate which rose from 1/10 sec to 8/10 sec and then fell off to 1 again. Each of six birds was watched through three of these cycles on each day and the presence or absence of a number of behaviour patterns in every 15-sec period was recorded.

Fig. 4 shows the results for the four birds which were influenced by the input; as far as could be seen the other two ignored it. The birds tended to be active when the tones were on and preening and song, if they changed at all, tended to occur when the tones were off. The results on preening, however, should not be stressed on the basis of only two hours of observation, as they depend a great deal on whether or not a long bout occurred. The feeding results were very varied; one bird tended to feed as locomotion was rising, one as it was falling, one as it was rising and as it was falling, and one at any point during it. This confirms the point made earlier and by Slater (1974b) that, although feeding does not occur when a bird is inactive, its relationship with locomotion is a loose one.

Three of the birds which were tested in this way were, on the following two days, tested for locomotion for 6 h, with the tape-recorder being played on the first but not on the second day. The autocorrelations obtained are shown in Fig. 5. Bird 6 showed a weak 15—18-min cycle without the tones, but a clear 18—21-min one with them. In the other two birds the main cycle of locomotion was at about twice the length of the tone cycle whether or not the tones were switched on, though in bird 14 there was a suggestion of

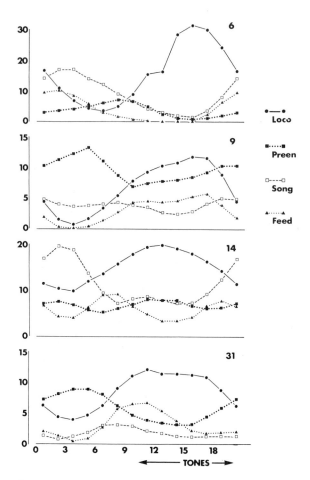

Fig. 4. The distribution of four behaviour patterns as shown by four birds during a 19.5-min tone cycle (described in the text). Each bird was observed for six cycles and the presence or absence of a behaviour pattern was noted in every 15-sec time unit. The results were summed over 1.5-min periods, thus giving a maximum score of 36. The figures plotted have been smoothed to remove minor fluctuations. (This involved adding to each figure, 2/3 of each of those immediately before and after it and 1/3 of those two before and two after it, then dividing this sum by 3.) The plots for each individual are labelled with its code number.

a weak 18—21-min component coming in when the tones were present. Unfortunately the amount of observational data on these two birds is insufficient to allow analysis of the results presented in Fig. 4 on the assumption that they were showing a 39-min cycle rather than 19.5-min one. Presumably the quite marked changes in locomotion which they show in Fig. 4 are because this behaviour peaked at the end of every second tone cycle.

In summary, one, and probably two, birds synchronised their locomotion

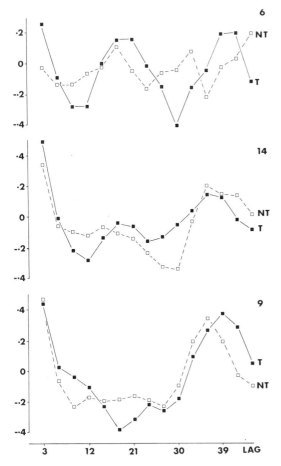

Fig. 5. Autocorrelations of locomotion with lags up to 45 min for three of the birds shown in Fig. 4, based on 6 h data with the tones being played (T) and 6 h on the following day without the tones (NT).

with the cyclical input, two birds showed cycles of twice the length of the tone cycle, and two birds appeared to ignore it altogether. That these experiments were not more successful may have been partly due to the choice of a rather short cycle length and partly due to the choice of stimulus. They did, however, suggest that cycles of external stimulation may synchronise and standardise in length the cycles of locomotion shown by zebra finches as long as their length is not too different from that of the cycle normally shown by the individual.

More systematic experiments have now been started using cycles of varying light intensity. Three birds have so far been tested and their results are shown in Fig. 6. All of these had been kept for 3 months on a 12-h day

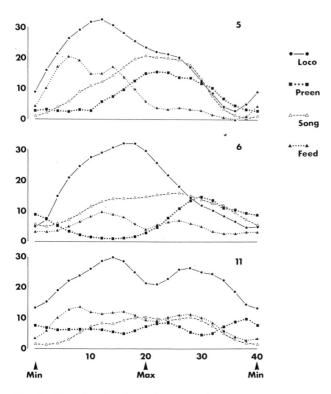

Fig. 6. The distribution of four behaviour patterns as shown by three birds during a 40-min light cycle (described in the text). Each bird was observed for ten cycles and the presence or absence of a behaviour pattern was noted in every 30-sec time unit. The results were summed over 2-min periods, giving a maximum possible score of 40, and smoothed using the procedure described in the legend to Fig. 4.

during which the light level in their cages rose in a linear manner from about 3 lux to about 52 lux and then fell again on a 40-min cycle repeated 18 times during the day. The low point of the light cycle was adequate for them to be observed on closed circuit television without special equipment, and birds were several times seen to fly around at this point indicating that they were not being forced into inactivity. The data for each bird in Fig. 6 are based on ten 40-min watches carried out, as nearly as possible, on consecutive days at the same time.

The behaviour of these birds was strikingly predictable from day to day. Bird 5 provides a good example. It was invariably inactive when the light intensity was at its low point but would quickly start to move around the cage, taking a long meal within a few minutes and, on some occasions, a shorter meal a little while later. Despite the fact that the data are smoothed, two peaks of feeding show up in Fig. 6 indicating that the timing of these meals was very consistent. The bird would remain active throughout the first half of the cycle, but as its activity began to decline it showed increasing amounts of singing and then preening. This preening continued when it became inactive about 10 min before the end of the cycle; for the last few minutes of the cycle it sat drowsily and showed none of the four behaviours plotted in Fig. 6 (the rise seen in locomotion and feeding at the end of the cycle results from the smoothing procedure which is described in the legend to Fig. 4). Bird 6 shows a similar general pattern, with locomotion and feeding being followed by song, then preening and then inactivity, except that the feeding is more spread out in this bird. Interestingly, this bird also followed the 20-min cycle used in the last experiment very closely (see Fig. 5). Bird 11 is more problematic, with a trough in locomotion and a slight peak of preening in the middle of the cycle as well as at the end. Its behaviour was much less well synchronised with the imposed cycle than was that of the other two: although it was usually inactive when the lights were low, it sometimes started to move about at once when they began to rise and on other occasions remained still for over 10 min. Nevertheless, there is a suggestion from the data that it was fitting approximately two behavioural cycles into one of the cycles of light intensity; this is interesting in view of the fact that, when its locomotion had been studied 2 years earlier, it had been found to show an 18-min cycle.

Discussion

The results reported in this section are largely preliminary and, as such, based on rather few birds. But they do seem to point to short-term cycles as an important and neglected influence on animal behaviour. Both of the experiments carried out so far suggest that birds can be induced to synchronise their locomotion cycles with those of imposed stimuli unless the period of these diverges too strongly from that which they normally show. Some birds show one cycle of behaviour for every two imposed cycles, and in one case there was a suggestion that the bird showed two cycles within each imposed one. One bird which was used in both experiments synchronised its behaviour with both the 20-min and the 40-min stimulus cycles.

The long-term aim of these studies is to try to tease apart the causal factors affecting different behaviours using cycles of different stimuli and lengths. If it proves possible to synchronise different behaviours with different imposed cycle lengths, this will provide evidence that the cyclical factors affecting the different behaviours are separate from one another. The clearest evidence that this can be achieved comes from data on the interrelation between locomotion and feeding. Present evidence suggests that these two behaviours are controlled by cycles of different lengths which are only loosely coupled to one another. While feeding occurs mainly during brief breaks in locomotion, it may occur at any stage during a period of activity and there are strong individual differences in the time of its occurrence in relation to the activity cycle. An additional piece of evidence comes from a pilot experiment in which six birds were put on a regime in which the lights in their cages were off for 5 min and on, at a constant level, for 25 min in each half-hour. This synchronised feeding which tended to occur in the first 5 min after the lights came on ($\chi_r^2 = 11.1$, $P < 0.05$, using a Friedman test between the five successive 5-min blocks when the lights were on). No significant effects were discovered in seven other behaviour patterns examined, including locomotion.

In contrast to feeding, some other behaviours do seem to fit into the cycles of locomotion at specific points. Both sequence analysis on birds in constant conditions (Slater and Ollason, 1972), and the results of birds which synchronised well with imposed cycles (bird 6 in Fig. 4; birds 5 and 6 in Fig. 6) suggest that this is true of song, preening and resting. It is possible that the oscillations in these behaviours are separate but synchronised so that further experiments will show that they can be dissociated from each other. An alternative, and at present more likely, hypothesis is that the probability of each of them is affected by the value of a single underlying variable. Although this might be called arousal, this word has many meanings: of these, that to which the variable proposed here would be most akin would be that termed activation by Andrew (1974). On this model locomotion would be most prevalent when the variable was high, song when it was rather lower, preening when lower still and resting at the bottom of the continuum.

A merit of this activation model is that it fits not only data on short-term cycles but also that on daily changes in behaviour as well (Ollason and Slater, 1973). Zebra finches are considerably more active in the morning than in the afternoon. Not surprisingly, sitting still, both alert and drowsily, rises during the course of the day, and there is also a slight increase in preening. Song falls with locomotion. These changes can be accounted for by assuming activation to be lower in the afternoon than in the morning; thus behaviours high on the continuum decrease and those low down increase. As before, feeding fails to fit into the model as its level was found to remain fairly constant during the period of the day in which the other behaviours were changing most rapidly.

If an activation model is to fit the facts already known about the temporal pattern of zebra finch behaviour, the variable underlying it must have certain characteristics. Its effects on behaviour must be to set probabilities rather than to determine which act should occur. Zebra finches do not go through a fixed sequence from locomotion to song to preening to rest, and there is considerable temporal overlap between these categories. It is possible for a bird which is moving round the cage to suddenly stop and rest or preen; likewise a resting bird may suddenly sing or move off. These types of sequence are, however, unusual. More common are alternations between adjacent behaviours on the postulated continuum: locomotion with song, song with preening and preening with resting. The level of the variable is thus seen as making certain behaviours more likely than others, rather than dictating exactly which should be shown. More precise determination of which behaviours are shown when, will, of course, depend on factors specific to them as well as on activation level.

A second point is that activation level need not vary in a cyclical manner. It would be hard to justify postulating this type of influence were it not for the fact that some birds show cycles of behaviour. Nevertheless, activation may be an important factor in the behaviour of other individuals, but one which varies in a non-cyclical manner, making it more difficult to detect. Finally, it should be mentioned that even in those birds which do show cycles, activation cannot be thought of as varying sinusoidally. The decline from locomotion, through song and preening, to resting is a slow process but the change from resting to locomotion tends to take place rather abruptly, with little, if any, song and preening in between. This can be seen clearly in the plots shown in Fig. 6. The level of a single variable underlying these changes in behaviour would thus have to follow a saw-tooth pattern, rather like a relaxation oscillation, running slowly down and then resetting rapidly.

There have been many attempts to explain changes in behaviour in terms of a single arousal-like variable, and in some cases efforts have been made to relate this to measures of neurophysiological arousal, but often without much success (Hinde, 1970; Andrew, 1974). The present results do no more than suggest the existence of such a variable but, if this can be confirmed, it will be an important task to discover its neural or endocrine basis and how this influences the behaviour patterns which are affected by it.

CONCLUSION

The work discussed in this article has been largely descriptive. While this approach may raise only a limited number of possibilities for the causation of the patterning of behaviour discovered, it is difficult to choose between these without more extensive experimental work. It is, nonetheless, certain at this stage that many different mechanisms underlie the temporal pattern of behaviour so that no single analysis can hope to elucidate all of them. At the simplest, one behaviour pattern may stimulate another in a more or less reflex-like manner; for example, it is presumably for this reason that bill-wiping follows drinking in zebra finches. But in most of the cases which have been discussed in this paper the mechanisms involved are more complicated, with each behaviour being influenced by a number of factors, both internal and external, specific and non-specific, which change with time and as a result of feedback from performance. These causal factors must, presumably, combine to give the behaviour some rating, which must be compared with those for all other behaviours before a decision is taken as to which act will occur. How all this is done is a dauntingly complex problem. Behavioural work can define it, and perhaps scratch the surface of it, but its solution lies not only inside the animal but also a good distance in the future.

Acknowledgements

I am grateful to Mr. Peter Holden for collecting some of the data shown in Table I, and to Professor Richard Andrew and Mr. Peter Clifton for making helpful comments on an earlier draft of this paper. The research reported here is supported by a grant from the Science Research Council.

REFERENCES

Andrew, R.J. (1956) Normal and irrelevant toilet behaviour in *Emberiza* spp. *Brit. J. Anim. Behav.*, 4, 85—91.
Andrew, R.J. (1974) Arousal and the causation of behviour. *Behaviour*, 51, 135—165.
Baerends, G.P., Brouwer, R. and Waterbolk, H.Tj. (1955) Ethological studies on *Lebistes reticulatus* (Peters): I. An analysis of the male courtship pattern. *Behaviour*, 8, 249—334.
Crook, J.H. (1961) The basis of flock organisation in birds. In W.H. Thorpe and O.L. Zangwill (Eds.), *Current Problems in Animal Behaviour*. Cambridge University Press, Cambridge, pp. 125—149.
Duncan, I.J.H. and Wood-Gush, D.G.M. (1972) An analysis of displacement preening in the domestic fowl. *Anim. Behav.*, 20, 68—71.
Duncan, I.J.H., Horne, A.R., Hughes, B.O. and Wood-Gush, D.G.M. (1970) The pattern of food intake in female Brown Leghorn fowls as recorded in a Skinner box. *Anim. Behav.*, 18, 245—255.
Fentress, J.C. (1972) Development and patterning of movement sequences in inbred mice. In J.A. Kiger (Ed.), *The Biology of Behavior*. Oregon State University Press, Corvallis, Oreg., pp. 83—131.
Fentress, J.C. (1973) Specific and nonspecific factors in the causation of behavior. In P.P.G. Bateson and P.H. Klopfer (Eds.), *Perspectives in Ethology*. Plenum Press, New York, pp. 155—224.
Heiligenberg, W. (1973) Random processes describing the occurrence of behavioural patterns in cichlid fish. *Anim. Behav.*, 21, 169—182.
Hinde, R.A. (1970) *Animal Behaviour, 2nd Ed.* McGraw-Hill, New York.
Iersel, J.J.A. van and Bol, A.C.A. (1958) Preening of two tern species. A study on displacement activities. *Behaviour*, 13, 1—88.
Lemon, R.E. and Chatfield, C. (1973) Organization of song of rose-breasted grosbeaks. *Anim. Behav.*, 21, 28—44.
McFarland, D.J. (1966) On the causal and functional significance of displacement activities. *Z. Tierpsychol.*, 23, 217—235.
McFarland, D.J. (1969) Mechanisms of behavioural disinhibition. *Anim. Behav.*, 17, 238—242.
McFarland, D.J. (1970) Adjunctive behaviour in feeding and drinking situations. *Rev. Comp. Animal*, 4, 64—73.
Ollason, J.C. and Slater, P.J.B. (1973) Changes in the behaviour of the male zebra finch during a twelve-hour day. *Anim. Behav.*, 21, 191—196.

Richter, C.P. (1927) Animal behavior and internal drives. *Quart. Rev. Biol.*, 2, 307—342.

Rowell, C.H.F. (1961) Displacement grooming in the chaffinch. *Anim. Behav.*, 9, 38—63.

Sevenster, P. (1961) A causal analysis of a displacement acitivity (fanning in *Gasterosteus aculeatus* L.). *Behaviour*, Suppl. 9, 1—170.

Slater, P.J.B. (1973) Describing sequences of behavior. In P.P.G. Bateson and P.H. Klopfer (Eds.), *Perspectives in Ethology*. Plenum Press, New York, pp. 131—153.

Slater, P.J.B. (1974a) Bouts and gaps in the behaviour of zebra finches, with special reference to preening. *Rev. Comp. Animal*, 8, 47—61.

Slater, P.J.B. (1974b) The temporal pattern of feeding in the zebra finch. *Anim. Behav.*, 22, 506—515.

Slater, P.J.B. and Ollason, J.C. (1972) The temporal pattern of behaviour in isolated male zebra finches: transition analysis. *Behaviour*, 42, 248—269.

Sollberger, A. (1965) *Biological Rhythm Research*. Elsevier, Amsterdam.

Tinbergen, N. (1951) *The Study of Instinct*. Oxford University Press, London.

Tinbergen, N. (1952) Derived activities: their causation, biological significance, origin and emancipation during evolution. *Quart. Rev. Biol.*, 27, 1—32.

Weiss, B., Laties, V.G., Seigel, L. and Goldstein, D. (1966) A computer analysis of serial interactions in spaced responding. *J. exp. Anal. Behav.*, 9, 619—626.

Wells, G.P. (1950) Spontaneous activity cycles in polychaete worms. *Symp. Soc. exp. Biol.*, 4, 127—142.

3

NEST CONSTRUCTION BY THE DOMESTIC HEN: SOME COMPARATIVE AND PHYSIOLOGICAL CONSIDERATIONS

D.G.M. WOOD-GUSH

Agricultural Research Council's Poultry Research Centre, Edinburgh (Great Britain)

In this chapter it is proposed to describe and discuss nesting building by the domestic hen and her behaviour up to and including oviposition, and to compare it with what is known about closely related species and with some other species which construct similar sorts of nests. Although there are many general descriptions of nests of different species few of these are quantitative so that assessment of the stimuli governing nest site selection is impossible and few contain descriptions of the motor patterns involved in the construction of the nest or accurate timing of the behavioural patterns in relation to ovarian events.

THE NESTING BUILDING BY THE DOMESTIC HEN

Biological considerations

The use of the domestic hen for a study of nest-building might seem to be a rather odd choice of species, but the hen possesses several advantages not usually considered: more is known about its physiology than is the case of any other bird, and domestication has bestowed another interesting dimension to aid the understanding of the physiological factors underlying its behaviour. For domestication constitutes a process of micro-evolution in which some changes are occurring rapidly before us as in the case of quick-breeding species such as the domestic fowl. In examining its behaviour one is sometimes aware of how some behaviour patterns which seem functionally useless can be explained by consideration of their evolutionary history and how, presumably through relaxed selection, the components of other behaviour patterns stand out under domestic conditions and show their relationships in a way which would be much more difficult in a wild species which could only survive if all these components were coordinated into a seemingly unitary behaviour pattern. In this respect the nesting behaviour of the domestic fowl constitutes a 'partially dissected' behaviour pattern with some components still very similar to other gallinaceous species and with some elements uncoordinated with others and seemingly non-functional.

Economic considerations

Furthermore, the subject of nesting behaviour in the hen has considerable economic implications. In the context of British agriculture, which has a larger production than that of Canada or of Australia and New Zealand combined, the poultry industry is one of the largest animal enterprises; its income amounting to £ 347 million in 1971—72. Revenue from eggs constitutes a major part of this sum and so a small increase in the egg production or in the efficiency of harvesting eggs involves very large sums of money.

Among the practical factors connected with the nesting behaviour of the hen is its close physiological relationship with ovulation. A study by Wood-Gush and Gilbert (1970) revealed that many hens may, on occasion, show nesting behaviour without oviposition and that in these cases the hen has ovulated and the egg has not entered the oviduct but dropped into the body cavity and been reabsorbed there. Thus nesting frequency is a better index of ovarian activity than is egg production, and the study pin-points a source of egg loss which may involve 5—20% of the hen's total ovulations. Moreover, nesting behaviour has at least two other agricultural applications. First, although the majority of commercial birds are kept in cages, some very large commercial egg suppliers, as opposed to stock suppliers, keep their hens on deep litter and in their case failure to lay in the nests leads to dirty eggs and expensive time-consuming egg collection. Furthermore, some pedigree birds used as the genetic foundation of the stocks used by the commercial egg producer are often kept in pens in which individual records are needed so that knowledge about nest selection and usage is required to prevent wastage through laying on the floor. Second, in the case of birds kept in battery cages there is the welfare problem of how birds adapt their behaviour to a very restricted environment (Wood-Gush, 1972).

In addition to the general aims outlined earlier, the experiments reported here were directed towards specific questions. Does a hen use the same nest for successive eggs within a sequence? If so, does she elaborate on the previous nest? If she does use the same nest for successive eggs and sequences of eggs, is it the site that governs her usage of the same nest or the nest itself and its contents? Observations were also made on early incubation behaviour and these, together with some preliminary results from a hormone experiment, are discussed in relation to ovulation rates in the hen.

Material and methods

Ten hens from a modern commercial strain of Rhode Island Red breed origin were used for the descriptive studies on nest-building and early incuba-

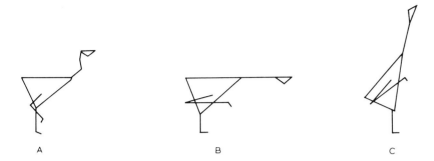

Fig. 1. Nest site selection. A: posture of a hen when normally walking and not motivated to nest. B: posture when fixating on nest site at onset of nesting behaviour. C: posture when 'examining' the walls of the pen.

tion behaviour, while the hormone work involved six Brown Leghorn hens. The Rhode Island Red females had all been kept in pens without trap-nests and selected as potential nest-builders by their behaviour at times near to oviposition. Once selected, they were kept in individual cages and transferred singly to a pen for observations on their nesting behaviour. The pens were 2.4 m square with solid walls up to a height of 1.3 m and wire mesh above. They were well lit and had litter consisting of wood-shavings on the floor. On the litter were scattered a number of white shaft feathers. Food and water were available in the pens and home cages, and the birds were kept on a constant 14-h day. Observations were made from the aisle outside the pen using a tape recorder which did not seem to disturb the birds. They were palpated each morning to test for the presence of an egg and a bird that was to be observed was put into the observation pen, if possible, several hours before she was due to lay, and for the nest construction observations, she was watched until half-an-hour after laying.

Nest site selection

The onset of nest site selection under these conditions was often heralded by vigorous wing-flapping and the bird became restless and gave the nesting call. Very often she 'examined' the walls keeping the keel up very high and appearing to wish to fly away (see Fig. 1). As time passed she tended to pay more attention to the corners of the pen and on these occasions tended to fixate on a corner with her neck in a horizontal position and held straight as in Fig. 1. At this time as she walked she raised her feet high and gave the impression of moving with stealth. At the beginning of nest site selection, the hen may stretch her neck and fixate only for a few seconds at a time and in some birds full fixation takes place only when she is very near the nest.

Nest construction

Having chosen a corner, the hen generally half crouches in the corner, and rotates herself, pushing her feet out sideways. She then lowers herself so that the keel forms an angle with the floor so that the hind part of the body is raised and the chest touches the ground. One circular movement of this sort amounting to a turn of 90° or more is noted as a rotation in Table I and two smaller rotations following upon one another without any intervening behaviour were counted as one rotation. When stationary the hen picks up litter or feathers and either throws them over one shoulder or places the litter or feather on her back. At other times she drops it at her chest or rakes the material towards her; one of these movements is recorded as a gathering in Table I.

It can be seen in the Table that the time taken to build the nest varied considerably between birds and for the building of different nests by the same bird, but most of the short times are due to the birds already being motivated by the time they were put into the pen, and therefore these observations are incomplete.

TABLE I

Summary of motor patterns used and time taken to build nests by seven hens

Bird	No. of gatherings		No. of rotations	Time taken	Remarks
	On nest	Off nest			
PM	513	157	117	2 h 47 min	
	574	61	201	4 h 15 min	Two nests built
	302	138	60	1 h 20 min	
PB1	153	4	72	1 h 42 min	
	140	2	15	46 min	
	120	2	5	1 h 18 min	
B	144	62	0	32 min	
	166	79	22	1 h	
RP	197	13	23	54 min	Some dust-bathing
	416	83	9	53 min	
R	89	27	18	24 min	Made two nests
	29	3	3	1 h 17 min	
W	190	13	45	1 h 15 min	
	199	31	23	58 min	
	168	59	2	39 min	
B1	196	59	47	1 h 48 min	
	107	29	11	45 min	
	223	7	12	1 h 9 min	

On one occasion dust-bathing occurred during a nest-building period, but generally nest-building was interrupted only by short forays off the nest when the hen might gather more nest material. When gathering off the nest the nesting material, either litter or a feather, is usually thrown over one shoulder but sometimes is placed on the back in which case it is carried back to the nest where it drops off during a rotation. Never was the hen seen to carry anything in her beak nor was she seen to deliberately remove material from her back. Gathering in many cases continues after oviposition for several minutes. At this time a few hens sometimes partially destroyed their nests by standing facing the nest and gathering material already on it and throwing it over a shoulder away from the nest. Two hens are noted in Table I as having built two nests for a single oviposition and others have also been seen to do this.

A finished nest is shown in Fig. 2. It can be seen to be round with a rim of litter caused by the outward movements of the legs during a rotation and by the chest during the same movement. The feathers on the rim in the nest were dropped into position during rotation and gathering movements.

The use of the nest for subsequent eggs

In the case of a wild bird, unless it is parasitic, successive eggs of a clutch will be laid within the same nest, and although domestic hens tend to use the same trap-nest (Wood-Gush, 1954) this experiment was designed to find out whether they would do so when given the chance to construct nests, and if

Fig. 2. A nest built by a hen from wood shavings and feathers.

they did use the same nest whether they would add nesting material to it. Finally the question was raised as to whether the hen was attracted to the site or the nest for later eggs within the sequence.

The same hens and pen were used as hitherto. Each hen was allowed to lay two successive eggs in the pen and as nests formed in the litter are very easily broken the position of each nest was plotted on a map of the pen. Half-an-hour after laying the hen was returned to her home cage until the next day for the laying of the second egg. The positions of the nests used on successive days were compared and if the degree of overlap was 80% or more they were deemed to be one and the same nest. This allowed for moving of the pliable walls, which might easily be shifted during rotations, and for the error of measuring the position of the nest, for judging the exact start of the inner rim of a wall made of litter added a further possible error. If a hen used the same nest for a pair of successive eggs she was tested for other pairs of ovipositions until she showed that she had used the same nests for laying six successive pairs of eggs which meant that she had shown a probability of 1/32 of using the same nest for ovipositing the eggs of a clutch or sequence. She was discontinued when she used two different nests for two succesive eggs. Furthermore, in order to test whether her choice of the nest would be influenced by the presence of an egg in the nest, she was tested for half the observations with the first egg of a pair left in the nest and for half the observations with the first egg removed. The order of this arrangement was randomised, some birds, therefore, having the egg removed for the first pair of observations and some having it present.

For the birds that tended to use the same nest for the second eggs as for the first egg, the question was then asked whether the hen was really using the same nest or whether she was using the same site. To answer this, a board

TABLE II

Use of previous days nest: + shows its use

Bird	With egg	Without egg
PM	— —	+
PB1*	+++	+++
R	+ —	++
W*	+++	+++
B1	— + —	—
RP*	+++	+++
B*	+++	+++
M*	+++	+++
MW*	+++	+++
MB1	+	++ —

* Shows $P < 0.05$.

41

TABLE III

Addition of feathers to previous nest

Bird	Added		Not added	
	Nest		Nest	
	With egg	Without egg	With egg	Without egg
PB1	+	++	—	—
W	++	++	—	—
RP	++	+	—	— —
B	++	+++	—	
M	+	+++	—	
MW	+	+++	— —	
Total	9	14	7	4

about a metre square was placed under the litter in the corner usually used by the hen, and after she had made the nest for the first egg and laid in it, she was, as usual, removed from the pen to her home cage half-an-hour after laying and the nest, resting on the hidden board was moved 50 cm away from the corner and litter placed in the position where the board had been. The next day the hen was returned to the pen and allowed to lay where she chose.

Results

As can be seen in Table II, a number of birds tended to use the same nest for both ovipositions in a sequence, and did so for six pairs of observations ($P < 0.04$). The presence of the first egg of a pair did not affect the behaviour of the birds. Also it can be seen (Table III) that they sometimes added nesting material (feathers) to the nest for the second oviposition of a pair. Fig. 3 shows the nest used for two successive ovipositions within a sequence in which there is only slight change in the position of the nest walls over the two days. Fig. 4 shows an example of two different nests being used and one in which the egg from the first oviposition was rolled by the hen to the new nest.

Of the six birds which were tested to see whether they were attracted to the nest or the site, only one oviposited in her old nest on its new site, 50 cm from its former site. The other five all built nests on the old site and two of the five rolled the previous day's egg to the second nest. One of these birds was allowed to lay in the nest three times before it was moved and she rolled two of the three eggs to the new nest on the old site.

42

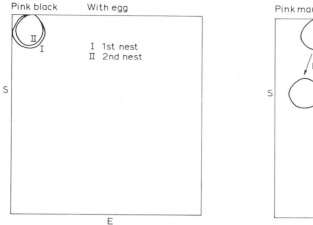
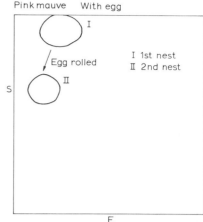

Fig. 3. An example of the same nest site being used for two consecutive eggs within a sequence or clutch.

Fig. 4. An example of two different sites being used for two consecutive eggs within a sequence or clutch.

The behaviour of the domestic hen in relation to the egg

All but two of the hens used in this study showed interest in their eggs after oviposition and some did so even before laying. These hens after laying usually rolled the egg to a position in which it would be in contact with the breast when she sat. Sometimes in order to do this the head was completely turned so that the comb was on the ground and set at right-angles to the neck. At other times, if the egg was in front of the hen it was rolled towards the breast by placing the bill over the egg. Most hens did this to the previous day's egg when they entered an existing nest, and some went through the complicated motions suggestive of egg-rolling shortly before laying in the absence of the previous egg. On one occasion the egg was rolled 1.5 m, and on another two eggs were each rolled 50 cm.

Some physiological considerations

Some of the physiological factors connected with nesting behaviour have been studied in birds kept under semi-intensive conditions in which the hens live in pens equipped with trap-nests. The behaviour under these conditions has been described in detail by Wood-Gush (1963): the hen before laying shows restlessness and gives a particular type of call which can be produced in young immature females by oestradiol monobenzoate (Wood-Gush and Gilbert, 1969). She then examines a number of nests, enters one and sits.

Some time after oviposition she usually cackles and stands waiting to be released from the trap-nest. Nest-examination, nest-entry and sitting are under the control of the postovulatory follicle (Wood-Gush and Gilbert, 1964) and as these behaviour patterns can be reproduced in ovariectomised hens by means of oestrogen and progesterone injections, it seems likely that the postovulatory follicle acts hormonally. A striking feature of the behaviour is that the stimulus for it generally occurs more than 24 h before the behaviour is seen, for the ovulation giving rise to the postovulatory follicle is generally rather more than 24 h before oviposition, and excision of the follicle shortly after ovulation destroys or disrupts the following nesting.

In these experiments, the birds which were accustomed to trap-nests were not seen to construct nests. However, a notable exception was seen: a Brown Leghorn hen that had been ovariectomised as an adult hen with much experience in the use of trap-nests constructed a nest. At the time she was seen to build the nest she had been receiving subcutaneous injections of oestradiol benzoate at an average dosage of 2 mg/day for 18 weeks and for the 4 weeks immediately before had received additional daily injections of 2 mg progesterone. Two weeks after the observation she was killed and the postmortem examination revealed no sign of original ovarian tissue. There was a small piece of right gonad present, about 5 mm long, and the oviduct was well oestrogenised.

Unfortunately the behaviour was completely unexpected and so no quantification was made; but for 2 h 10 min she performed the nest-building

Fig. 5. The nest built by a bilaterally ovariectomised hen treated with oestrogen and progesterone.

movements described earlier with rather more gathering off the nest than was seen in the birds described earlier. The nest built by her is shown in Fig. 5 and can be compared with the nest in Fig. 2 built by a normal hen; no major differences can be seen. The experiment has not been repeated as yet.

THE NESTING BEHAVIOUR OF OTHER GALLINACEOUS BIRDS

This subject has attracted little attention apart from brief descriptions of the nests in the general ornithological literature. One of the few studies to rise above the anecdotal is that by Jenkins *et al.* (1963) on red grouse nests, and although it contains little information on nesting behaviour, it does discuss some factors involved in nest site selection. One hundred and sixty-three nests were studied, nearly all were in heather but occasional ones were in other vegetation. Generally they were sheltered and overhung by foliage, but 17% were open above. More than three-quarters were on hard, well-drained ground, and about two-thirds were on flat, as opposed to sloping, ground. One was found on a farm road used by tractors and another was just 20 cm off a road used by motor traffic; eggs were hatched from both of these. The nests studied were mostly shallow scrapes with shallow linings of pieces of moorland vegetation. In a later paper Watson and Jenkins (1964) state that the hen may make nesting scrapes for up to a fortnight before laying and that cocks may also make scrapes. The authors state that one captive grouse hen laid eggs in a nest without a cock being present. However they do not state whether she could hear or see a cock in the vicinity. The hen usually covers its eggs with vegetation when the clutch nears completion. However, this behaviour is variable; some nests being so covered that the eggs are invisible while in others the eggs are left uncovered. The motor patterns used are not described.

In the case of the ptarmigan, *Lagopus mutus*, a little more information is available (Watson, 1972) on the motor patterns. The cock starts making scrapes 2 weeks before the hen lays. One male was seen to make 15 scrapes in an hour at noon, sitting in some for a few seconds or minutes. The hen of this pair apparently did not sit in any of the scrapes and laid 6 days later. Generally, nearer oviposition the hen makes scrapes and sits in them for several minutes at a time with the cock alongside, and in this respect the behaviour is more reminiscent of the domestic fowl under range conditions in which the cock may lead the hen to nest sites. The nest is a natural hollow which is scantily lined with vegetation or a few feathers. The nests are usually in the open, sheltered by a large stone and the scrape may be enlarged during egg-laying and incubation. After laying the hen covers her eggs partly or completely with loose grass, lichen, moss or other vegetation (Watson, 1972). She does this shortly after laying an egg or while still sitting, by placing the material in the nest on each side of her (MacDonald, 1970), and in this respect the behaviour has some resemblance to some of the gathering movements of the domestic hen.

The laying by japanese quail hens, *Coturnix coturnix japonica*, which is a domesticated species, in open range conditions has been very briefly described. Stevens (1961) states that the hen does very little nest construction. She lays her first egg in a shallow, scratched-out depression and adds a little dead vegetation at each oviposition. Feathers or down were not seen by him in the nests. In another population described by Rothstein (1967), no nests were found and the hens merely laid their eggs in three areas. Some examination of these sites was seen, but he does not describe the behaviour in any detail.

DISCUSSION

From the experiments on hens it can be seen that some modern strains of domestic hens at least do indulge in long periods of nest-building and build nests which are used by some birds for successive eggs within a sequence.

Some of the movements used by the hen in constructing her nest are similar to some used by other birds whose nests are primarily scrapes. Hinde (1958) describes a movement in the domestic canary which he calls turning and which is very similar to what I have called rotating. In his stimulating and thorough study of the black-headed gull, Beer (1963) described sideways building which is very similar to gathering on the nest described earlier. In making or increasing the hollow for its nest the black-headed gull performs backward scraping movements of the feet unlike those of the domestic hen which are made outwards, but further observations on the hen on different types of ground are necessary before this dissimilarity is accepted. In both species, however, the chest is kept down and the hind part of the body raised. The black-headed gull, unlike the canary and many other species, does not have a discrete period of nest-building but like the domestic hen is liable to indulge in nest-building over a long period of the reproductive cycle.

In one experiment Beer (1963) was able to transfer nests of black-headed gulls from one position to another in stages without upsetting the birds. However, a strict comparison with the experiments described above is difficult for he does not say how far he moved the nest at any one time.

In the literature there appears to be no information on nest-building by hens on open range, and McBride *et al.* (1969) writing on the Queensland feral domestic fowl gives little information on nesting behaviour except to say that the cock might escort the hen to the potential nest sites after she had given the nesting call. Having accepted a nest site the hen stands within 1 m of the nest cackling and throwing twigs and leaves onto her back, these fall off when she sits and this completes the simple nest. While sitting she might throw twigs toward the nest and on her back. Kruijt (1964) in his comprehensive study of the social behaviour of the Burmese red jungle fowl,

(*Gallus gallus spadiceus*), gives no information on their nest site selection behaviour or on the construction of the nest and Collias and Collias (1967) confine themselves to the description of one nest.

The literature indicates that the nests of other gallinaceous birds are no more elaborate than those of the domestic hen and that nest material may be added after oviposition and the eggs covered. From the very brief description of nest-building by the ptarmigan it seems that the motor patterns involved are likely to be the same as in the domestic hen. In this group of species with poorly constructed nests it would be of interest to know more about nest site selection. From the literature it is obvious that some nest sites are not well hidden or are maladaptive in other respects; one would therefore like to know more about this component of nesting behaviour, how much variation there is in the time taken over nest site selection and in the intensity of the behaviour shown by individuals in different species, and the stimuli governing nest site selection in specific cases. In addition to learning more about the motor patterns involved in nest construction it would be of interest to know the timing of events in relation to oviposition and ovulation, as well as the relative dependence on stimulation from external stimuli such as the mate and features of the nest.

Finally, in considering the behaviour of the domestic hen in comparison to other related species it would be necessary to study her nesting behaviour under a wide variety of environments.

Two points of particular interest arise in the nesting behaviour of the hen. The first concerns the nest site. Its selection takes a relatively long time and once selected, the site appears to be more important than the nest itself. In a group such as the Pheasianidae, in which the nest is of scanty construction and relatively open to predators, success in hatching may be largely dependent on the cryptic quality of the nest and, although bizarre choices may be found, the choice of site may be of such importance that when the nest is moved the bird chooses to return to the site rather than the nest.

The second interesting feature of nest-building behaviour in the domestic hen is the fact that this behaviour can be independent of the cock, and it would be of interest to know how widespread this is in gallinaceous birds. In birds such as the ring dove female there is a step-wise interaction between external stimuli and succeeding hormones (Lehrman, 1965), but in the domestic fowl it seems that given sufficient oestrogen and progesterone the behaviour will appear and no step-wise interactions are necessary. In fact some elements of nest-building may be seen in the case of certain birds even in battery cages, where relevant stimuli from the nest are minimal. It would be interesting to know whether the relative independence from the cock in nest-building is to be found in other gallinaceous females. Occasional nest-building by ring dove females in the presence of castrated males which do not show the normal behaviour has been seen (Erickson, personal communication) and so it is possible that the propensity to build nests with a weaker external stimulus than is usually thought necessary may be fairly widespread, and not a mere aberration due to domestication.

The observations on the ovariectomised hen suggest that nest-building is dependent on progesterone and oestrogen, and this is supported by the fact that they can induce nest-examination and nest-entry in birds having trap-nests at their disposal (Wood-Gush and Gilbert, 1973). However, it is as yet not certain whether the behaviour is dependent on the synergistic action of the two hormones or whether one alone would suffice, nor is it known whether the behaviour pattern differs in any quantitative way from that seen in intact birds. A few intact Brown Leghorns have been observed and their behaviour has been less complete than that of intact birds described here, and the ovariectomised Brown Leghorn, for their nests are very rudimentary compared to them. Therefore, it seems as if the hormones at the dosage given exaggerated the normal behaviour of the Brown Leghorn bird and indeed this bird had been under continual observation for a year and had not shown any nest-building behaviour in the pens nor was it certain how much nest site selection behaviour had been shown. A number of questions, therefore, have yet to be answered.

In comparing the hen with wild gallinaceous birds, the timing of nesting behaviour in relation to ovulation would be of interest. In the hen, nesting behaviour, including nest-construction, occurs in the few hours before oviposition even in the case of the first egg of a clutch. However, in the red grouse and possibly in the ptarmigan, it seems that nesting behaviour occurs days, rather than hours, before oviposition of the first egg of a clutch. The question then arises as to whether the same hormones play the same roles in the two species or whether the behaviour is governed differently. In young pullets approaching their first oviposition, nest calling, which is under oestrogen control, is sometimes heard several days before the first ovulation, and this suggests that oestrogen may possibly play a larger role in nesting behaviour in the other species, but it does not preclude the possibility that progesterone is released earlier relative to oviposition in these species than in the domestic hen.

The presence of egg-rolling behaviour may have important implications for egg production for if it is equated with early incubation behaviour and if incubation in the domestic hen is caused by prolactin, as indicated by Riddle et al. (1935), then the possibility that prolactin is present at the time of oviposition in this strain of birds must be considered. In mammals prolactin release may be caused by oxytocin (Cowey, 1969), and while arginine-vasotocin and oxytocin are probably the hen's posterior pituitary hormones (Gilbert, 1971b), oxytocin levels are highest just before laying (Sturkie and Lin, 1966), and thus it is possible for prolactin to be present at the time that egg-rolling is seen. On the other hand, early incubation in the hen may, like that of the ring dove, be under progesterone control. Using radioimmuno-assay techniques Cunningham and Furr (1972) studied the progesterone lev-

48

els in birds of two strains including the one used in this study and found a peak at 4—7 h before ovulation. This was confirmed by Wilson and Sharp (1973) who found a peak in luteinising hormone levels in three birds of this breed 4—5 h before the oviposition of the previously ovulated egg. Allowing a 2-h lapse between progesterone activity and luteinising hormone release (Gilbert, 1971a) this would indicate a peak of progesterone at a similar time. Hence a progesterone peak is present in the circulating blood at about the right time for it to trigger egg-rolling.

Egg-rolling behaviour has been found in two of three strains of laying birds kept at the Poultry Research Centre, and it is of practical interest to note that both these strains are relatively poor layers compared to the third strain in which the egg-rolling behaviour has not been found up to now. If prolactin does indeed control the behaviour then it might account for the relatively poor rate of egg production in these two strains for it suppresses gonadal activity (Lofts and Marshall, 1956; Lake and Furr, 1971). On the other hand, if the egg-rolling behaviour is due to progesterone then the difference between the strains may be due to:

(1) Neural threshold differences to progesterone.

(2) Differences in circulating progesterone levels.

(3) Differences in timing of the progesterone peaks in relation to the ovulatory cycle.

(4) Absence of the necessary neural pathways.

Since egg-rolling is associated with poor producers, at least at this laboratory, this may indicate that the output of gonadotrophins is faulty in some way, and hence also steroid production.

REFERENCES

Beer, C.G. (1963) Incubation and nest building behaviour of black-headed gulls III. The pre-laying period. *Behaviour*, 21, 13—77.

Collias, N.E. and Collias, E.C. (1967) A field study of the red jungle fowl in North-Central India. *Condor*, 69, 360—386.

Cowey, A.T. (1969) General hormonal factors involved in lactogenesis. In M. Reynolds and S.T. Folley (Eds.) *Lactogenesis: The Initiation of Milk Secretion at Parturition*. University of Pennsylvania Press, Philadelphia, Pa., pp. 157—169.

Cunningham, F.J. and Furr, B.J.A. (1972) Plasma levels of luteinizing hormone and progesterone during the ovulatory cycle of the hen. In B.M. Freeman and P.E. Lake (Eds.), *Egg Formation and Production Poultry Science Symposium 8*. British Poultry Science Ltd., Edinburgh, pp. 51—64.

Gilbert, A.B. (1971a) Control of ovulation. In D.J. Bell and B.M. Freeman (Eds.), *Physiology and Biochemistry of the Domestic Fowl*. Academic Press, New York, pp. 1225—1236.

Gilbert, A.B. (1971b) Transport of the egg through the oviduct and oviposition. In D.J. Bell and B.M. Freeman (Eds.), *Physiology and Biochemistry of the Domestic Fowl*. Academic Press, New York, pp. 1345—1350.

Hinde, R.A. (1958) The nest-building behaviour of domesticated canaries. *Proc. zool. Soc. Lond.*, 131, 1—48.

Jenkins, D., Watson, A. and Miller, G.R. (1963) Population studies on red grouse *Lagopus lagopus scotius* (Lath), in North-East Scotland. *J. Anim. Ecol.*, 32, 317—376.

Kruijt, J.P. (1964) *Ontogeny of Social Behaviour in Burmese Red Jungle fowl (Gallus gallus spadiceus)*. Brill, Leiden pp. 1—196.

Lake, P.E. and Furr, B.J.A. (1971) The endocrine testes in reproduction. In D.J. Bell and B.M. Freeman (Eds.), *Physiology and Biochemistry of the Domestic Fowl*. Academic Press, New York, pp. 1469—1484.

Lehrman, D.S. (1965). Interaction between internal and external environments in the regulations of the reproductive cycle of the ring dove. In F.A. Beach, (Ed.), *Sex and Behaviour*. Wiley, New York, pp. 355—380.

Lofts, B. and Marshall, A.J. (1956) The effects of prolactin administration on the internal rhythm of reproduction in male birds. *J. Endocr.*, 13, 101—106.

MacDonald, S.D. (1970) The breeding behaviour of the rock ptarmigan. *Living Bird*, 9, 195—238.

McBride, G., Parer, I.P. and Foenander, F. (1969) The social organization and behaviour of the feral domestic fowl. *Anim. Behav., Monogr.*, 2, 127—181.

Riddle, O., Bates, R.W. and Lahr, E.L. (1935) Prolactin induces broodiness in fowl. *Amer. J. Physiol.*, 111, 352—360.

Rothstein, R. (1967) Some observations on the nesting behaviour of japanese quail (*Coturnix coturnix japonica*) in pseudo-natural conditions. *Poultry Sci.*, 46, 260—262.

Stevens, V.C. (1961) Experimental study of nesting by coturnix quail. *J. Wildlife Management*, 25, 99—101.

Sturkie, P.D. and Lin, Y. (1966) Release of vasotocin and oviposition in the hen. *J. Endocr.*, 35, 325—326.

Watson, A. (1972) The behaviour of the ptarmigan. *Brit. Birds*, 65, 6—26 and 93—117.

Watson, A. and Jenkins, D. (1964) Notes on the behaviour of the red grouse. *Brit. Birds*, 57, 137—165.

Wilson, S.C. and Sharp, P.J. (1973) Variations in plasma LH levels during the ovulatory cycle of the hen, *Gallus domesticus. J. Reprod. Fert.*, 35, 561—564.

Wood-Gush, D.G.M. (1954) Observations on the nesting habits of Brown Leghorn hens. *Section Paper 10th World Poultry Congress, Edinburgh*, Dept. Agric., Scotland, pp. 187—192.

Wood-Gush, D.G.M. (1963) The control of the nesting behaviour of the domestic hen. I. The role of the oviduct. *Anim. Behav.*, 11, 293—299.

Wood-Gush, D.G.M. (1972) Strain differences in response to suboptimal stimuli in the fowl. *Anim. Behav.*, 20, 72—76.

Wood-Gush, D.G.M. and Gilbert, A.B. (1964) The control of the nesting behaviour of the domestic hen. II. The role of the ovary. *Anim. Behav.*, 12, 451—453.

Wood-Gush, D.G.M. and Gilbert, A.B. (1969) Oestrogen and the pre-laying behaviour of the domestic hen. *Anim. Behav.*, 17, 586—589.

Wood-Gush, D.G.M. and Gilbert, A.B. (1970) The rate of egg loss through internal laying. *Brit. Poultry Sci.*, 11, 161—164.

Wood-Gush, D.G.M. and Gilbert, A.B. (1973) Some hormones involved in the nesting behaviour of hens. *Anim. Behav.*, 21, 98—103.

4

INTEGRATION OF GONADOTROPHIN AND STEROID SECRETION, SPERMATOGENESIS AND BEHAVIOUR IN THE REPRODUCTIVE CYCLE OF MALE PIGEON SPECIES

R.K. MURTON and N.J. WESTWOOD

Institute of Terrestrial Ecology, N.E.R.C., Monks Wood Experimental Station, Huntingdon (Great Britain)

INTRODUCTION

Population gene pools contain an enormous amount of potential variability representing the past average fitness of the species. Sexual selection allows a male or female to choose a mate which is demonstrably the fittest under the existing ecological conditions and thereby to short-cut the more slowly acting effects of natural selection. Zahavi (1975) has formulated the 'handicap principle' which imagines that the secondary sexual handicap carried by an individual is a measure of its capacity to cope with its environment. However, there are genetical objections to this hypothesis as presently formulated (O'Donald, 1975) and the processes of sexual selection still remain obscure. It would be thought that a male bird which can afford to wear and display brilliantly coloured plumage must be very competent at avoiding predation; a bowerbird (Fam. Ptilonorhynchidae) which is able to collect large quantities of display objects, yet is still able to find food for itself, portrays evidence that it is a particularly efficient individual under existing conditions and so must be an ideal choice for any female. In like manner other behaviours related to mate selection are to be regarded as conventions which provide qualitative measures of the immediate fitness of the individual. Unfortunately, we know very little about the correlations that presumably exist between display plumage and other attributes contributing to fitness and the exact genetical mechanisms involved. They may help to explain the close interdependence of the hormone secretions which regulate gametogenesis and those mediating reproductive behaviour and the intimate morphological association of endocrine and germinal tissues.

Timing is the essence of successful reproduction. The physiological state of the sexes must be integrated for fertilisation of the egg and for parental duties and breeding must be accomplished at the most advantageous season ecologically. For the majority of avian species the seasonal variation in the daily photoperiod provides the most reliable proximate regulator of breeding periodicity. A profound dependence of all birds on the photoperiod follows from the fact that the alternation of day and night functions as a 24 h controlling oscillator, or *zeitgeber* which entrains endogenous rhythms of cell and body metabolism, feeding and activity cycles (Aschoff, 1965; Menaker, 1971; Pavlidis, 1973). It is now clear that the photoperiodic regulation of moult, migration and reproductive cycles is also achieved by the appropriate phasing of endogenous circadian and circannual rhythms of endocrine function and of tissue sensitivity (Lofts *et al.*, 1970; Follett, 1973; Meier and Dusseau, 1973; Murton, 1975). Thus the mediation of gametogenesis and of reproductive behaviour has to be considered in terms of circadian and circannual periodicities whose phase relationships are always an adaptive function of the environmental photoperiod.

REGULATION OF BREEDING PERIODICITY

Circadian rhythms of photosensitivity

The classical experiments of Hamner (1963 and 1964) demonstrated the involvement of circadian rhythm mechanisms in the seasonal testicular enlargement of the house finch, *Carpodacus mexicanus*, Hamner used ahemeral light cycles, *i.e.* cycles not equal to 24 h, in which the subject was first entrained to a short period of light (6 h) and then kept in constant dark during which time any circadian oscillators could free-run. The length of the dark period was subsequently varied in different groups to give cycle lengths ranging from 12 h to 72 h. Only cycles of 12, 36 and 60-h duration initiated spermatogenesis, and not cycles of 24, 48 or 72 h suggesting that the photo-induction of gonad growth is a periodic phenomenon which can only be induced at specific times during the dark period. This hypothesis was further tested by subjecting groups of birds to the previously non-inductive cycles of 24, 48 and 72 h but this time 1 h of light was given as a pulse at different times during the dark period. Fig. 1 summarises both of these experiments and shows how they are consistent with a rhythm of photosensitivity in which an oscillator is above a threshold during limited phases of the light cycle. So-called resonance experiments of this kind have also been performed with white-crowned sparrows, *Zonotrichia leucophrys gambelii* (Farner, 1965), juncos, *Junco hyemalis* (Wolfson, 1966) and japanese quail, *Coturnix coturnix japonica* (Follett and Sharp, 1969). In addition, it has been shown in quail that there is a periodic release of avian immunoreactive luteinising hormone (IR-LH Follett, see p. 66 below) under such lighting regimes (Follett and Davies, 1975).

Resonance experiments are strong evidence for the involvement of a circadian rhythm mechanism in avian photostimulated gonad growth. They

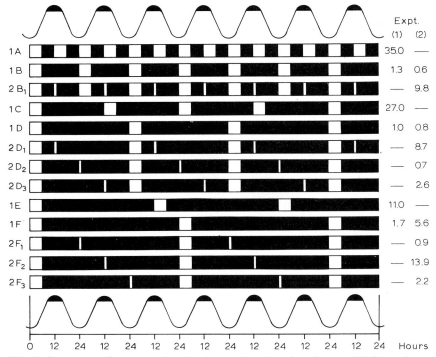

Fig. 1. Summary of two resonance type experiments using ahemeral light cycles performed by Hamner (1965) in which house finches were pulsed with bursts of light. In Experiment 1 pulses of 6-h light were combined with different intervals of dark (1A—1F). For Experiment 2, the light schedules 1B, 1D and 1F, which previously failed to evoke testicular growth were used as controls for experiments in which a further 1-h light pulse was given to interrupt the dark period ($2B_1$, $2D_1$, $2D_2$ . . . etc.). Mean testicular weights are given to the right of the light schedules which are shown as extending over seven days. The experiments actually lasted 20—54 days. An interpretation of the results is presented in terms of a light-sensitive oscillator which free-runs throughout the dark periods and whose sensitive phase is engaged by only certain of the light pulses.

suggest that the Bünning (1960) model of the photoperiodic clock is a valid concept. The explicit development of this hypothesis by Pittendrigh and Minis (1964 and 1971) is depicted in Fig. 2 to show how light supposedly performs a dual role. The strength of the day—night oscillator (*zeitgeber*) depends on the intensity and duration of the light period and this determines the phase of the light-sensitive circadian oscillator in the animal. Photoperiodic induction can occur if the photoinducible phase of the light-sensitive oscillator, imagined to be that part of an oscillator above some threshold, coincides with daylight. Since light is postulated to have these two simultaneous actions on any entrained oscillator there are methodological problems in distinguishing phasing and inductive effects.

Another class of experiments has involved interrupted or skeleton light

54

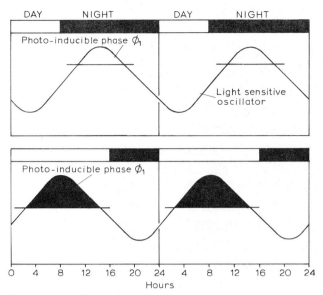

DAY   NIGHT   DAY   NIGHT

Photo-inducible phase ϕ_1

Light sensitive
oscillator

Photo-inducible phase ϕ_1

0   4   8   12 16 20 24   4   8   12 16 20 24
Hours

Fig. 2. A modification of the Pittendrigh model of the insect photoperiodic clock which has been used to explain light-induced gonad development in birds. The daylength (intensity of light) determines the strength of the day—night oscillator or *zeitgeber*. This is depicted as having a direct effect on the level (threshold) of the entrained oscillation and hence its frequency. The phase angle assumed during entrainment is a function of the frequencies of the controlling and controlled oscillators. When light coincides with the photoinducible phase a response can occur. In fact, current hypotheses consider that there is more than one light-sensitive driving oscillator, perhaps a coupled oscillator system, to which the oscillators controlling gonadotrophin secretion are in turn coupled, directly or indirectly. The system presumably allows the hormone rhythms to assume a varying phase relationship depending on the light intensity of the *zeitgeber*.

schedules given in a 24-h cycle (Fig. 3). Such experiments do not prove the involvement of circadian rhythm mechanisms in the control of gonad development, but they have been considered to define the limited phases of sensitivity to light within each 24-h cycle which are indicated by the resonance experiments just described and predicted by the Princeton model of the photoperiodic clock (Fig. 2). Fig. 3 refers to some rather limited experimental results obtained for the collared dove, *Streptopelia decaocto*, and, in addition, to some statistically valid data pertinent to house sparrows, *Passer domesticus*, to facilitate a discussion of the principles involved. The collared dove is an Indo-Himalayan species which underwent a remarkable range spread during the present century to reach N. Europe and Britain in the early 1950s (Fisher, 1953; Hudson, 1965). In Britain, it is mostly a commensal of man and is currently expanding into some urban habitats where it lives alongside the feral pigeon, *Columba livia* var. In India, the collared dove is the ecological equivalent of the African collared doves, including *Streptopelia*

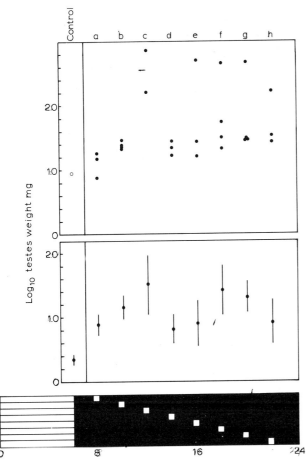

Fig. 3. Testicular development in photosensitive collared doves (upper panel) and house sparrows (lower panel) exposed for 7 and 20 days respectively to different asymmetric skeleton light schedules. The control condition at the start of the experiments is shown in the left-hand panel. House sparrow data based on Murton and Westwood, 1974.

roseogrisea from which the so-called Barbary dove* or ring dove, a much favoured laboratory subject mentioned again below, is a domesticated descendant. The collared doves contributing to Fig. 3 were wild caught from Ellesmere Port, Cheshire, in late December and on being laparotomised were found to have regressed testes. They were held for 27 days on the skeleton

* Whereas the authors correctly point out that the specific name of 'risoria' for the Barbary dove is not acceptable for this reason, it is widely used in the psychological literature, and for this reason we have not made the use of the specific name 'roseogrisea' a uniform practice throughout the volume. (Eds.)

56

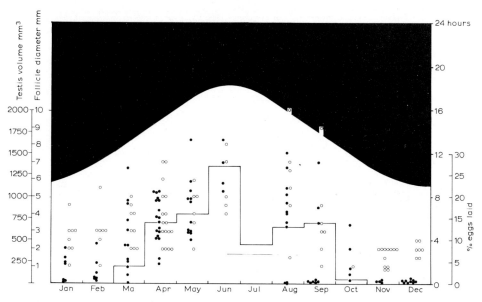

Fig. 4. The breeding season of the collared dove in England. The gonad cycle is based on birds collected at Ellesmere Port in 1970—71, with testicular volume (●) or diameter of largest follicle (○) being plotted for individual birds. Records of percentage of eggs laid throughout the year are based on a free-living population in Cambridgeshire in 1971—72. Total number of eggs examined was 187.

light schedules with the results depicted in Fig. 3. Fig. 4 details the gonad cycle of wild collared doves in Britain in relation to the egg-laying season and the seasonal variation in the day—night *zeitgeber*. The gonads appear to re-crudesce into breeding condition between February and March under a photoperiod of about 12 h (sunrise to sunset plus civil twilight). A skeleton simulation of a 12-h photoperiod was very effective in stimulating the only two subjects so tested, whereas 8-h and 10-h skeleton simulations were not very effective (Fig. 3).

Fig. 3 makes it clear that it was the distribution of the light pulses and not the total daylength which was important in stimulating photoinduced gonad growth. Similar results have now been obtained with several other species: house sparrow (Menaker, 1965; Murton et al., 1970; Murton and Westwood, 1974), tree sparrow, *Passer montanus* (Lofts and Lam, 1973), white-crowned sparrow (Farner, 1965), house finch (Hamner, 1968), japanese quail (Follett and Sharp, 1969) and wood pigeon, *Columba palumbus* (Murton, 1975). Moreover, using this type of experiment with greenfinches, *Carduelis chloris*, it was shown that immunoreactive LH, as measured by radioimmunoassay in the Bagshawe system (see p. 64 below), attained peak plasma titres under a light schedule that did not result in spermatogenetic development, although the interstitial Leydig cells were markedly stimulated;

spermatogenesis occurred with a different skeleton under which IR-LH levels were only slightly elevated. These results were considered to be evidence that different gonadotrophins (LH and follicle stimulating hormone (FSH)) were secreted at different phases of a circadian rhythm, perhaps being controlled by distinct circadian oscillators (Murton *et al.*, 1969 and 1970a). An alternative interpretation could be that separate components of a gonadotrophin complex are synthesised at different phases of a circadian rhythm.

Recently, it has been shown that the detailed daily activity rhythm of perch-hopping in the greenfinch and chaffinch, *Fringilla coelebs*, is best explained on the basis of two major components being involved in the pacemakers of the controlling circadian rhythms (Daan, 1975). Indeed, there are several reasons for supposing that a system of two coupled oscillators having an opposite dependence of their frequency on light intensity is likely (Pittendrigh, 1958 and 1974). If this is so, it becomes conceivable that LH and FSH could be linked to such a coupled pacemaker oscillator so that their phase relationships alter according to *zeitgeber* strength. There is evidence that testis growth in photoresponsive subjects is stimulated when the end of the daily activity cycle is pulsed by light (Hamner and Enright, 1967). Menaker and Eskin (1967) used green light to phase the daily activity rhythm of house sparrows without causing the induction of gonad growth. They then pulsed different phases of the rhythm with white light and only a pulse given at the end of the activity cycle resulted in gonad development. Gonad growth in photoresponsive subjects might depend primarily on FSH and the phase relationship with LH might not be too critical. This would partly explain why gonad recrudescence can occur under a wide range of photoperiods. But the complete seasonal reproductive cycle also involves the phenomenon of refractoriness to otherwise stimulatory daylengths and the range of photoperiods that will break refractoriness by allowing rehabilitation of the testis interstitium is much more critical (Hamner, 1968; Murton *et al.*, 1970b; Murton and Westwood, 1974).

Many species which have been photostimulated by skeleton light schedules exhibit a bimodal response in gonad growth of the kind seen in Fig. 3. This does not mean that there are two phases of photosensitivity in each 24-h cycle, as some workers have assumed. This is because such skeletons only simulate complete photoperiods up to the point where the duration between the beginning of the first pulse and end of the second is in the order of 14 to 16 h (Pittendrigh and Minis, 1964; Menaker, 1965). With light pulses simulating longer durations a phase jump occurs in the daily activity cycle and entrainment is to the start of the short light pulse. Pavlidis (1973) gives a mathematical explanation in terms of oscillator theory and Fig. 5 shows how the data in Fig. 3 should be interpreted. Considering the house sparrow records it can be shown that an 18-h skeleton is as effective as a 12-h skeleton because sparrows entrain to the start of the short light pulse thereby making an effective skeleton photoperiod of 12.5 h. This leaves us to decide between

58

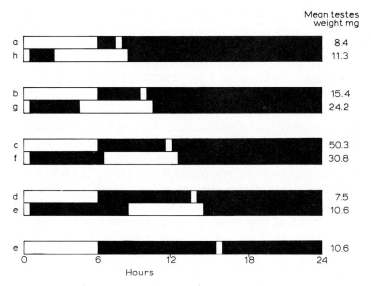

Fig. 5. Asymmetric skeleton schedules as used in Fig. 3 replotted to show probable manner of entrainment by house sparrows and collared doves. The beginning of the short light pulse serves as 'dawn' when the pulse is used to simulate complete photoperiods which are of about 14 h or longer duration. The mean testicular weights of the house sparrows are given to the right of the figure. Schedule e should have produced a better result than observed if entrainment had been to the start of the short light pulse, and so it is assumed that the cycle was interpreted as a long day.

two further interpretations. First, do skeleton photoperiods of 12 h or 18 h (read as 12.5 h) have a similar photostimulatory effect because they delimit approximately the same daylength and hence allow near enough the same phase relationship to develop between entrained hormone rhythms? Or second, do the two photoperiods provide a similar inductive effect, in spite of the long light pulse being at the start of the cycle in one case and at the end in the other? A continuous 14-h photoperiod is definitely stimulatory for house sparrows and the observation that a 14-h skeleton is ineffective does suggest that an inductive effect is important. Similar reasoning could be applied to the collared dove data.

Circannual rhythms in breeding cycles

Birds possess endogenous rhythms of moult, migratory disposition and reproductive capacity which persist under constant conditions with a period length of up to 12 months (Gwinner, 1967, 1971 and 1973; Schwab, 1971). Such periodicities have been termed circannual although they may free-run under certain constant light conditions (*e.g.* 12-h light—12-h dark (LD12:12)) with a period length which is considerably less than 12 months; this is strong evidence that an endogenous timing mechanism is involved. Furthermore, circannual rhythms, like circadian rhythms, are entrainable by the seasonal photoperiodic cycle within a limited range of *zeitgeber* frequencies and they assume a phase angle to the *zeitgeber* which is a function of the endogenous oscillator frequency and that of the *zeitgeber* (Gwinner, 1973; Goss, 1969a and b; Goss and Rosen, 1973). In the starling, *Sturnus vulgaris*, a constant photoperiod of LD11:13 prolongs the period of active gametogenesis while long photoperiods (LD13:11 and above) rapidly force the hypophyseal—gonad axis through a cycle of activity into a lasting state of photorefractoriness during which time the gonad remains regressed and inactive (Hamner, 1971; Schwab, 1971). A photoperiod close to LD12:12 uniquely enables a periodicity of gonad growth and involution followed by further cycles having a period length of about nine months (*cf.* the breeding periodicity of the wideawake tern, *Sterna fuscata*, on Ascension Island (Chapin, 1954; Ashmole, 1963)). In the collared dove there was little variation in period length of the breeding cycle between subjects held for 12 months on a range of constant photoperiods. However, under constant very low light intensity of under 0.01 lux, and under LD8:16 the gonad cycle was a bit shorter than under longer photoperiods or constant light of over 200 lux (see Fig. 6). The birds used for these experiments were again taken from the wild in Cheshire in November and December.

There is evidence that circannual rhythms are somehow related to circadian rhythms for starlings having long periods of circadian activity exhibited longer circannual periodicities in their gonad and moult cycles and *vice versa* (Gwinner, 1973). The correlation was not however, proportional. Pigeons and doves tend to have long breeding seasons compared with other birds which implies that their oscillator systems must be relatively phase stable under a wide range of seasonal photoperiods (Murton, 1975). This may account for the observation that their breeding rhythms, as shown in Fig. 6, are little altered under a wide range of photoperiods, in contrast to the situation noted above for the starling. Since a complete gonad cycle can be expressed under almost constant dark in the collared dove it is evident that photoinduction is not an absolute necessity for stimulation of the neurohypophyseal mechanism.

In Britain the gonads of collared doves recrudesce into a functional condition in March and regress again in September so the cycle lasts about seven months (Fig. 4). When doves were taken from the field in July and held on LD16:8 photoperiods until December their testes remained functional until November and then spontaneously regressed (Murton, 1975). These birds, therefore, had enlarged gonads for nine months, equivalent to the cycle length expressed under medium-to-long constant photoperiods (*cf.* Fig. 6). We conclude that the birds have an endogenous rhythm which is expressed under constant light conditions which allows the oscillator system to free-run. The endogenous rhythm is entrained by the seasonal light cycle in England

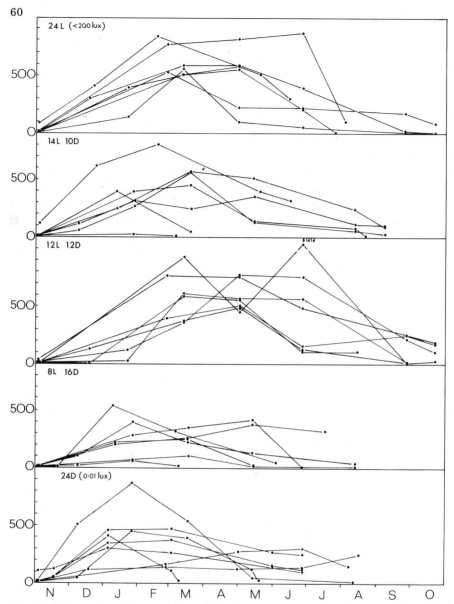

Fig. 6. Changes in testicular volume as measured by repeated laparotomy of individual collared doves held on the constant photoperiodic regimes indicated for 12 months. The data are consistent with the hypothesis that an increase in light intensity increases the frequency of the entrained oscillation, so that the level becomes raised relative to some threshold.

for this has an amplitude and acts as a controlling oscillator whereas constant light conditions provide no amplitude, that is, the *zeitgeber* is effectively damped. The response of the doves to skeleton light schedules suggests that the daily photoperiod is important for the entrainment process, that is, we must suspect that any circannual periodicity and its entrainment is compounded in some way from circadian rhythm mechanisms and the mechanism by which these are entrained. This is a topic for further research. In passing it may be suggested that the gonad cycle of the collared dove in Britain is not yet well adapted to ecological conditions. Genotypes having the capacity to remain in a reproductive condition for longer are probably being selected and we suspect that this explains the variability we note in our data (*e.g.* Figs. 3, 4 and 6). A similar process has been identified in the feral pigeon (Murton *et al.*, 1972 and 1973).

In many respects the gonad cycles of free-living feral pigeons in Britain resemble those of the collared doves, although breeding is possible over a longer period (Murton *et al.*, 1973). The testes gradually regress from a maximum size which is achieved in April, as Fig. 7 shows. We assume that the hormones necessary for gonad maintenance are gradually phased into a non-functional temporal relationship so that eventually in August or September a short season of refractoriness results. The feral pigeon is a convenient study animal in that birds taken into captivity from the field can be induced to undergo a full behaviour cycle leading to egg production (Murton *et al.*, 1969a). Fig. 7 shows how the time elapsing between initial pairing under experimental conditions — at which stage the males would perform the

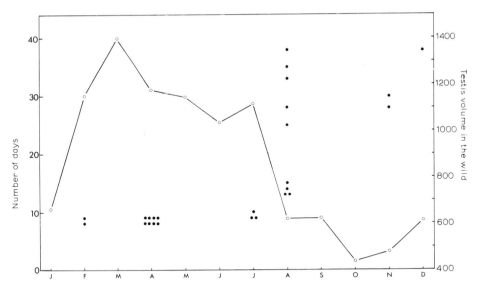

Fig. 7. Seasonal changes (●) in the number of days elapsing between the point at which individual male feral pigeons were introduced to females, when all delivered the bowing display, and egg-laying. All males had testes measuring more than 700 cu. mm when they were initially paired. Seasonal change (○) in mean testis volume of free-living feral pigeons is superimposed as a graph based on data in Murton *et al.*, 1972.

bowing display — until egg-laying increased as the season progressed. A similar lengthening of the courtship cycle has been noted in the canary with advance of the season (Hinde, 1962 and 1967). Such results are explicable if an appropriate phasing of gonadotrophin and gonadal hormone secretion is necessary for the expression of reproductive behaviour as well as for gametogenesis. The next section considers evidence for the direct involvement of gonadotrophins in the behaviour cycle of the feral pigeon.

ENDOCRINE AND BEHAVIOUR CHANGES DURING THE BREEDING CYCLE OF MALE PIGEONS

When a reproductively active male feral pigeon is introduced to a receptive female conspecific, he usually begins a sequence of courtship displays which stimulate the female to accept mounting and copulation and thereafter to participate in behaviour leading to nest-building (Fabricius and Jansson, 1963; Murton et al., 1969a). Exceptions occur because in some free-living populations the birds practice negative assortive mating so that certain phenotypes are reluctant to pair with each other, this also applying to captive subjects (Murton et al., 1973). It is now well established for the domesticated Barbary dove, Streptopelia roseogrisea var., primarily through the efforts of Lehrman and his colleagues, that reproductive development depends on the interaction of external stimuli — which include direct stimuli from the physical environment and also appropriate behaviour from the mate — and internal physiological changes (see Lehrman, 1964 and 1965 for summaries). Participation in nest-building behaviour initiates hormone secretion in the female which leads successively to oviduct development, ovulation, and oviposition, whereupon both male and female are prepared to sit on the eggs (Lehrman, 1958; Lehrman and Brody, 1961; Bruder and Lehrman, 1967). Similarly, the interaction of internal and external factors, including participation in appropriate behaviours, stimulate nest-building behaviour, oviduct growth and brood patch development in domesticated canaries (White and Hinde, 1968; see also Hinde, 1967 for summary). Although most studies to date have been concerned with the behavioural—endocrine correlates which can be established in the female it would be expected that the male's neuroendocrine apparatus should also reflect the behavioural changes which are manifested during the breeding cycle. This section investigates this topic.

Subjects were adult feral pigeons originating from a study area at Salford, Manchester. All birds were sexed and their gonads measured during an exploratory laparotomy after which those whose organs were fully recrudesced in size were held separately indoors in visual, but not auditory, isolation on a daily photoperiod of LD16:8. Each cage measured $51 \times 46 \times 41$ cm and, by removing an opaque sliding partition, could be connected with a second (and third) thereby making a double (treble) compartment. Two or three days after laparotomy a male and a female were introduced to each other by removing the partition and approximately half-an-hour was allowed for the male to begin performing the bowing display. As soon as this happened the birds were separated. Non-responsive males were tested on one or sometimes two subsequent occasions with different females, but if the males remained inactive they were discarded. We worked, therefore, with sexually active males which were known to be capable of giving the bowing display and with females which had elicited such a display from a male. Thus once the bowing display was obtained the pair were allocated to each other but kept separate. At this stage the largest follicles of the females measured 3—5 mm in diameter, and so they were not in a regressed condition (under 1.0 mm).

Two days after obtaining a preliminary display bout the cage partitions were removed and the pairs were allowed continuous contact in the presence of nest bowls and nesting material; they were checked three or four times per day to establish the stage to which their displays had progressed. As a full description of the behavioural cycle of the feral pigeon has already been given only brief notes are necessary here (Fabricius and Jansson, 1963; Murton *et al.*, 1969). Subjects were sacrificed between 1100 and 1300 h at different stages of the cycle as follows:

(1) The bowing display was again obtained whereupon the males were immediately killed and the females returned to stock (six pairs). These subjects defined the condition of all subjects at the start of the experiment and so they served as controls. The bowing display involves sexual and aggressive components (see discussion).

(2) Six pairs were killed after reaching the stage of nest-demonstration, which occurred one or two days after initial pairing. During the nest-demonstration display the male crouches on a potential nest site and with lowered head and vibrating wings emits a soft call-note, the nest call. Nest-demonstration is a submissive display which helps the female to overcome her fear of the male and to approach close to him. She comes to nibble at the male's head and neck feathers and he responds with mutual caressing movements. After the initial bowing phase and during the early stages of nest-demonstration, the pair indulge in much courtship feeding, this display sometimes being followed by copulation. These activities decline in frequency as nest-demonstration and caressing develop and become consolidated and the female assumes a more dominant role. After four to six days, but longer later in the breeding season (see above), the female is sufficiently assertive to push the male from the nest site. This causes an increase in aggressive behaviour by the male and he may again give the bowing display. He is now stimulated to collect nesting material which he brings back to the female; she alone or both birds may build the twigs into the nest structure and the birds enter the stage defined at (3).

(3) On the first day that twigs were added to the nest-bowl they were removed and the birds allowed a further 24 h before being killed, *i.e.* the

subjects were killed at the same time on the following day when more nest material had accumulated (six pairs).

(4) Pigeons are determinate layers producing two eggs per clutch over a two to three day period. Males and females were killed the day after the first egg was laid and before the second egg appeared (six pairs).

(5) Four pairs (two other pairs were overlooked and hatched their eggs) were sacrificed after incubating their eggs for 14 days; the full incubation period is 17.5 days.

(6) Seven pairs hatched their eggs and were killed when their young were 10—11 days old.

(7) A final group also hatched their eggs. When the squabs were 7—9 days old they were killed and the adults were left together for a further 14 days whereupon eight males were killed. These are considered as birds that could begin a new cycle and they have been included in some figures below as unpaired males not yet stimulated by a female.

Gonadotrophin secretion

Immediately subjects were killed their gonads and epididymides were preserved and later prepared for histological study. Blood samples were collected with edetic acid (EDTA at 2 mg/ml) as anticoagulant, centrifuged at 2500 r.p.m. and the plasma stored at −15°C. Half of each sample was assayed for 'LH' activity in an HCG/human LH radioimmunoassay (Wilde *et al.*, 1967) using serial dilutions of the Second International Reference Preparation of HCG and human gonadotrophin (HMG) and purified luteinising hormone (LH-Hartree) as standard. Previous studies indicated that avian plasma samples and pituitary extracts cross-react in this system (Bagshawe *et al.*, 1968) while the specificity of the assay has been investigated in relation to purified preparations of mammalian LH and FSH (Bagshawe *et al.*, 1966). Recovery of $[^{125}I]$HCG was initially recorded as a percentage of that recovered in the absence of unlabelled antigen and expressed as IU/ml. The remaining half of each plasma sample was kindly assayed by Dr. B. Follett in a radioimmuno- assay system developed against avian (chicken) luteinising hormone (Follett *et al.*, 1972). This assay has been made possible by the separation and purifica- tion of two avian gonadotrophic factors, an 'LH' and an 'FSH', from chicken pituitaries (Stockell-Hartree and Cunningham, 1969; Scanes and Follett, 1972). The LH fraction possesses strong biological activity in the rat ovarian ascorbic acid depletion assay but none in the Steelman—Pohley rat FSH bioassay (Furr and Cunningham, 1970).

Considering the IR-LH (Bagshawe) assays (Fig. 8) it is evident that titres were generally low at the start of the behaviour cycle. The sensitivity of the assay is such that recovery percentages greater than 80% (equivalent to plasma concentrations less than 0.022 IU/ml and depicted as open circles in Fig. 8) fall outside the straight line region of the dose—response graph and cannot

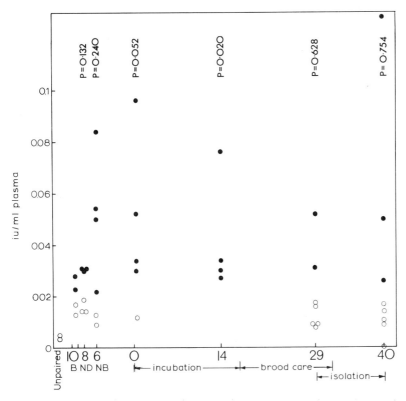

Fig. 8. Plasma titres of immunoreactive luteinising hormone as measured in the Bagshawe assay system during the reproductive cycle of individual male feral pigeons. The abscissa gives the day and behaviour stage reached in relation to egg-laying equivalent to day 0. B, bowing; ND, nest-demonstration; NB, nest-building. Open circles: points outside straight-line region of dose—response graph (see text). Probability values are Mann—Whitney U-tests comparing each stage with the bowing and unpaired birds as controls.

be distinguished except in terms of a crude scoring system. Since this invalidates *t*-test assumptions for some of the readings, a non-parametric Mann—Whitney U-test (two-tailed) is used to test for statistical significance in difference between IR-LH readings at the various behaviour stages compared with those obtained at the bowing phase of the cycle (details given in Fig. 8). The results are not sufficiently well defined to determine the precise point at which there was a general increase in immuno-LH secretion but it seems likely that this was not until about the time of nest-building. Certainly by the time of egg-laying there had been an increase in this gonadotrophin activity. The variability between the males in any one group may well have depended on genuine individual differences in the level of circulating hormone, that is, the normal temporal pattern of secretion could fluctuate around a higher mean in some birds than in others. It is also possible that there were variations in

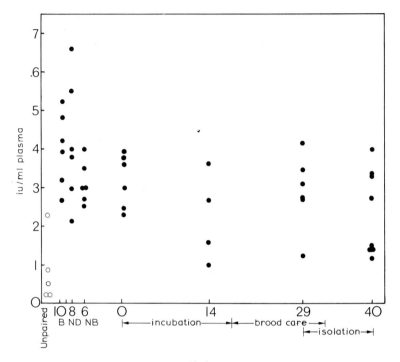

Fig. 9. Plasma titres of immunoreactive luteinising hormone as measured in the Follett assay system during the reproductive cycle of individual male feral pigeons. The abscissa gives the day and behaviour stage reached in relation to egg-laying. Abbreviations as for Fig. 8, except that open circles refer to unpaired males.

the temporal phasing of hormone release in different individuals. Ideally it would be desirable to compare individuals whose plasmas have been assayed at regular intervals throughout the 24-h but this was beyond our technical capacity.

The IR-LH (Follett) assays produced different results from those obtained in the Bagshawe system (Fig. 9). Indeed, there was no correlation between the two sets of assay results for the same plasma samples, so it is clear that different immunoreactive substances were being measured. That detected by the Follett system exhibited a marked elevation at pairing and the highest titres recorded in individuals were at the bowing and nest-demonstration stages of the cycle (Fig. 9), whereas this was when the Bagshawe assays indicated minimal titres (Fig. 8). In short, plasma immunoreactivity as measured in the Bagshawe system increased later in the cycle than did IR-LH (Follett) and corresponded temporally with the appearance of new Leydig cells. It might, in view of this and previous results (Murton et al., 1969) be more appropriate to refer to the Bagshawe immunoreactive substance as an interstitial cell stimulating hormone (ICSH). IR-LH (Follett) appeared to be more prominent at that stage of the cycle when Leydig cells were clearly secreting and were disappearing, and when the epididymal columnar epithelium, supposedly dependent on androgen secretion, was maximally stimulated.

Histology of testes and epididymides

Changes in testicular cytology during the courtship, incubation and brooding stages were apparent from the counts of germ cells summarised in Fig. 10; this figure would become too confused if standard deviations

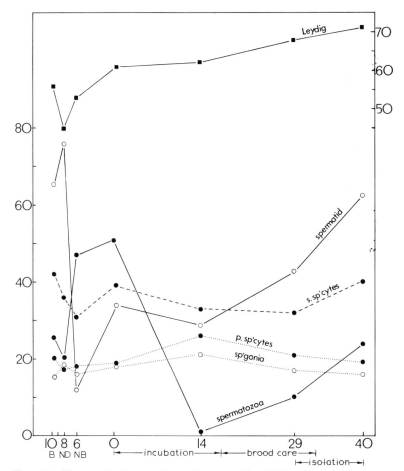

Fig. 10. Changes in the number of germ cells and Leydig cells in segments of the testes of feral pigeons during the reproductive cycle. For method of counting cells and error terms see footnote to Appendix 1. The abscissa gives the day and behaviour stage reached when samples were taken. Abbreviations as for Fig. 8.

were given so these are detailed in Appendix 1. When males were first paired and were prepared to give the bowing display their testes contained relatively large numbers of spermatids and few spermatozoa. Doubtless many sperm that had already been produced had passed from the testes tubules via the rete testes and were already stored in the sperm sacs at the distal end of the vasa deferentia. Following pairing there was a distinct wave of spermateleosis between the nest-demonstration and nest-building stages, that is, over a period of two days in the present subjects. As spermatids were rapidly metamorphosed a peak in numbers of spermatozoa resulted between the nest-building and egg-laying stages (Fig. 10). The maturation of spermatids coincided with a depletion of mature Leydig cells and followed after the highest plasma IR-LH (Follett) titres were noted. After egg-laying spermatozoa were released from the enfolding Sertoli cells and disappeared and very few remained in the tubules after the males had taken their share of incubation for 14 days; presumably the spermatozoa had migrated to the epididymides and vasa deferentia. There were probably no new spermatids produced during the period of incubation and indeed there was little germ cell activity until the eggs hatched, whereupon there was a steady build-up of spermatids to the numbers noted in birds at the beginning of the cycle.

The temporal variations in spermatogenetic activity were reflected in the overall histological appearance of the testis tubules when viewed in low power cross-section. At pairing the tubules had a cluttered untidy appearance because sperm free in the lumina were partly intermixed with other types of germ cell. In contrast, at egg-laying the tubules presented a much more ordered appearance with the various germinal stages organised into well-defined cellular associations, displaying a clear radial coordination. Evidently an earlier generation of germ cells had matured into spermatozoa and these had then become intermixed with subsequent generations of cells. Following pairing spermatogenesis had been slowed or stopped, coincident with a partial clearance of material accumulated in the tubule lumina, and it then started afresh giving rise to an ordered sequence of new germ cells.

There were changes in the appearance and number of interstitial Leydig cells and their nuclei during the cycle (Table I). The growth and maturation of the Leydig cells appears as a continuous process so that it is not possible to make definitive criteria for separating old and young cells. Mature cells, which are typically large with large nuclei containing little chromatin, could readily be distinguished from small new cells with small dark elliptical nuclei and much chromatin, but intermediate categories could only arbitrarily be categorised. Cells with pycnotic nuclei which had reached the end phase of their secretory cycle could also be recognised. Mature Leydig cells were steadily depleted during the preincubation behaviour phase during which time the number with pycnotic nuclei increased. Although some new Leydig cells were probably formed at this time, large numbers of juvenile interstitial cells did not appear until the onset of incubation. The new generation of

TABLE I

Condition of Leydig cells of the pigeon testis during the reproductive cycle

To count Leydig cells, an interstitial space was centred at random and all Leydig cell nuclei were counted. The boundaries of the tubules were then followed to the next space and so on until ten spaces had been checked. This technique ensured that cognizance was taken of those Leydig cells which were squashed between tubules as well as those positioned in the interstices.

		Number* ± S.D. of Leydig cells which were:				Number of phagocytes in interstitium
		Juvenile	Mature	Pycnotic	Total	
Bowing	(6)	22 ± 13	28 ± 11	7 ± 3	57 ± 17	13 ± 10
Nest-demonstration	(6)	15 ± 10	20 ± 9	10 ± 4	45 ± 18	10 ± 11
Nest-building	(6)	18 ± 6	15 ± 6	20 ± 8	53 ± 12	16 ± 18
After first egg	(6)	17 ± 10	17 ± 9	27 ± 17	61 ± 31	11 ± 7
After 14 days incubation	(4)	28 ± 11	16 ± 12	18 ± 3	62 ± 21	11 ± 9
Brooding young (10—11 days)	(7)	20 ± 8	26 ± 26	23 ± 7	69 ± 35	10 ± 6
After removed from young for 14 days	(8)	25 ± 8	16 ± 8	31 ± 10	72 ± 18	23 ± 23

* The means are based on the total number of Leydig cells recorded per bird, the number of individuals being given in brackets.

Leydig cells arising during incubation matured and possibly became secretory throughout and after the period of brood care for there was an increase in pycnotic nuclei during and after brooding. By the time the birds had been isolated from their young for 14 days large numbers of macrophages had invaded exhausted areas of interstitium (Table I). Pigeons can lay eggs and begin a new cycle while they still have three-quarters grown young.

Development of the columnar lining cells of the epididymides supposedly depends on androgen secretion (Witschi, 1945). Jones (1970) has shown that the height of these epididymal lining cells is at a maximum in California quail *Lophortyx californicus* when their interstitial tissue is losing lipids and cholesterol, as these precursor materials are apparently rapidly converted to androgen. In the pigeons there was a marked increase in the height of these cells, and in the diameter of the epididymal duct, soon after pairing (Fig. 11). Maximum development of the epididymides was noted at the nest-demonstration and nest-building stages and thereafter the organ regressed to the time of egg-hatching, remaining small until the adults were separated from their young. There was a reasonably close temporal correlation between the degree of epididymal development and evidence of secretory activity by the inter-

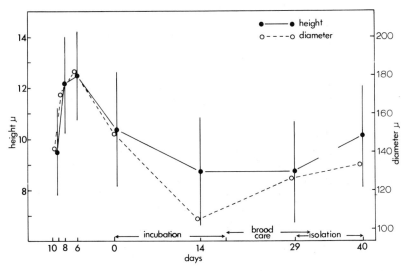

Fig. 11. Changes in diameter of the epididymis (dashed line) and in the height of the columnar lining cells (solid line) during the reproductive cycle of the feral pigeon. The abscissa conventions are as in Fig. 8.

stitium. As the columnar epithelium enlarged, presumably under the influence of androgen release, spermatids were seen to mature into spermatozoa (Figs. 10 and 11). There is some evidence that androgens mediate the process of spermateleosis (Lofts and Murton, 1973). Spermiation, like ovulation, is supposed to require a burst of LH activity; a wave of spermiation occurred during the period of incubation, when titres of IR-LH (Bagshawe) were elevated but not so obviously those of IR-LH (Follett).

Different pairs took variable times to complete the preincubation behaviour cycle. If the plasma gonadotrophin content or histological condition was plotted strictly on a time scale the clear patterns discussed above vanished. Evidently, the behaviour stage achieved provided a better indication of endocrine and histological conditions than the time elapsing from pairing. It appeared that involvement in particular behaviours influenced the internal condition of the birds. Thus, pairing immediately caused an increase in IR-LH (Follett) followed by increased secretory activity in the Leydig cells and development of the epididymides.

EFFECT OF EXOGENOUS HORMONES ON PIGEON BEHAVIOUR

The bowing display of the pigeon has a sexual—aggressive motivation and functions mostly as a ritualised appeasement. It combines crouched submissive postures and head-lowering at the termination of the downward bow, with tail-fanning movements reminiscent of braking. The aggressive component is typified by the male advancing at the females with a high-stepping gait, his head held upright. Ambivalence is apparent in the display for if the male catches the object of his advances he may attempt to mount and copulate, especially if the chased bird crouches, or he may seize the retreating bird by its feathers and attempt to fight or he may himself turn and flee. Nest-demonstration is a more submissive display which attracts the female to approach closely to the future nest site and it culminates in nest-building and egg-laying (see p. 63 above). There is much evidence to suggest that high endogenous androgen titres are involved at the beginning of the pigeon reproductive cycle since injections of testosterone propionate also mediate bowing behaviour in Barbary doves (Erickson and Lehrman, 1964; Erickson *et al.*, 1967) and *Columba livia* (Murton *et al.*, 1969) and this hormone can reestablish male-type behaviour in castrates (Beach, 1948; Collias, 1950; Erickson and Lehrman, 1964). In precise studies by Hutchison (1970a and b), castration completely eliminated bowing but chasing behaviour and nest-soliciting continued at a diminished frequency. Moreover, using hypothalamic micro-implants of testosterone propionate, Hutchison showed that high-diffusion implants resulted in maximum bowing behaviour whereas low-diffusion implants induced nest-solicitation in the virtual absence of chasing and the complete absence of bowing. 'Androgen-type' displays (bowing, driving, mounting and copulation) are stimulated when a male becomes paired and the high plasma androgen concentrations thought to exist at this stage could account for the peak development of the columnar epithelium of the epididymides which occurs two days later.

Further evidence that androgen titres may decline during the preincubation behaviour cycle comes from examining the biosynthetic capacity of testis extracts incubated *in vitro* with tritiated pregnenolone as the isotopically labelled precursor steroid (Lofts *et al.*, unpublished). Fig. 12 illustrates how the capacity to synthesise testosterone from pregnenolone declines markedly between initial pairing and bowing and egg-laying, recovering somewhat by the time of egg-hatch, and that the biosynthetic pathway tends to be blocked at the production of 17-α-hydroxyprogesterone at egg-laying. It is of interest that progesterone apparently provides the endocrine stimulus which induces male and female Barbary doves to incubate their eggs (Lehrman, 1958 and 1963; Lehrman *et al.*, 1961).

The stimulus encouraging androgen secretion at the start of the reproductive cycle presumably acts via hypothalamic—endocrine control mechanisms and not neural pathways. Conceivably pairing stimulates a gonadotrophin secretion which at least then facilitates the expression of androgen dependent mechanisms. Exogenous mammalian NIH.FSH, but not exogenous mammalian NIH.LH, increases the driving and aggressive components of the bowing display, particularly the sexual—aggressive components, while testosterone propionate enables a more natural sequence with bowing progressing to mounting and copulation (Murton *et al.*, 1969a). Exogenous oestradiol

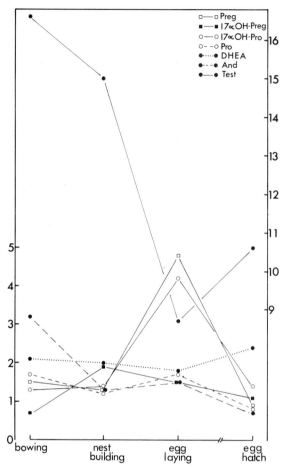

Fig. 12. Capacity of testicular extracts of feral pigeons to biosynthesise steroids from tritiated pregnenolone supplied as a labelled precursor according to the behaviour stage reached in the reproductive cycle. Radioactive steroids were counted by a liquid scintillation spectrophotometer after thin layer chromatography and presented as a percentage conversion index =

$$\frac{\text{radioactivity of each steroid sample}}{\text{total radioactivity of precursors}} \times 100$$

Abbreviations in order refer to pregnenolone, 17-α-hydroxypregnenolone, 17α-hydroxyprogesterone, progesterone, dehydroxyandrosterone, androstenedione and testosterone.

benzoate causes a profound stimulation of nest-demonstration behaviour under a wide range of conditions (Vowles and Harwood, 1966; Murton et al., 1969; Hutchison, 1970a and b). This might imply that oestrogen functions in the normal behaviour cycle of the pigeon or that it is effective because it has an inhibitory effect on FSH—androgen secretion. In this context progesterone is also of interest because Erickson *et al.*, (1967) discovered that male castrates, which were induced to give the bowing display with androgen injections, gave only nest-demonstration if progesterone was combined in the treatment. In the next section we consider the endocrine and histological effects of oestrogen or progesterone treatment and then consider the behaviour of oestrogenised subjects given FSH or androgen replacement therapy.

Endocrine and histological effects of oestrogen and progesterone injection

Subjects were 27 male feral pigeons taken from an outside holding pen in August, which by exploratory laparotomy were all shown to have enlarged testes of a size consistent with an ability to reproduce. Each bird was held in an individual cage in an indoor bird room under an LD16:8 photoperiod. Controls received ten sham injections of normal saline solution over a period of 14 days. Another 15 birds were given ten injections of 2 mg of oestradiol benzoate in 0.4 ml arachis oil on different days during a 14-day period, *i.e.* a total dose of 20 mg. Similarly, the remaining six subjects were given ten injections of 2 mg progesterone over a period of 14 days. Injections were given between 09.00 and 10.00 h. On the morning of the fifteenth day following initial injection the six controls, together with six progesterone-treated and six oestrogen-treated subjects were sacrificed for plasma assay and histological examination, about 1 h after injection of the steroids. This left nine oestrogen-treated subjects which were kept in their individual cages without further experimental interference. Fourteen days after the cessation of oestrogen injections five of these birds were killed (and the others laparotomised) and 14 days later (28 days from the termination of injections) the remaining four birds were killed. The rate at which spermatogenetic activity resumed following oestrogen treatment was studied in these birds but their plasma was not assayed.

There was a small increase in gonad weight and volume of the controls after transfer to LD16:8 (Fig. 13). Progesterone inhibited this increase while oestradiol benzoate caused a drastic reduction in testicular size and tubule diameter (Fig. 13). The histological appearance of the testes from oestrogenised birds was distinctive. Most tubules contained only a resting ring of spermatogonia with a few degenerating primary spermatocytes (Fig. 14). None of the individuals had an intact layer of primary spermatocytes and few of these cells were advanced beyond the meiotic stage of prophase. Moreover, the cells, other than spermatogonia, were morphologically and topographically abnormal and contained vacuoles where the cell cytoplasm had contracted from the nucleus. As in the case of the domestic rat (Lacy and Lofts, 1965), spermatogonia were not reduced in number and it was the primary spermatocytes which appeared to be most sensitive to oestrogen treatment for they were depleted in number and few survived to pachytene.

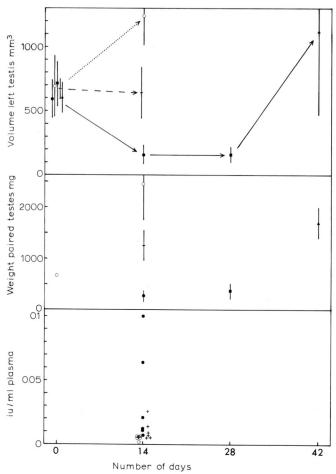

Fig. 13. Effect of injecting feral pigeons with 20 mg oestradiol benzoate or 20 mg proges-terone over 14 days. Mean testes volumes refer to all subjects studied, these were measured by laparotomy. Testicular weight was only obtained at autopsy and so the controls were different birds from those treated. This also applied to the plasma IR-LH titres which were measured by the Bagshawe radioimmunoassay method. Vertical lines indicate standard errors. Open symbols are for controls, solid symbols for oestrogen-treated birds and crosses for progesterone-treated subjects.

The interstitial areas were enlarged and of unusual appearance (Fig. 14). The mature Leydig cell generation could be distinguished because the cell nuclei were large, with slightly crenulated boundary membranes, and they contained little chromatin. These cells were aggregated as nests of cells in which the inner and adjacent cell walls had disintegrated, but not those on the outside, so that 2—12 Leydig nuclei were encapsulated within one membrane (Fig. 14). In addition, a new generation of juvenile Leydig cells with small spindle-shaped nuclei had been formed and large numbers of these cells could be seen as a ring surrounding the tubules, apparently being proliferated from fibroblasts in the basement membrane of the testis tubules. Further from the tubule membrane, the cells and their nuclei became bigger and rounder and the nuclei contained numerous dark chromatic granules. There were virtually no lipids inside the testis tubules as revealed by Sudan black B, but the interstitium was occluded with amorphous deposits. These occurred as a dense mass in the cytoplasm of the old Leydig generation so that the encapsulated aggregations noted in the wax sections above appeared as black spheres in the gelatin prepared material (Fig. 14). In addition, the juvenile Leydig cells contained numerous sudanophilic droplets.

In the oestrogen-treated subjects allowed a recovery period of 14 days, the tubules had expanded, mitotic figures could be distinguished in the spermatogonia, new healthy primary spermatocytes were present in increased numbers and a few secondary spermatocytes had been proliferated. The encapsulated Leydig cells had been invaded by capillaries and were in the process of being phagocytosed, while the nuclei of the new generation of interstitial cells had enlarged and the cells appeared to be secretory. The sudanophilic lipid deposits were much reduced and, in one specimen, absent altogether; vacuoles had appeared in the Leydig cytoplasm. After 28 days the processes observed above were still evident. Tubules were approaching their fully expanded diameter and spermatozoa were now present. The encapsulated Leydig cells had mostly been lost, but an occasional empty capillary or one with one or two phagocytes could be found; exceptionally a capillary packed with phagocytes was evident. The second generation of Leydig cells had matured, become depleted of lipid and the nuclei of many of these cells were pycnotic, indicating that the cells were at the end of their secretory phase; many cells were disintegrating or had disappeared and macrophage activity was evident.

Compared with the oestrogen-treated birds, those given progesterone exhibited a more variable histological appearance, for in some tubules a few spermatozoa could be distinguished, while in others germ cells beyond the secondary spermatocyte stage were not present (Fig. 15). The tubules were not much regressed in size and numerous meiotic figures were apparent in the primary spermatocytes ranging from early prophase to metaphase and all seemed normal in appearance. Essentially, development beyond the stage of secondary spermatocytes had been inhibited and these germ cells were accumulating in the tubule lumina. The interstitial cells did not differ in appearance from the control situation, judging by the wax embedded material, but gelatin sections revealed an increased deposition of sudanophilic lipid in the Leydig cytoplasm. Appreciably more lipid was also visible in the Sertoli cell cytoplasm than was present in control or oestrogen treated subjects (Fig. 15). This lipid was apparent as droplets located near the periphery of the tubule and sometimes the droplets had coalesced to form large globules.

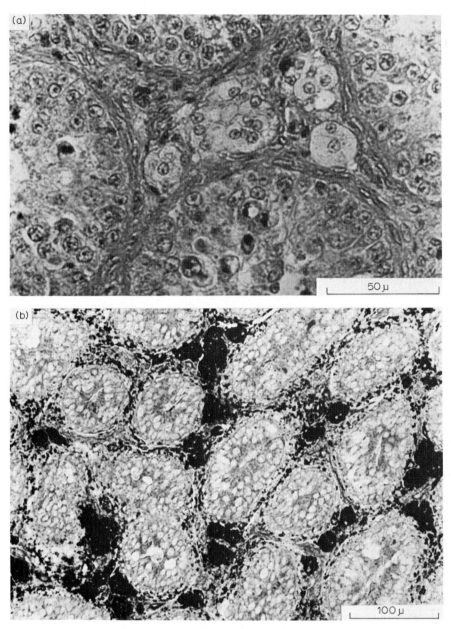

Fig. 14. Transverse section of testis of *Columba livia* following ten injections of 2 mg oestradiol benzoate. a: the tubules contain only degenerating primary spermatocytes and a ring of spermatogonia. The old Leydig cells are encapsulated while a new generation of Leydig cells are differentiating from fibroblasts sited in the basement membrane of the tubules. Fixed in Bouin's solution and stained iron haematoxylin and orange G. b: sudano-philic lipoidal material (black stain) is concentrated in the encapsulated Leydig cells as solid areas while small lipid droplets are also associated with the juvenile generation of interstitial cells. Note the absence of lipid in the testis tubule. Fixed in formol—saline, postchromed by Baker's method and coloured with Sudan black B.

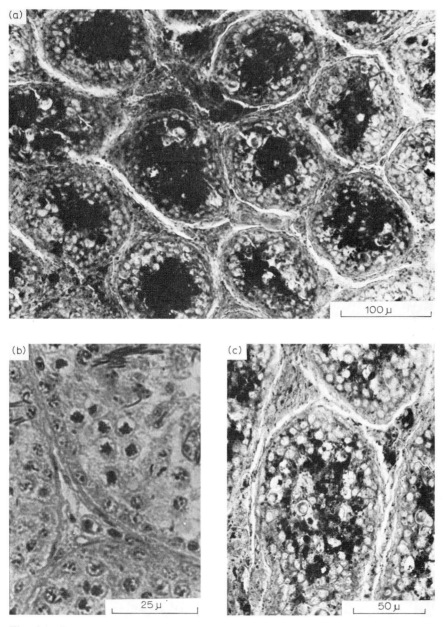

Fig. 15. Transverse section of testis of *Columba livia*. a: gelatin section from bird given ten injections of 2 mg progesterone to show sudanophilic lipoidal material (black) inside testis tubules and also as small droplets in the Leydig cell cytoplasm. Baker's method with Sudan black B. b: wax section from subject treated as (a) to show that there has not been a proliferation of juvenile Leydig cells and there is no indication of the older generation of cells becoming encapsulated. The tubules still contain spermatids. Bouin fixation, iron haematoxylin and Orange G. c: gelatin section from bird given 15 × 1 mg injections of testosterone propionate to compare with (a) and to show a similar build-up of Sertoli cell lipids within the testis tubules. Contrast with Fig. 14b. Sudan black B.

Oestrogen stimulated a significant (Mann—Whitney U-test) elevation in plasma levels of IR-LH (Bagshawe) in all but one of the subjects over the control situation, whereas progesterone produced somewhat variable results, which did not differ significantly from the control birds (Fig. 13). Although it is well established that oestrogen stimulates LH secretion in mammals, it has not so far been certain whether a similar response occurs in birds. In the female mammal oestrogenic hormone supposedly reduces the synthesis and release of FSH. The same appears to be true in the male rat for FSH treatment immediately restores spermatogenetic development, in particular meiotic divisions of the primary spermatocytes (Lacy and Lofts, 1965). It is possible that the suppression of spermatogenesis in the present studies also depended on an inhibition of FSH synthesis or release. The massive steatogenesis of the interstitium was presumably consequent on the accumulation of unsaturated precursor steroidogenic material under the influence of endogenous LH. Two distinct processes were recorded: (a) the accumulation of lipid in the mature Leydig cells as if a releasing factor was absent and (b) the proliferation of new Leydig cells which also became lipoidal. It is to be noted that spermatogenesis was not maintained by high endogenous titres of IR-LH (Bagshawe).

The observations of Erickson et al., (1967) suggest that progesterone selectively insulates the CNS against the action of testosterone. It is possible that progesterone more generally inhibits androgenic function for there is evidence that the maturation of germ cells from the stage of secondary spermatocyte to mature sperm depends on androgen and not FSH (Lofts, 1962; Lofts and Murton, 1973). The histological manifestation following progesterone treatment was an inhibition of development in those germ cells supposedly requiring androgen for maturation. It is not certain whether progesterone causes any suppression of FSH activity. Continued treatment with exogenous testosterone can result in lipid deposition in the tubules (Fig. 15), perhaps in consequence of an inhibition of FSH secretion by feedback mechanisms (Lofts et al., 1970).

Behaviour elicited by FSH or androgen following oestrogen treatment

Subjects were 12 adult male feral pigeons with enlarged testes withdrawn from a communal holding flight. The males were kept in isolation in the middle section of a battery of three cages (see p. 62). In each left-hand compartment was a reproductively active female (12 females) which two days after exploratory laparotomy was allowed access to the adjacent male for 15 min. In this way it was confirmed that the males would perform the bowing display. The males were now kept in isolation while being given pectoral injections of 10×2 mg oestradiol benzoate as above (p. 73). There were two rest days following the last injection and then the males were introduced to the females around 10.00 to 11.00 h by removal of the partition (day 1). All behaviour was recorded for 15 min whereupon a second reproductively active male, which had been held in the third compartment, was introduced to the female and first male. The behaviour of the three birds was recorded for a further 15 min and then all subjects were separated and confined in their own cages, in visual but not auditory isolation. The 12 second males were known to be capable of giving the bowing display, but they were switched from cage to cage so that a different male was presented to the 12 pairs in the experiments below. In the same way, following the experiment just described, the females were moved to different cages so that in subsequent tests the males experienced a new female and then a new male. In this way we hoped to avoid pair bonds being established and the behaviour cycle progressing to nest-demonstration. The next day (day 2) all the 12 test males were again laparotomised, which confirmed that oestrogen treatment had caused their testes to regress. After two days for recovery the males were divided into two groups.

Group 1. Six of the males were given 0.2 mg testosterone propionate in arachis oil at 10.00 h and after half-an-hour were introduced to a female (day 5). Behaviour was recorded for 15 min and then a second, but, as mentioned above, different male was introduced and behaviour recorded for another 15 min after which the three subjects were isolated in their own cages. Exactly the same procedure was followed during the next two days, *i.e.* morning testosterone injection followed by behaviour observations, each time involving different partners. On day 8 the experimental males were killed and their gonads removed for histological examination.

Group 2. On day 5 the remaining six males were given 0.2 mg NIH.FSH in normal saline. In exactly the same way as the testosterone-treated birds, they were reintroduced to females after half-an-hour and behaviour was noted for 15 min before a second male was introduced for a further 15 min. These tests continued on days 6 and 7 and test subjects were then sacrificed on day 8. The results consider the number of bouts of a particular behaviour pattern per 15-min of observation. When the second male was introduced some of the behaviour was directed at the female, some at the second male and some at no bird in particular. We distinguished these situations but did not keep a comparable record when only a single male and female were together.

When tested at the termination of oestrogen therapy, males displayed to females with bowing and some attack movements, even though their testes were regressed in size and on histological criteria (see above) possibly inactive. The combined frequency of bowing and attacking was not altered when a second male was introduced (Table II) but the subjects now delivered a higher proportion of attacks compared with bowing displays (Table III). Hence, the presence of a second male did not alter the quantity of display, but the quality was changed and it became mostly directed at the rival male. It should be remembered that the experimental males had only 15-min ex-

TABLE II

Number of bowing and aggressive displays given by oestrogen-treated males when tested with a female or a female plus a second male, and according to subsequent hormone treatment

Ref. no. of ♂	Number of display bouts given by each male per 15 min					
	After oestrogen therapy		Injected with androgen following oestrogen therapy		injected with FSH following oestrogen therapy	
	♀ only	♀ + second ♂	♀ only	♀ + second ♂	♀ only	♀ + second ♂
B 930	16	16	0	0		
G 292	0	1	0	0		
G 296	1	10	0	20		
G 288	17	16	0	13		
211	11	42	4	36		
258	18	14	16	18		
213	14	17			6	7
721	0	0			6	16
256	10	5			18	8
729	14	36			29	30
726	0	20			0	29
281	39	35			36	34

There is no significant difference between scores for oestrogenised males tested with a female or with a female plus second male using a Wilcoxon matched pairs signed-ranks test.

Androgen caused a significant decrease at 0.05 probability level in display frequency for males tested with females alone (randomisation test for matched pairs) but not when tested with a second male. Thus, there was a significant increase (at less than 0.05 level) in the frequency of bowing and aggressive displays when a second male was introduced to a male treated with androgen (randomisation test).

There was no significant change in display frequency in males treated with FSH whether they were tested with a female alone or female plus second male.

The apparent difference in frequency of bowing plus attack, when the scores for the six testosterone-treated birds are compared with those given FSH, is not quite statistically significant ($P = 0.094$, Mann—Whitney U-test, 2-tailed).

perience of the females before the second male was introduced. That the males did not attack their females and perform more bowing to the rival males appeared to depend on the behaviour of the sexes; females usually retreated and adopted submissive postures, whereas rival males exhibited more aggression. As Table III shows, a total of 212 distinct display bouts were recorded for the 12 males when they were tested for 15 min with a female and another male. In toto 113 of these display bouts were directed at

TABLE III

Percentage of bowing display bouts or attacks made by oestrogen-treated males to female partners alone or with the addition of a male rival, and according to subsequent hormone treatment

	Only ♀ present	With ♀ plus second ♂		Displays given to ♂	Total§ displays
		Displays given to ♀			
Oestrogen only n = 12					
% attack	23	53		82	40
% bowing	77	47		18	60
Total bouts*	140	17		96	212
χ^2	11.290		7.221		
	P<0.001		P<0.01		
Oestrogen followed by testosterone n = 6					
% attack	11	65		91	45
% bowing	89	35		9	55
Total bouts**	109	20		149	315
χ^2	40.641		10.623		
	P<0.001		P<0.01		
Oestrogen followed by FSH n = 6					
% attack	31	97		90	39
% bowing	69	3		10	61
Total bouts**	245	35		143	379
χ^2	4.338		2.002		
	P<0.05		n.s.		

* Observations are for one 15-min period per bird = 12 × 15 bird-min in total.
** Observations are for three 15-min periods per bird covering three days = 6 × 15 × 3 bird-min.
§ These totals include displays which were not directed at the ♀ or second ♂ so that columns 2 and 3 do not summate to equal column 4.
The first column of χ^2 values compare frequency of attack and bowing between column 1 and column 4. The second column of χ^2 values compare frequency of attack and bowing between column 2 and column 3.

the other male or female so that 99 (47%) were not directed at any bird in particular. In some earlier experiments (Murton et al., 1969) six untreated intact males delivered the bowing display during 35 out of 48 two-min observation periods (72%) and they attacked during 14 (28%). This ratio of bowing to attack does not differ from that noted in the oestrogen-treated males

(χ^2 = 0.038; n.s.); the total quantity of display cannot be judged from these records since different recording techniques were adopted.

Subsequent treatment of oestrogenised subjects with testosterone resulted in a decrease in the total frequency of display (Table II); only records collected on days 1 and 5 are detailed, so that single observation periods of 15 min can be compared for each bird. This is not the case in Table III in terms of the total number of bouts, because records covering three days of observation have been totalled, *i.e.* all observations recorded on days 5, 6 and 7. Compared with the situation immediately following oestrogen treatment, testosterone increased the proportion of bowing to attacking when only females were present (89% bows against 77%; χ^2 = 5.913; $P < 0.05$) but not when another male was present (55% bows against 60%; χ^2 = 1.284; n.s.). With the introduction of the second male, relatively little display was directed at the female, and the proportion not directed at either bird again amounted to 47%.

The total incidence of bowing and attacking combined per individual was not reduced following FSH treatment, unlike the situation pertaining when testosterone was used (Table II). In males given FSH the ratio of bowing to attacking while only a female was present was not significantly altered compared with the situation following oestrogen treatment (77% bows against 69%; χ^2 = 2.670; n.s.) (Table III). However, following FSH injection, the presence of a second male resulted in much aggression and test males now failed to discriminate their own mates from the rival. Nevertheless, many of their displays were not directed at either the female or rival male and when the total number of display bouts was considered the ratio of bowing to attack was the same as it had been just after oestrogen administration (61% bows against 60%; χ^2 = 0.098; n.s.) (Table III). With FSH, the proportion of displays not directed at either the second male or the female was 53%.

Behaviour of males with partially regressed testes

Many feral pigeons are relatively insensitive to changes in daylength imposed at inappropriate stages during the normal gonad cycle. However, a proportion of birds do exhibit testicular regression in response to a decrease in daylength (Murton *et al.*, 1973). For this experiment, males caught in a study area at Salford, Manchester, during May and June 1970, were transferred to the laboratory and following an exploratory laparotomy subjected to an LD8:16 daily photoperiod. After 37 days the testes of seven of the birds had regressed (Fig. 16) and these subjects were now used for the experiment described here. Our rationale was that the testes had decreased in size because gonadotrophin secretion had been inhibited (or become inappropriately phased) under the short day regimes and we sought to test the response of such males to a reproductively active female. So, as in the experiments just described, two days after the second laparotomy the males were allowed

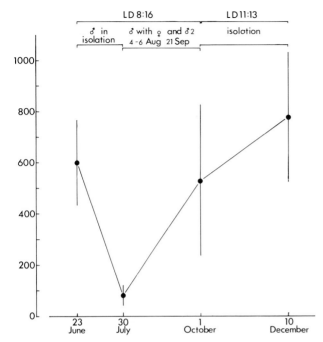

Fig. 16. Changes in mean volume of left testis of certain male feral pigeons which were transferred from outdoor aviaries to the artificial light regimes depicted beginning 23 June. Vertical bars indicate standard errors.

access for 15 min to a reproductively active female and then a second male was introduced and behaviour recorded for a further 15 min. This process was repeated during the next two days, whereupon subjects were now left in isolation on LD8:16 for a further 46 days before being tested again over the next three days. They were laparotomised again before being left in isolation on LD11:13 before being killed 70 days later (Fig. 16).

Fig. 16 shows how testis size decreased in response to a reduction in the daily photoperiod, but how, following the behaviour tests, testis size subsequently increased in spite of the fact that the photoperiod was not altered. It is possible that participation in courtship behaviour stimulated gonadotrophin secretion which led to some testicular recrudescence. Some further, but slight, increase in testis size was noted when the birds were kept under an 11-h photoperiod following the cessation of the behaviour tests. The combined incidence of bowing and attack displays during the first 15-min test session gave scores of 24, 15, 10, 9, 2, 0 and 0, there being no significant difference from the comparable scores for the subjects treated with oestrogen (as given in Table II). Similarly, the number of these displays delivered in the next 15 min, when a second male was present, did not differ significantly from that given by oestrogen-treated birds in the presence of a second male.

84

TABLE IV

Percentage of bowing display bouts and attacks, made by males with regressed testes to female partners alone or with the addition of a rival male

Observations are total for three 15-min periods per bird covering three days = 3 $\times$ 15 $\times$ 7 bird-min.

	Only ♀ present	With ♀ plus second ♂		
		Displays given given to ♀	Displays given given to ♂	Total displays
4—5 August n = 7				
% attack	44	92	95	44
% bowing	56	8	5	56
Total bouts	187	93	118	400
	χ^2 = 0.006*		n.s.**	
21 September n = 7				
% attack	29	100	83	29
% bowing	71	0	17	71
Total bouts	49	17	29	126
	χ^2 = 0.003*		n.s.**	
χ^2 comparing August and September results				
	3.756			8.903

*Comparing columns 1 and 4
**Comparing columns 2 and 3

However, compared with subjects given oestrogen treatment, birds with small testes exhibited a higher proportion of attack to bowing displays when tested with females only (*cf.* Tables III and IV) (χ^2 = 15.538; P < 0.001). In contrast, there was no difference when a second male was present (χ^2 = 0.658; n.s.).

After a month, during which time the testes increased in size, the behaviour of the males changed so that they now produced a higher proportion of bowing displays and the frequency of bowing was not altered when a second male was present (Table IV). The normal pattern of testicular recrudescence involves spermatogenesis under the influence of gonadotrophin secretion and androgen seems to be important during the later stages (Lofts and Murton, 1973). It seems reasonable to assume that growth of the testes was associated with a rising plasma androgen concentration and that this caused the change in behaviour in favour of bowing.

CONCLUSIONS

The results in Table III make it clear that it is important to consider the context in which displays are given in evaluating the mediating properties of hormones. Evidently, testosterone favours the sexual—appeasement component of the introductory courtship display, *i.e.* bowing, whereas FSH encourages the aggressive attacking and chasing component. Moreover, testosterone allows the male to distinguish and behave differently towards his female than to a rival male, whereas if affected by FSH a male is predisposed to attack all intruding pigeons indiscriminately. Free-living pigeons initially attack all intruders but later come to react selectively to the most persistent female. Castration does not immediately eliminate chasing and attacking (Hutchison, 1970a and b) and conceivably this is because FSH remains effective. The behaviour evoked with exogenous FSH and testosterone during the present experiments reinforces the view that at the beginning of pair formation endogenous titres of FSH are elevated and cause aggressive components of the courtship display to predominate (as mentioned, exogenous mammalian LH will not cause the increase in aggressive behaviour which is noted after injection of mammalian FSH). The gonadotrophin which we designate as FSH could be implicated in the release of testosterone from the Leydig cells. It has to be postulated that behavioural stimuli from the potential mate somehow stimulate the Leydig cells and facilitate the development of the epididymides. The progression of the preincubation behavioural sequence then appears to involve a diminution of FSH-type aggressive behaviour as well as of androgenic behaviour. This can be achieved artificially by treatment with exogenous oestrogen which we suggest acts by inhibiting the release of endogenous FSH. Clearly, gonadotrophins affect reproductive behaviour directly and their involvement cannot be neglected when the influence of gonadal steroids is under investigation. Since gonadotrophin secretion is regulated by circadian and circannual rhythm mechanisms these have to be considered in studies of behaviour. It has been realised for a long time that there are diurnal rhythms in bird behaviours, the general restriction of song to dawn and dusk providing a well known example. It seems likely that the hormones underlying behaviour will be shown to assume threshold plasma concentrations during limited phases within the daily 24-h *zeitgeber* and phase-shift effects will be recognised as increasingly important in comparative ethology. It will be interesting to discover whether any behaviour patterns show evidence of a circannual periodicity.

REFERENCES

Aschoff, J. (Ed.) (1965) *Circadian Clocks. Proc. Feldafing Summer School.* North-Holland, Amsterdam

Ashmole, N.P. (1963) The biology of the wideawake or sooty tern *Sterna fuscata* on Ascension Island, *Ibis*, 103b, 297—364.

Bagshawe, K.D., Wilde, C.E. and Orr, A.H. (1966) Radioimmunoassay for human chorionic gonadotrophin and luteinizing hormone, *Lancet*, i, 1118—1121.

Bagshawe, K.D., Orr, A.H. and Godden, J. (1968) Cross-reaction in radioimmunoassay between human chorionic gonadotrophin and plasma from various species. *J. Endocr.*, 42, 513—518.

Beach, F.A. (1948) *Hormones and Behaviour*. Harper and Bros., New York.

Bruder, R.H. and Lehrman, D.S. (1967) Role of the mate in the elicitation of hormone-induced incubation behaviour in the ring dove. *J. comp. physiol. Psychol.*, 63, 382—384.

Bünning, E. (1960) Circadian rhythms and time-measurement in photoperiodism. *Cold Spr. Harb. Symp. quant. Biol.*, 25, 249—256.

Chapin, J.P. (1954) The calendar of Wideawake Fair. *Auk*, 71, 1—15.

Collias, N.E. (1950) Hormones and behaviour with special reference to birds and the mechanisms of hormone action. In E.S. Gordon (Ed.), *A Symposium on Steroid Hormones*. University of Wisconsin Press, Madison, Wisc., pp. 277—329.

Daan, S. (1975) Light intensity and the timing of daily activity in finches. *Ibis*, in press.

Erickson, C.J. and Lehrman, D.S. (1964) Effect of castration of male ring doves upon ovarian activity of females. *J. comp. physiol. Psychol.*, 58, 164—166.

Erickson, C.J., Bruder, R.H., Komisaruk, B.R. and Lehrman, D.S. (1967) Selective inhibition by progesterone of androgen-induced behaviour in male ring doves (*Streptopelia risoria*). *Endocrinology*, 81, 39—45.

Fabricius, E. and Jansson, A.-M. (1963) Laboratory observations on the reproductive behaviour of the pigeon (*Columba livia*) during the pre-incubation phase of the breeding cycle. *Anim. Behav.*, 11, 534—547.

Farner, D.S. (1965) Circadian systems in the photoperiodic responses of vertebrates. In J. Aschoff (Ed.), *Circadian Clocks. Proc. Feldafing Summer School*. North-Holland, Amsterdam, pp. 357—369.

Fisher, J. (1953) The collared turtle dove in Europe. *Brit. Birds*, 46, 153—181.

Follett, B.K. (1973) Circadian rhythms and photoperiodic time measurement in birds. *J. Reprod. Fert.*, Suppl. 19, 5—18.

Follett, B.K. and Davies, D.T. (1975) The neuroendocrine control of reproduction. *Symp. zool. Soc. Lond.*, in press.

Follett, B.K. and Sharp, P.J. (1969) Circadian rhythmicity in photoperiodically induced gonadotrophin release and gonadal growth in quail. *Nature (Lond.)*, 223, 968—971.

Follett, B.K., Scanes, C.G. and Cunningham, F.J. (1972) A radioimmunoassay for avian luteinizing hormone. *J. Endocr.*, 52, 359—378.

Furr, B.J.A. and Cunningham, F.J. (1970) The biological assay of chicken pituitary gonadotrophins. *Brit. Poultry Sci.*, 11, 7—13.

Goss, R.J. (1969a) Photoperiodic control of antler cycles in deer. I. Phase shift and frequency changes. *J. exp. Zool.*, 170, 311—324.

Goss, R.J. (1969b) Photoperiodic control of antler cycles in deer. II. Alterations in amplitude. *J. exp. Zool.*, 171, 223—234.

Goss, R.J. and Rosen, J.K. (1973) The effect of latitude and photoperiod on the growth of antlers. *J. Reprod. Fert.*, Suppl. 19, 111—118.

Gwinner, E. (1967) Circannuale Periodik der Mauser und der Zugunruhe bei einem Vogel. *Naturwissenschaften*, 54, 447.

Gwinner, E. (1971) A comparative study of circannual rhythms in warblers. In M. Menaker (Ed.), *Biochronometry*. Nat. Acad. Sci., Washington, D.C., pp. 405—426.

Gwinner, E. (1973) Circannual rhythms in birds: their interaction with circadian rhythms and environmental photoperiod. *J. Reprod. Fert.*, Suppl. 19, 51—65.

Hamner, W.M. (1963) Diurnal rhythm and photoperiodism in testicular recrudescence of the house finch. *Science*, 142, 1294—1295.

Hamner, W.M. (1964) Circadian control of photoperiodism in the house finch demonstrated by interrupted-night experiments. *Nature (Lond.)*, 203, 1400—1401.

Hamner, W.M. (1968) The photorefractory period of the house finch. *Ecology*, 49, 211—228.

Hamner, W.M. (1971) On seeking an alternative to the endogenous reproductive rhythm hypothesis in birds. In M. Manaker (Ed.), *Biochronometry*. Nat. Acad. Sci., Washington D.C., pp. 448—461.

Hamner, W.M. and Enright, J.T. (1967) Relationships between photoperiodism and circadian rhythms of activity in the house finch. *J. exp. Biol.*, 46, 43—61.

Hinde, R.A. (1962) Temporal relations of brood patch development in domesticated canaries. *Ibis*, 104, 90—97.

Hinde, R.A. (1967) Aspects of the control of avian reproductive development within the breeding season. *Proc. Int. Orn. Congr.*, 14, 135—154.

Hudson, R. (1965) The spread of the collared dove in Britain and Ireland. *Brit. Birds*, 58, 105—139.

Hutchison, J.B. (1970a) Influence of gonadal hormones on the hypothalamic integration of courtship behaviour in the Barbary dove. *J. Reprod. Fert.*, Suppl. 11, 15—41.

Hutchison, J.B. (1970b) Differential effects of testosterone and oestradiol on male courtship in Barbary doves (*Streptopelia risoria*). *Anim. Behav.*, 18, 41—51.

Jones, R.E. (1970) Effects of season and gonadotrophin on testicular interstitial cells of California quail. *Auk*, 87, 729—737.

Lacy, D. and Lofts, B. (1965) Studies on the structure and function of the mammalian testis. I. Cytological and histochemical observations after continuous treatment with oestrogenic hormone and the effects of FSH and LH. *Proc. roy. Soc. B*, 162, 188—197.

Lehrman, D.S. (1958) Induction of broodiness by participation in courtship and nest-building in the ring dove (*Streptopelia risoria*). *J. comp. physiol. Psychol.*, 51, 32—36.

Lehrman, D.S. (1963) On the initiation of incubation behaviour in doves. *Anim. Behav.*, 11, 433—438.

Lehrman, D.S. (1964) Control of behaviour cycles in reproduction. In W. Etkin (Ed.), *Social Behaviour and Organisation among Vertebrates*, University of Chicago Press, Chicago, Ill., pp. 143—166.

Lehrman, D.S. (1965) Interaction between internal and external environments in the regulation of the reproductive cycle of the ring dove. In F.A. Beach (Ed.), *Sex and Behaviour*. Wiley, New York, pp. 355—380.

Lehrman, D.S. and Brody, P.N. (1961) Does prolactin induce incubation behaviour in the ring dove? *J. Endocr.*, 22, 269—275.

Lehrman, D.S., Brody, P.N. and Wortis, R.P. (1961) The presence of the mate and of nesting material as stimuli for the development of incubation behaviour and for gonadotrophin secretion in the ring dove (*Streptopelia risoria*). *Endocrinology*, 68, 507—516.

Lofts, B. (1962) The effects of exogenous androgen on the testicular cycle of the weaver-finch *Quelea quelea*. *Gen. comp. Endocr.*, 2, 394—406.

Lofts, B. and Lam, W.L. (1973) Circadian regulation of gonadotrophin secretion. *J. Reprod. Fert.*, Suppl. 19, 19—34.

Lofts, B. and Murton, R.K. (1973) Reproduction in birds. In D.S. Farner and J. King (Eds.), *Avian Biology, Vol. 3*. Academic Press, New York, pp. 1—107.

Lofts, B., Follett, B.K. and Murton, R.K. (1970) Temporal changes in the pituitary—gonad axis. *Mem. Soc. Endocrinol.*, 18, 545—575.

Meier, A.H. and Dusseau, J.W. (1973) Daily entrainment of the photoinducible phases for photostimulation of the reproductive system in the sparrows, *Zonotrichia albicollis* and *Passer domesticus*. *Biol. Reprod.*, 8, 400—410.

Menaker, M. (1965) Circadian rhythms and photoperiodism in *Passer domesticus*. In J. Aschoff (Ed.), *Circadian Clocks. Proc. Feldafing Summer School*. North-Holland, Amsterdam, pp. 385—395.

Menaker, M. (1971) Rhythms, reproduction and photoreception. *Biol. Reprod.*, 4, 295—308.

Menaker, M. and Eskin, A. (1967) Circadian clock in photoperiodic time measurement: a test of the Bünning hypothesis. *Science*, 157, 1182—1185.

Murton, R.K. (1975) Ecological adaptation in avian reproductive physiology. *Symp. zool. Soc. Lond.*, in press.

Murton, R.K. and Westwood, N.J. (1974) An investigation of photo-refractoriness in the house sparrow by artificial photoperiods. *Ibis*, 116, 298—313.

Murton, R.K., Bagshawe, K.D. and Lofts, B. (1969) The circadian basis of specific gonadotrophin release in relation to avian spermatogenesis. *J. Endocr.*, 45, 311—312.

Murton, R.K., Thearle, R.J.P. and Lofts, B. (1969) The endocrine basis of breeding behaviour in the feral pigeon (*Columba livia*): I. Effects of exogenous hormones on the pre-incubation behaviour of intact males. *Anim. Behav.*, 17, 286—306.

Murton, R.K., Lofts, B. and Orr, A.H. (1970) The significance of circadian based photosensitivity in the house sparrow *Passer domesticus*. *Ibis*, 112, 448—456.

Murton, R.K., Lofts, B. and Westwood, N.J. (1970a) The circadian basis of photoperiodically controlled spermatogenesis in the greenfinch *Chloris chloris*. *J. Zool.*, 161, 125—136.

Murton, R.K., Lofts, B. and Westwood, N.J. (1970b) Manipulation of photorefractoriness in the house sparrow *Passer domesticus* by circadian light regimes. *Gen. comp. Endocrinol.*, 14, 107—113.

Murton, R.K., Thearle, R.J.P. and Thompson, J. (1972) Ecological studies of the feral pigeon *Columba livia* var. I. Population, breeding biology and methods of control. *J. appl. Ecol.*, 9, 835—874.

Murton, R.K., Westwood, N.J. and Thearle, R.J.P. (1973) Polymorphism and the evolution of a continuous breeding in the pigeon, *Columba livia*. *J. Reprod. Fert.*, Suppl. 19, 561—575.

O'Donald,P. (1975) *J. theoret. Biol.*, in press.

Pavlidis, T. (1973) *Biological Oscillators: their Mathematical Analysis*. Academic Press, New York.

Pittendrigh, C.S. (1958) Perspectives in the study of biological clocks. In A.A. Buzzati-Traverso (Ed.), *Symposium on Perspectives in Marine Biology*. Univ. California Press, Berkeley, Calif., pp. 239—268.

Pittendrigh, C.S. (1974) Circadian oscillations in cells and the circadian organization of multicellular systems. In F.A. Schmitt (Ed.), *The Neurosciences: Third Study Program*. pp. 437—458.

Pittendrigh, C.S. and Minis, D.H. (1964) The entrainment of circadian oscillations by light and their role as photoperiodic clocks. *Amer. Naturalist*, 98, 261—294.

Pittendrigh, C.S. and Minis, D.H. (1971) The photoperiodic time measurement in *Pectinophora gossypiella* and its relation to the circadian system in that species. In M. Menaker (Ed.), *Biochronometry*. Nat. Acad. Sci., Washington, D.C., pp. 212—247.

Scanes, C.G. and Follett, B.K. (1972) Fractionation and assay of chicken pituitary hormones. *Brit. Poultry Sci.*, 13, 603—610.

Schwab, R.G. (1971) Circannian testicular periodicity in the European starling in the absence of photoperiodic change. In M. Menaker (Ed.), *Biochronometry*. Nat. Acad. Sci., Washington D.C., pp. 428—445.

Stockell-Hartree, A. and Cunningham, F.J. (1969) Purification of chicken pituitary follicle-stimulating hormone and luteinizing hormone. *J. Endocr.*, 43, 609—616.

Vowles, D.M. and Harwood, D. (1966) The effect of exogenous hormones on aggressive and defensive behaviour in the ring dove (*Streptopelia risoria*). *J. Endocr.*, 36, 35—51.

White, S.J. and Hinde, R.A. (1968) Temporal relations of brood patch development, nest-building and egg-laying in domesticated canaries. *J. Zool.*, 155, 145—155.

Wilde, C.E., Orr, A.H. and Bagshawe, K.D. (1967) A sensitive radioimmunoassay for human chorionic gonadotrophin and luteinizing hormone. *J. Endocr.*, 37, 23—35.

Witschi, E. (1945) Quantitative studies on the seasonal development of the deferent ducts in passerine birds. *J. exp. Zool.*, 100, 549—564.

Wolfson, A. (1966) Environmental and neuroendocrine regulation of annual gonadal cycles and migratory behaviour in birds. *Recent Progr. Horm. Res.*, 22, 177—244.

Zahavi, A. (1975) Sexual selection — a selection for a handicap. *Anim. Behav.*, 23, 237.

APPENDIX 1

Mean number of germ-cells in the testis tubules of feral pigeons

Means are for the number of individual birds given in brackets.

	Spermatogonia	Primary spermato-cytes	Secondary spermato-cytes	Spermatids	Spermatozoa
Bowing					
(6)	15.3 ± 2.8	20.3 ± 3.6	42.0 ± 12.6	65.5 ± 22.5	25.8 ± 8.2
Nest-demonstration					
(6)	18.5 ± 4.5	17.3 ± 5.3	35.7 ± 10.0	76.0 ± 20.4	20.5 ± 15.3
Nest-building					
(6)	15.8 ± 1.3	18.2 ± 5.3	30.7 ± 9.4	12.2 ± 4.3	47.3 ± 15.8
After first egg					
(6)	17.7 ± 2.3	18.7 ± 5.0	39.3 ± 7.5	33.8 ± 12.2	50.7 ± 21.6
After 14 days incubation					
(4)	21.0 ± 1.8	26.3 ± 3.1	33.3 ± 5.1	29.3 ± 25.5	1.3 ± 1.9
Brooding young (10—11 days)					
(7)	16.9 ± 1.6	21.0 ± 5.6	32.3 ± 11.2	42.7 ± 23.4	10.4 ± 11.0
After removed from young 14 days					
(8)	15.8 ± 2.8	19.4 ± 2.8	40.4 ± 6.0	62.5 ± 17.7	23.9 ± 10.6

Notes

The numbers of germ cells were counted along a line transecting the widest section of ten tubules per individual to give a total germ cell count. The means ± S.D. for 4 to 8 individuals according to group are detailed, not the means per single tubule.

The increase in primary spermatocytes after 14 days of incubation was significant (t_8 = 2.97; P = 0.02—0.01) as was the increase in numbers of spermatogonia (t_8 = 2.54; P = 0.05—0.02).

The increase in numbers of spermatids from nest-building to egg-laying was significant (t_{10} = 4.15; $P < 0.002$) but the apparent increases following the onset of brooding and then removal from the young were not significant.

5

THE HORMONAL BASIS OF COOPERATIVE NEST-BUILDING

CARL J. ERICKSON and MARIE CHRISTINE MARTINEZ-VARGAS

Psychology Department, Duke University, N.C. and Laboratories for Reproductive Biology, The University of North Carolina (U.S.A.)

Socially coordinated behaviour is a prominent feature of vertebrate reproduction, but there is an extraordinary variety of complexity and duration to be found in the sexual interactions of the many species. In some aquatic forms, for example, it is sufficient for the male and female to release their gametes in the same place at about the same time. Among the mammals, internal fertilization necessitates more extensive behavioural coordination, but viviparity and the restriction of lactation to the female place limits upon the parental contribution of the male once insemination has occurred. Hence interaction of the sexes during the parental phases of the mammalian reproductive cycle are often limited to brief episodes when protecting the young or, as in a few carnivores, when the two parents provide food for the weaned offspring.

Among birds, on the other hand, there are no critical anatomical or physiological constraints linking the nurture of the embryo or the care of the young to either the male or the female once the eggs have been laid. Without these constraints, the coordinated reproductive activities of the two sexes are likely to reflect more clearly than in other animals the great variety of ecological demands to be found in the many habitats. Under most circumstances the chance of reproductive success is enhanced if both parents care for the young, and perhaps it is for this reason more than any other that the great majority of avian species are monogamous (Lack, 1968).

In doves and pigeons the behavioural contribution of the male to the total reproductive effort approaches that of the female. Not only does the male share in the care and feeding of the young after hatching, but, by analogy with mammals, he shares the demands of gestation with the female through his incubation of the eggs; and he shares in the preparation of a supportive environment for the developing embryo through his participation in nest-construction. Because the male and the female both make significant contributions to all phases of the reproductive cycle, it is tempting to conclude that the underlying physiological processes are similar in the two sexes.

Although to some extent this may indeed be the case, it should be noted that courtship, nest-building, incubation, and parental care are functional labels for behaviour patterns that may differ considerably between male and female. In the ring dove (*Streptopelia risoria*), for example, the activities of the male and female tend to be complementary rather than merely additive. The male gathers the nesting material, and the female uses it to construct the nest; the male incubates the eggs during the afternoon hours, and the female sits on them throughout the remainder of the day. Moreover, there is some evidence from our own observations suggesting that there are some differences in the times of day during which the two sexes feed their young. How is this division of labour established? When attempting a physiological analysis of the breeding behaviour of the Columbidae, both the similarities and the differences in behaviour must be kept in the foreground.

THE NEST-BUILDING BEHAVIOUR OF THE RING DOVE

As our first step toward an understanding of the behavioural complementarity of breeding birds we have examined the nest-construction of ring doves. The characteristic structure of this species can hardly be considered a monument of avian architecture. It is a rough, loosely-woven nest lacking the elegance of many other avian fabrications. The selection of nesting materials and their formation into a receptacle for the eggs seems hardly more than random. Nonetheless, this activity has several characteristics warranting attention.

When a male and female dove are placed together in a laboratory cage, nest-building does not begin at once. Instead, the partners spend two or three days in active display. Only by the third or fourth day does nest-building become a prominent feature of the behaviour of the pair. Fig. 1 depicts the average nest-building activity of six pairs of doves. Each pair was placed, in turn, into a large cage in an isolated room and left undisturbed until the eggs were laid. Each day the pairs were observed during four 54-min periods. The first period began at 08.00 h when the lights in the cage were automatically switched on; the second began at 09.00 h, the third at 15.00 h and the fourth at 20.00 h. (The lights were automatically switched off at 22.00 h.) Because the interval between introduction and egg-laying varies from pair to pair, the data were averaged for only the first four days after introduction and also for the four days preceding egg-laying. It is apparent that twig-gathering by the male increases steadily from the second or third day and reaches a peak about two days prior to egg-laying. Moreover, this activity varies markedly throughout each day. In this study we found that it reached a peak between the second and seventh hour of each day. More recently, Erickson and Hutchison (in preparation) have found that the daily peak occurs about the third or fourth hour. In any case it is clear that there is a striking diurnal rhythm to twig-gathering as well as a dramatic overall increase in nest-building activity from the second or third day onward.

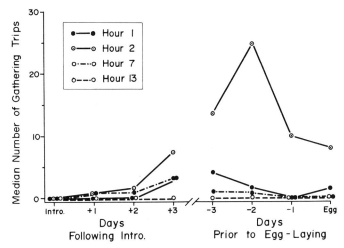

Fig. 1. Twig-gathering by male ring doves. Behaviour is depicted from introduction to egg-laying at four different times of day. (Reproduced by permission of Martinez-Vargas and Erickson, 1973.)

Why does nest-building not begin for several days? The delay could possibly be explained in two different, but not mutually exclusive, ways. The most obvious hypothesis involved an explanation in terms of nest-building hormones. Erickson (1970) has shown that male displays stimulate ovarian activity in the female. This increased ovarian activity is reflected in a sharp rise in plasma levels of oestradiol (Korenbrot *et al.*, 1974) as can be seen in Fig. 2. Silver *et al.*, (1974) have now traced a rise in plasma progesterone in the female as well. Thus it seems very likely that female participation in nest-building is linked to this rise in oestrogen and progesterone. Conversely, it seems possible that a reciprocal effect could account for the gradual rise in twig-gathering by the male. Specifically, stimuli from the female might create a hormonal state within the male which would be conducive to twig-gathering and nest-construction. This hormone state could be associated with an increase in androgen levels beyond those required for display, or they could involve other hormones such as oestrogens or progestogens of gonadal or adrenal origin. It should be noted that this hypothesis carries the assumption that the male is initially unprepared hormonally to gather twigs and that he only begins this activity when the appropriate nest-building hormones are secreted in response to several days of interaction with the female. An alternate hypothesis may be suggested, however. It is quite possible that twig-gathering by the male is more directly influenced by the emerging behaviour of his female partner than by any changes in his hormone state. Some years ago Miller (1965), reporting her studies of dove nest-building, suggested that ". . . the male's behaviour is 'shaped' by the female's behaviour (p. 55)". Our own extended observations tend to support this view.

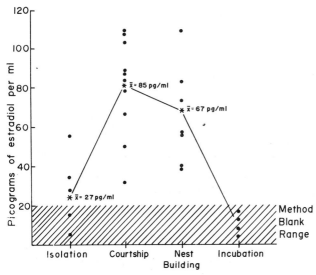

Fig. 2. Oestradiol levels in plasma taken from individual female ring doves during various phases of the breeding cycle. Each point is the average of duplicate determinations for a single individual. The method blank was determined by assaying 2.0 ml column eluate. (Reproduced by permission of Korenbrot *et al.*, 1974.)

When a male and female are first introduced to one another, the male chases the female and often pecks at her quite aggressively. This phase passes rapidly, however, and is replaced by a period of mutual display lasting several days. The predominant behaviour of this second phase has been termed 'nest-soliciting', and it is a prominent activity of both male and female. Fig. 3 portrays a male engaged in this nest-soliciting display. Except for the first day, nest-soliciting occurs almost exclusively in the immediate area where the nest will be built. The display has three principal features: (1) the head is held low — below the level of the feet if the dove is perched — and the tail is held high; (2) the wings are shaken or vibrated in a steady rhythm several times each second, and (3) the bird emits a characteristic vocalization, termed a 'nest-coo', every few seconds while in this position. Uninterrupted nest-soliciting bouts may vary in duration from a few seconds to an hour or more. This activity seems to be very important in the establishment of the nest site and in the commencement of nest-construction itself.

Although nest-soliciting is performed by both the male and the female, its typical distribution in time differs in the two sexes. In Fig. 4 we have presented the median nest-soliciting performances of six males. The display has been portrayed in terms of two measures, the wing-flipping component and the vocalization. Also included is an indication of the time spent by these males at the nest site. Three features should be noted: (1) nearly all of the male's time at the nest site is spent in nest-soliciting; (2) nest-soliciting decreases on

Fig. 3. Male ring dove performing the nest-soliciting display at the nest site.

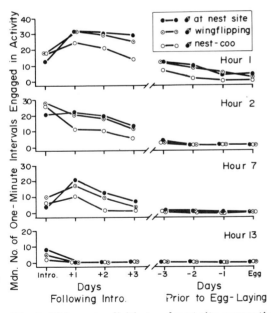

Fig. 4. Male nest-soliciting and nest site occupation. Behaviour is depicted from introduction to egg-laying at four different times of the day. Nest-soliciting is measured separately in terms of wing-flipping and nest-cooing. (Reproduced by permission of Martinez-Vargas and Erickson, 1973.)

96

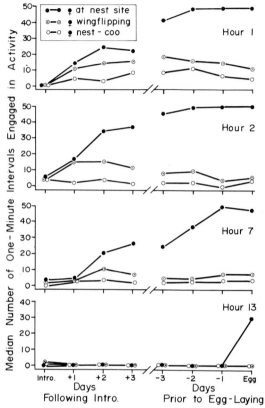

Fig. 5. Female nest-soliciting and nest site occupation. Behaviour is depicted from introduction to egg-laying at four different times of the day. Nest-soliciting is measured separately in terms of wing-flipping and nest-cooing. (Reproduced by permission of Martinez-Vargas and Erickson, 1973.)

each successive day after the first or second day, and (3) nest-soliciting declines throughout each day. In Fig. 5 we see the changes in female performance over the same period. In contrast to the behaviour of the male (1) the time spent by the female at the nest increases steadily each day from introduction to egg-laying; (2) the nest-soliciting fills a decreasing percentage of her time at the nest, and (3) initially, nest-soliciting increases each day and declines only slightly as egg-laying approaches.

While Figs. 4 and 5 provide an interesting comparison of male and female behaviour, a grasp of the full relevance of these behavioural changes requires the following brief description of the interactions of the breeding partners.

On his first day with a female, the male may nest-solicit often, but the female may show little response. On the second day his nest-soliciting seems to attract the female, and she may join him in the nest site and nest-solicit next to him. If the male leaves the site at this stage, however, the female is likely to abandon it as well. By the third or fourth day an important change seems to take place. As on previous days, the female joins the male at the nest site; but when he leaves, there is a high probability that she will remain behind. It seemed significant to us that it was at approximately this time that the male began to select twigs and carry them to the nest site. Moreover, it appeared entirely possible that the male had been physiologically prepared to gather twigs from the day of introduction to the female but that the behaviour emerged only when the female had become attached to the nest area. If females could be induced to develop this behaviour immediately upon introduction to a male, evidence for this relationship might be provided. It was this notion that formed the basis of a second study.

As indicated earlier, the ovary of the female ring dove secretes increasing amounts of oestradiol and progesterone during the few days immediately following introduction to a courting male (Korenbrot et al., 1974; Silver et al., 1974). It seems likely that the increasing attachment of the female to the nest site is linked to the rising titres of these steroid hormones. If this is indeed the case, treatment of females with substantial amounts of exogenous oestrogen and progesterone prior to their introduction to male partners might be expected to induce their attachment to the nest site within hours or even minutes. Such behaviour might, in turn, stimulate the males to gather twigs and deliver them to the treated females.

In our study (Martinez-Vargas and Erickson, 1973) 12 females were given intramuscular injections of diethylstilboestrol (a synthetic oestrogen) and progesterone for five days prior to their introduction to male partners. Each was then placed in a large cage with a supply of nesting material and an elevated shelf containing a shallow wooden box for nest-construction. Twelve control females were treated similarly but received only the oil vehicle without hormone.

The differences between groups were apparent shortly after introduction to the test cages. Within an hour or two most of the hormone-treated females were spending much of their time in the nest site, and by the end of the first day nine of the 12 males paired with them had delivered some nesting material (Fig. 6). Six of the 12 pairs built substantial nests on this first day In contrast, the control females spent little time at the nest site, and only three males of this group carried any nest material. Moreover, in only one case was the material delivered in an amount that was sufficient to form a structure resembling a nest.

These observations suggest, then, that the hormonal condition of the female and her hormone-dependent behaviour are important determinants of the twig-gathering exhibited by the male. It appears that the establishment of a special hormonal state within the male is not a requirement for his participation in nest-building. Does this mean, however, that the male's hormones are unnecessary for his participation? Although the twig-gathering

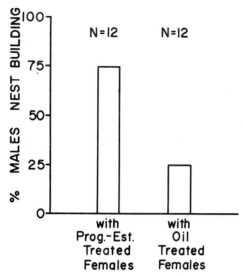

Fig. 6. Twig-gathering by males mated with progesterone—oestrogen-treated females or mated with oil-treated control females. (Reproduced by permission of Martinez-Vargas and Erickson, 1973.)

of the male may require no special nest-building hormone(s), the possibility remained that hormones are nonetheless crucial to this behaviour. Moreover, those hormones which support his nest-soliciting as well as other of his displays (Erickson, 1970) could be the same as those which promote his selection and delivery of nesting material to the female at the nest site. It was this question which formed the basis of a study by Martinez-Vargas (1974).

As a preliminary to her investigation, Martinez-Vargas matched three groups of males according to their twig-gathering activities when exposed to females that had been treated with oestrogen and progesterone. She then castrated all males and kept them in visual isolation from other birds for three weeks. At the end of this period they were tested with untreated females for nest-soliciting and other gonad-dependent displays. Absence of these displays in the males was accepted as an indication of the effectiveness of castration. Following this test 33 castrated males were assigned to three groups of equal size. Those of one group were given daily injections of testosterone propionate, males of a second group were treated with oestradiol benzoate, and males in the third group were given injections of the oil vehicle as a control. At the completion of three weeks of treatment (six weeks after castration) the males were tested for their nest-building behaviour with females that had been pretreated with oestrogen and progesterone. Fig. 7 shows that castrated males receiving no steroid hormone performed little or no twig-gathering. It should be noted that these males failed to select and deliver nesting material in spite of the fact that female stimulus animals often

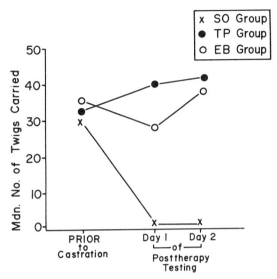

Fig. 7. Twig-gathering performances of castrated males treated either with oestradiol benzoate (o), testosterone propionate (•), or sesame oil (×). Males were paired with females that had been treated with oestradiol and progesterone. (Reproduced by permission of Martinez-Vargas, 1974.)

spent long periods at the nest site. In marked contrast, castrated males given either testosterone propionate or oestradiol benzoate responded with active twig-gathering, much like the males of the previous study. Viewed together, these studies show that gonadal secretions are very important in the nest-building activities of both the male and the female. Social stimulation is of fundamental significance as well, but the way in which this social stimulation influences nest-construction behaviour differs in the two sexes. The male's principal influence upon the female can be characterized as a 'primer' effect (Wilson and Bossert, 1963). His displays stimulate female ovarian activity resulting in a rise in oestrogen and progesterone. These steroid hormones in the female, in turn, promote the emergence of those specific behaviour patterns which directly elicit twig-gathering in the male. The responsiveness of the male is dependent upon his own gonadal hormones, however, and without them the behaviour of the female has little effect in inducing his participation.

Although the above summary describes the principal influences upon the nest-building behaviour of the male and female, it is possible that other aspects of the interaction have an additional facilitating effect on this behaviour. For example, the female may 'prime' in the male the secretion of non-gonadal hormones and (when he is not castrated) gonadal hormones that could further promote his selection and delivery of nesting material. Conversely, the female may be further stimulated to remain at the nest site by the male's nest-soliciting and active engagement in twig-gathering.

LOCUS OF HORMONE ACTION

Several studies have indicated that the courtship displays of the male ring dove are directly influenced by the action of gonadal steroid hormones upon the brain. As Hutchison (1967, 1970b and 1971) and Barfield (1971) have demonstrated, hormone implants are generally most effective when placed in the preoptic region and anterior hypothalamus. In the study of Martinez-Vargas (1974) described above, systemic injection of castrated males with gonadal hormones was very effective in promoting twig-gathering behaviour. Could localized implantation of gonadal hormones in the brain produce a similar induction of twig-gathering behaviour if males were introduced to properly prepared females? Erickson and Hutchison (in preparation) have collected preliminary evidence suggesting that gonadal hormones do indeed support the nest-building activity of the male through their direct action upon the CNS. Using a procedure which somewhat parallels that of the Martinez-Vargas study, Erickson and Hutchison examined the response of castrated males to intracerebral implants of testosterone propionate. Prior to castration, males were tested for twig-gathering with females that had been treated with oestrogen and progesterone. As in the initial study of Martinez-Vargas and Erickson (1973) these males readily engaged in nest-building. In fact, they were much more active nest-builders than the males of previous studies. This was apparently due to the introduction as nest-building material of split cane, the strong flexible material used in the weaving of seats for chairs. Six to eight weeks after castration, however, very little cane was gathered by the males. It should be noted that a few birds continued to carry substantial amounts of nesting material long after castration, a consideration to which we will return later. But in the great majority of cases, little or no gathering was being performed when the males were prepared to receive their intracerebral implants of testosterone propionate approximately two months after castration. This reduction in behaviour was completely reversed by implants of testosterone propionate in the preoptic region and anterior hypothalamus (Fig. 8). Generally, participation in nest-building was restored to a level at or above that exhibited prior to castration. In contrast, implants in palaeostriatal or most archistriatal regions produced no discernible rise in twig-gathering behaviour.

In our initial studies of breeding ring doves we noted that nest-building was largely confined to the hours of late morning and early afternoon (Fig. 1). In the Erickson and Hutchison study the pairs with effective brain implants of testosterone propionate exhibited this same daily rhythm of behaviour. This indicates that the diurnal changes in nest-building activity are not dependent upon underlying changes in testicular secretion. It is possible

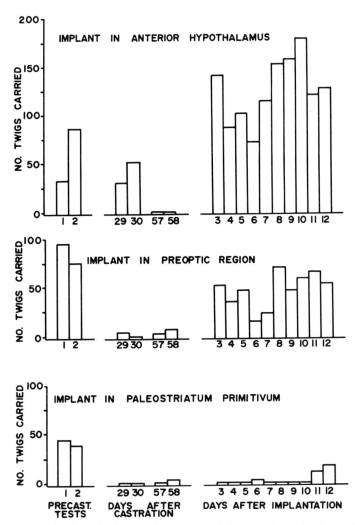

Fig. 8. Twig-gathering performances of three males prior to castration, following castration, and following intracerebral implantation with testosterone propionate.

that they are influenced by gonadal rhythms in the stimulus females since these females were not gonadectomized. On the other hand, these females were being given large amounts of oestradiol and progesterone, and it can be assumed that their normal patterns of ovarian activity were disrupted as a result. The most likely hypothesis is that daily rhythms reflect changes within the CNS itself and that these changes are independent of the patterns of steroid hormone secretion from the gonads. It should be further noted that the behaviour of implanted males was in no way discernibly different from that of normal nest-building males.

Although the locations of all testosterone propionate implants have not yet been determined, preliminary analysis indicates that there is considerable correspondence between the effective sites from which courtship and copulation are elicited (Hutchison, 1967, 1970b and 1971; Barfield, 1971) and those from which twig-gathering can be produced. Moreover, a recent autoradiographic analysis of testosterone uptake in the brain of the male ring dove (Martinez-Vargas *et al.*, 1974) suggests a further striking correspondence between those sites from which courtship and twig-gathering can be elicited and those basal areas of the brain in which [^{3}H]testosterone or its metabolites become highly concentrated.

OESTROGEN OR ANDROGEN?

Nest-building was induced in the male ring dove through intracranial implantation of testosterone propionate; oestrogens have not yet been applied by this route to determine their effects with respect to this behaviour. Martinez-Vargas (1974) found, however, that oestradiol benzoate, when given systemically, was as effective as similarly applied testosterone propionate in generating this behaviour. We are reasonably confident that if oestradiol were implanted into the anterior hypothalamus or preoptic region, it, too, would effectively induce twig-gathering behaviour in male ring doves.

Hutchison (1971) has implanted both testosterone propionate and oestradiol monobenzoate into the brains of castrated male doves and has found that the nest-soliciting display can be restored with either of these steroid hormones. Stern and Lehrman (1969) and Stern (1974) found that castrated male ring doves would not readily incubate eggs. Yet if systemic injections of either testosterone propionate or oestradiol benzoate were combined with progesterone, the males would then sit on the eggs. Progesterone alone was ineffective. In summary, nest-soliciting displays, twig-selection and delivery, and incubation behaviour of the male can be induced by either oestrogen or androgen application.

Although both classes of steroid hormones are very effective in supporting these behaviour patterns, several investigators have suggested that the oestrogens, rather than the androgens, may be the naturally effective agents controlling several of these behaviour patterns in male pigeons and doves. Lofts *et al.* (1968) noted that when male pigeons were treated with 22,25-diazacholesterol dihydrochloride, an avian antifertility agent, low doses damaged the Sertoli cells and there was an associated decrement in nest-soliciting (these authors designate the behaviour as 'nest-demonstration'). Higher doses damaged both Sertoli and Leydig cells, and the nest-soliciting and bow-cooing displays were both reduced. They suggested that nest-soliciting is normally controlled by oestrogen from the Sertoli cells and bow-cooing displays by androgen from the the Leydig cells; but they produced no direct evidence that oestrogen from the testis, if produced in the Sertoli cells, reaches the blood and brain. Hutchison (1970a) found that male Barbary doves which did not nest-solicit prior to castration did so more readily when treated with oestradiol monobenzoate than when given testosterone propionate. He suggested that oestrogen might be the normally effective hormone in the nest-soliciting display of the male dove.

Korenbrot et al. (1974) used a radioimmunoassay in an attempt to detect and measure oestradiol in the blood plasma of courting and nest-building male ring doves. They were unable to detect the hormone. The possibility remains, however, that blood samples were collected at times of the day when the hormone was not being actively secreted or that some other form of oestrogen was present but not detected. There is the further possibility that oestrogen is the effective agent in the nervous system but that it is not found in the blood because it is converted from androgen by the nervous system itself. For instance, Naftolin et al. (1972) and Vangala et al. (1973) have demonstrated that in the mammalian brain androstenedione, one of the androgens, can be aromatized to oestrone by tissue of the anterior hypothalamus and limbic system.

While it is clear that oestrogens, when given exogenously, can produce several behaviour patterns that are characteristic of the male ring dove's reproductive cycle, there is as yet no direct evidence indicating that oestrogens from any endogenous source enter the blood or brains of males doves or pigeons.

In contrast, testosterone propionate is very effective in stimulating all of the behaviour patterns that have been linked to oestrogens. Moreover, testosterone is present in the testes of male doves (Katongole and Hutchison, personal communication; Zarutskie et al., unpublished observations), and it is also found in the blood (Silver et al., personal communication). Testosterone also enters the brain (Stern, 1972) and the fact that the bow-coo display of the male can be elicited by localized brain implants of testosterone propionate but cannot be produced by oestrogen treatments — either in the form of brain implants or systemic injections — argues that androgen controls at least some behaviour patterns without prior conversion to oestrogen (Martinez-Vargas, 1973).

At present there is insufficient evidence to determine whether the androgens or the oestrogens are more directly involved in the control of male nest-soliciting, twig-gathering, or incubation behaviour. From our viewpoint, however, the evidence currently available favours a major role of the androgens.

THE RELATIONSHIP OF NEST-BUILDING BEHAVIOUR TO OTHER REPRODUCTIVE BEHAVIOUR PATTERNS OF THE MALE RING DOVE

As indicated earlier, the courtship displays, nest-building activities, and incubation behaviour of the male ring dove are all reduced or eliminated following gonadectomy; replacement therapy with gonadal hormones promotes their return. It appears, however, that these behaviour patterns differ in terms of their relationships to the hormones influencing them. When a male ring dove is castrated, his bow-coo and nest-soliciting displays exhibit a very rapid decline; and usually they are not elicited more than a week or ten days following gonadectomy. Moreover, it is relatively difficult to reinstate these behaviour patterns to precastration levels using either systemic injections or hypothalamic implants of testosterone propionate (Hutchison, 1970a and 1971). If this therapy is delayed more than 30 days after castration, attempts at reinstatement of the behaviour using these techniques become increasingly difficult, although the nest-soliciting display is more readily induced than is bow-cooing (Hutchison, 1969).

The participation of the male in nest-building appears to be quite differently affected by castration. Gonadectomy may result in only a slight decline of the behaviour during the first few weeks following surgery, and not until some 40 days or more have passed does this behaviour reach a low level in a majority of the animals. A few males may remain remarkably active for several months. Moreover, systemic injections (Martinez-Vargas, 1974) or brain implants (Erickson and Hutchison, in preparation) readily restore the behaviour to levels at or above those performed by the animals prior to gonadectomy. (See Fig. 8 for examples.) This restoration occurs even when replacement therapy is delayed more than two months after castration.

Incubation behaviour may resemble nest-building more than male courtship in this regard. Stern and Lehrman (1969) found that fewer progesterone-treated male ring doves incubated eggs when castrated than when intact. But fully one-third of the castrated males did incubate eggs one month after castration. Similarly, Silver *et al.* (1973) found that untreated castrated males incubated less than intact males when tested about one month following gonadectomy, but many were still exhibiting the behaviour at this time. Unfortunately, there have not yet been any longitudinal studies tracing the decline of incubation following castration, nor have there been studies to determine whether the sensitivity of systems underlying nest-building or incubation change as a function of time following removal of the testes.

THE POSSIBLE FUNCTIONAL SIGNIFICANCE OF VARIATIONS IN HORMONAL DEPENDENCE

It is possible that the variations described above simply reflect differences in the effectiveness of the stimulus conditions for the male in the various testing contexts. For example, twig-gathering might have been more readily produced because of the attractive nesting material provided or because the cage conditions for examining courtship display were in some way inadequate. On the other hand, the differences in behavioural decline following castration and the variations in response to replacement therapy may represent basic differences in physiological organization, and these differences could have functional significance of a fundamental nature.

It has been suggested (Trivers, 1972) that through their courtship, birds may communicate information to potential mates concerning their own promise as parents. Trivers argues that the reproductive cycle represents an ongoing investment in the offspring. This being the case, it is to the advantage of each partner to mate with individuals that will contribute optimally to successful reproduction. In many monogamous species, such as the doves and pigeons, the male and female both make essential contributions to the reproductive effort after copulation has occurred. As indicated above, the male and female share in the construction of the nest, incubation of eggs, and care of the squabs. Thus it would be advantageous for each animal to know at an early stage whether the mate being selected was physiologically prepared to carry through with the entire breeding sequence. Specifically, it would be of considerable advantage to a female ring dove to be able to detect those males that are physiologically prepared to engage in nest-building, incubation, and squab care. Because the physiological readiness of males may vary considerably with age and season, such selection by the female could be critical in determining her breeding success. It is, of course, advantageous to the male when he is physiologically prepared, to communicate his parental potential to the available females. The most likely means of communication is through courtship, since in the male ring dove there are no visible morphological changes that correspond with changes in his reproductive condition.

Courtship displays can only have this function if they do, in fact, predict parental readiness. Otherwise selection for female responsiveness to them would not have occurred. If courtship displays are predictive of parental activities, however, there must be an appropriate organization of underlying mechanisms linking them. In the male ring dove we have seen that courtship displays, nest-building activities, and incubation of the eggs are all promoted by the secretion of gonadal steroids. (The role of the gonads in squab care is unclear.) There is, then, some reason to believe that these activities share a common basis. But the fact that these various behaviour patterns are influenced by the same steroid hormones is insufficient evidence in itself to ensure that courtship is predictive of nest-building and incubation unless the sensitivities of the various androgen-dependent systems are organized in such a way that the later behaviour patterns necessarily follow upon the earlier ones. For example, if the systems mediating courtship displays were highly sensitive to androgen but the others were not, newly maturing males or adult males slowly increasing in their androgen output in the breeding season might readily court females but would be incapable of following through in the other stages of breeding. On the other hand, if the sensitivity of systems mediating the courtship displays is lower than that of the other behaviour patterns, the emergence of display is a virtual guarantee that hormone levels are adequate for the later-appearing behaviour patterns as well.

It should be noted that there is no reason to assume that the gonadal hormones have to be sustained throughout the breeding cycle. The studies by Stern and Lehrman (1969), Silver *et al.* (1973), Stern (1974), and Erickson and Hutchison (in preparation) indicate that even after the removal of the gonads and the presumed withdrawal of the hormone, the males are capable of nest-building and incubating eggs for several weeks or months. During the normal breeding these stages would be reached within a fortnight of initial contact between the male and female. It appears, though, that testosterone secretion does, in fact, rise during the period of active courtship and nest-building before falling as incubation begins (Silver *et al.*, personal communication). The significance of these changes is not clear, but they may represent a 'fail-safe' organization ensuring that the nervous system is sufficiently sensitized by the androgen at the beginning of the reproductive cycle. Such an organization would guarantee that the male is fully prepared to negotiate the successive phases of the breeding cycle successfully.

The notion that courtship may be predictive of behaviour in later stages of the reproductive cycle has recently received empirical support from Nisbet (1973) in his study of a colony of common terns (*Sterna hirundo*) on the Massachusetts coast. In this species it is the male that provides most of the food for the young during the period shortly after hatching. Moreover, there is a wide variation among the males in their capacity to locate and select the proper food for the chicks. Therefore, selection of a male mate with the appropriate paternal capabilities is of considerable consequence to the breeding success of the female. Females appear to select males as mates, not on the basis of their demonstrated parental behaviour, however, but on the basis of their courtship early in the breeding season. According to Nisbet, the courtship of the males passes through three phases. In the first phase, they carry fish throughout the colony while displaying to females; the females are sometimes fed the fish but only irregularly. In the second phase the males and females may be seen paired on the feeding grounds, and the male often feeds the female. During the third phase the female remains in the male's territory, and she is fed there by the male until the clutch of eggs is complete. In the group of terns under study, Nisbet found an appreciable variation among the pairs in the total weight of the clutch, in the amount of food fed to the chicks by the male, in the weight of the brood, and in the number of chicks fledged. There was also considerable variation among the pairs in the weight of the third egg and the weight of the third chick. (Common terns lay but three eggs, and it is this final egg that seems most to reflect the nutritive investment of the parents.) Most significant, however, was Nisbet's discovery that all of these variables were positively correlated with the amount of food fed to the female by the male during courtship. Thus, the male's courtship feeding was predictive of his later parental behaviour. As has been suggested for the ring dove, this indicates a link between the mechanisms controlling courtship and those influencing parental behaviour. But Nisbet's study raises the possibility of further relationships as well.

In some species courtship may represent more than the hormonal readiness of a male to engage in nest-building or parental care. Conceivably, the quality of the courtship could reflect the special capabilities or talents of the male vis-à-vis these later phases of the reproductive cycle. Many aspects of reproductive behaviour benefit from experience acquired in their performance. In the ring dove, for example, incubation behaviour and squab care are significantly affected by prior breeding experience (Lehrman and Wortis, 1960 and 1967). Thus, it would be a distinct advantage for individuals to communicate through courtship their personal talents for parental care. In the black-crowned night heron (*Nycticorax nycticorax*), for instance, stick-waving is a conspicuous component of the courtship display (Noble and Wurm, 1940). Is it possible that this activity provides a means of communicating the individual's special ability to find and successfully compete for scarce nesting material? At the present time we know of no clear evidence supporting such a suggestion, but the possibility exists that courtship could transfer information of this kind.

In several species of terns it is necessary for the male to feed small, easily digested types of food material to the newly hatched chicks. To the extent that fishing experience increases the success of adult terns in finding food for themselves (Ashmole and Tovar, 1968; Dunn, 1972), experience may be of considerable importance in the selection of appropriate food for the young as well. Buckley and Buckley (1972) noted that young royal terns (*Sterna* (*Thalasseus*) *maxima maxima*) occasionally refused food brought to them by their parents. Possibly this was because the food was too large or of the wrong type for ingestion. Whether the ability of the male to select and deliver the appropriate food depends upon experience or simply his own individual proclivities, his attractiveness to females could possibly depend, not only upon the vigour of his courtship as an indication of his general parental inclination, but upon the type of food he presents during courtship.

CONCLUSIONS

The sequence of courtship, nest-building, incubation of eggs, and parental care of young which characterizes the reproductive cycle of many avian species is customarily viewed as a chain of discrete phases. Commonly, the behavioural phases are assumed to be associated with a series of rather disparate physiological (*e.g.* hormonal) states. In the male dove, however, we have suggested that nest-building behaviour is influenced by the same gonadal steroids that support courtship and incubation behaviour. The fact that several phases of the reproductive cycle are affected by similar physiological factors may have considerable functional significance. We have suggested that because of these common physiological links, the breeding partners may predict the parental capacities of their mates.

108

REFERENCES

Ashmole, N.P. and Tovar, S.H. (1968) Prolonged parental care in royal terns and other birds. *Auk*, 85, 90—100.

Barfield, R.J. (1971) Activation of sexual and aggressive behavior by androgen implanted ·into the male ring dove brain. *Endocrinology*, 89, 1470—1476.

Buckley, F.G. and Buckley, P.A. (1972) The breeding ecology of royal terns *Sterna (Thalasseus) maxima maxima*. *Ibis*, 114, 344—359.

Dunn, E.K. (1972) Effect of age on the fishing ability of sandwich terns *Sterna sandvicensis*. *Ibis*, 114, 360—366. ⁄

Erickson, C.J. (1970) Induction of ovarian activity in female ring doves by androgen treatment of castrated males. *J. comp. physiol. Psychol.*, 71, 210—215.

Hutchison, J.B. (1967) Initiation of courtship by hypothalamic implants of testosterone propionate in castrated doves (*Streptopelia risoria*). *Nature (Lond.)*, 216, 591—592.

Hutchison, J.B. (1969) Changes in hypothalamic responsiveness to testosterone in male Barbary doves (*Streptopelia risoria*). *Nature (Lond.)*, 222, 176—177.

Hutchison, J.B. (1970a) Differential effects of testosterone and oestradiol on male courtship in Barbary doves (*Streptopelia risoria*). *Anim. Behav.*, 18, 41—51.

Hutchison, J.B. (1970b) Influence of gonadal hormones on the hypothalamic integration of courtship behaviour in the Barbary dove. *J. Reprod. Fert.*, Suppl., 11, 15—41.

Hutchison, J.B. (1971) Effects of hypothalamic implants of gonadal steroids on courtship behaviour in Barbary doves (*Streptopelia risoria*). *J. Endocr.*, 50, 97—113.

Korenbrot, C.C., Schomberg, D.W. and Erickson, C.J. (1974) Radioimmunoassay of plasma estradiol during the breeding cycle of ring doves (*Streptopelia risoria*). *Endocrinology*, 94, 1126—1132.

Lack, D. (1968) *Ecological Adaptations for Breeding in Birds*. Methuen, London.

Lehrman, D.S. and Wortis, R.P. (1960) Previous breeding experience and hormone-induced incubation behavior in the ring dove. *Science*, 132, 1667—1668.

Lehrman, D.S. and Wortis, R.P. (1967) Breeding experience and breeding efficiency in the ring dove. *Anim. Behav.*, 15, 223—228.

Lofts, B., Murton, R.K. and Thearle, R.J.P. (1968) The effects of 22,25-diazacholesterol dihydrochloride on the pigeon testis and on reproductive behaviour. *J. Reprod. Fert.*, 15, 145—148.

Martinez-Vargas, M.C. (1973) *The Induction of Nest Building in the Ring Dove (Streptopelia risoria): Hormonal and Social Factors*. Unpublished Ph.D. dissertation, Duke University, Durham, N.C.

Martinez-Vargas, M.C. (1974) Nest building in the ring dove (*Streptopelia risoria*): hormonal and social factors. *Behaviour*, 50, 123—151.

Martinez-Vargas, M.C. and Erickson, C.J. (1973) Social and hormonal determinants of nest building in the ring dove (*Streptopelia risoria*). *Behaviour*, 45, 12—37.

Martinez-Vargas, M.C., Sar, M. and Stumpf, W.E. (1974) Brain targets for androgens in the dove (*Streptopelia risoria*). *Amer. Zoologist*, (Abstract) 14, 1285.

Miller, S. (1965) *Nest-Building Activity in the Ring Dove (Streptopelia risoria)*. Unpublished Ph.D. dissertation, Rutgers University, New Brunswick, N.J.

Naftolin, F., Ryan, K.J. and Petro, Z. (1972) Aromatization of androstenedione by the anterior hypothalamus of adult male and female rats. *Endocrinology*, 90, 295—298,

Nisbet, I.C.T. (1973) Courtship-feeding, egg-size and breeding success in common terns. *Nature (Lond.)*, 241, 141—142.

Noble, G.K. and Wurm, M. (1940) The effect of testosterone propionate on the black-crowned night heron. *Endocrinology*, 26, 837—850.

Silver, R., Feder, H.H. and Lehrman, D.S. (1973) Situational and hormonal determinants of courtship, aggression, and incubation behavior in male ring doves (*Streptopelia risoria*). *Horm. Behav.*, 4, 163—172.

Silver, R., Reboulleau, C., Lehrman, D.S. and Feder, H.H. (1974) Radioimmunoassay of plasma progesterone during the reproductive cycle of male and female ring doves. (*Streptopelia risoria*). *Endocrinology*, 94, 1547—1554.

Stern, J.M. (1972) Androgen accumulation in hypothalamus and anterior pituitary of male ring doves; influence of steroid hormones. *Gen. comp. Endocr.*, 18, 439—449.

Stern, J.M. (1974) Estrogen facilitation of progesterone-induced incubation behavior in castrated male ring doves. *J. comp. physiol. Psychol.*, 87, 332—337.

Stern, J.M. and Lehrman, D.S. (1969) Role of testosterone in progesterone-induced incubation behaviour in male ring doves (*Streptopelia risoria*). *J. Endocr.*, 44, 13—22.

Trivers, R.L. (1972) Parental investment and sexual selection. In B. Campbell (Ed.), *Sexual Selection and the Descent of Man*. Aldine, Chicago, Ill., pp. 136—179.

Vangala, V., Reddy, R., Naftolin, F. and Ryan, K.J. (1973) Aromatization in the central nervous system of rabbits: effects of castration and hormone treatment. *Endocrinology*, 92, 589—594.

Wilson, E.O. and Bossert, W.H. (1963) Chemical communication among animals. *Recent Progr. Horm. Res.*, 19, 673—716.

6

TARGET CELLS FOR GONADAL STEROIDS IN THE BRAIN: STUDIES ON HORMONE BINDING AND METABOLISM

RICHARD E. ZIGMOND*

M.R.C. Neurochemical Pharmacology Unit, Department of Pharmacology, Medical School, University of Cambridge (Great Britain)

While evidence for the dependence of reproductive behaviour on gonadal hormones stems from the earliest observations of experimental endocrinology (Berthold, 1849)**, knowledge of the sites of action of these hormones has only recently been acquired. Part of the behavioural effects of oestrogens and androgens may result from actions on neurones in the spinal cord and sensory neurones innervating the genital area (Hart and Haugen, 1968; Komisaruk *et al.*, 1972; Kow and Pfaff, 1974). However, the most clearly demonstrated site of action of these hormones is the preoptic—hypothalamic region of the brain. The hypothesis that this area contains target cells for oestrogens and androgens is based on experiments in which these hormones were implanted into different parts of the nervous system in gonadectomised animals. These studies have shown that many of the behavioural effects produced by injections of oestrogens and androgens can also be elicited by implanting the hormones directly into the preoptic—hypothalamic area (see Hutchison, 1975 and Chapter 7 in this volume). (Fig. 1a.)

Another approach to the study of gonadal hormone action in the brain has been to gonadectomise animals and give them 'replacement therapy' using radioactive steroids, thus facilitating the subsequent localisation of the compounds in various parts of the body. When tritiated oestradiol, testosterone or progesterone was injected into a number of mammalian species, certain cells in the brain were found to concentrate and retain one or more of these compounds. Similar phenomena have been described in less detail in a number of avian species.

The distribution of cells in the brain which concentrate different steroids and the biochemical characteristics of the retention process have been re-

---

* Alfred P. Sloan Foundation Research Fellow.
** A translation of the original German article into English is included in the introduction to Harris (1955).

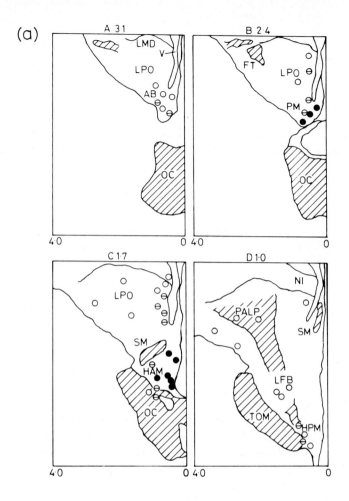

(a)

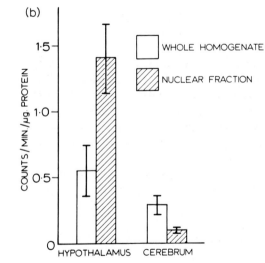

(b)

viewed recently (Zigmond, 1975). This chapter will be limited to a discussion of those aspects of gonadal steroid retention which are particularly relevant to avian behaviour. However, since by far the most complete information about steroid hormone retention in the brain exists for oestradiol in the rat, and since most of the main properties of this system seem to hold for other steroids and other species, the retention of oestradiol in the rat brain will be described in some detail.

LIFE HISTORY OF A STEROID HORMONE IN THE BRAIN

There seems to be no barrier to the entry of steroid hormones into the central nervous system. In all cases studied these hormones penetrated rapidly into the brain after a peripheral injection and the amount of hormone found in the brain was proportional to the plasma level over a wide range of concentrations (Laumas and Farooq, 1966; Eisenfeld, 1967; Raisinghani et al., 1968; McEwen and Pfaff, 1970; McEwen and Weiss, 1970; McEwen et al., 1970a). Thus the brain is in a position to monitor fluctuations in the circulating levels of these substances.

In terms of the initial interaction of steroid hormones with their target cells there is no evidence for specific binding sites on the cell membrane (King and Mainwaring, 1974). These compounds probably enter all cells in the body but are only retained in certain cells. This retention results from the presence in these cells of macromolecules having a high affinity for the steroid in question. In the case of oestradiol for instance, it has been shown that in the cytoplasm of certain brain cells, macromolecules exist which bind oestrogens.

Most studies of this binding have been carried out in vitro using 'cytosol' fractions prepared by disrupting brain cells by homogenisation and then re-

Fig. 1. a: location of intracerebral implants of testosterone proprionate in castrated male Barbary doves. (Reproduced by permission of Hutchison, 1971.) Each symbol denotes the position of the tip of the implant in the transverse section and the type of behaviour elicited (for sterotaxic coordinates see original paper). ●, denotes complete courtship with aggressive and nest-orientated behaviour. ⊖, denotes incomplete courtships, lacking in either aggressive or nest-orientated behaviour. ○, denotes no courtship, behaviour similar to a castrate. AB, area basalis; OC, optic chiasma; FT, tractus frontothalamicus; HAM, nucleus hypothalamicus anterior medialis; HPM, nucleus hypothalamicus posterior medialis; LFB, lateral forebrain bundle; LMD, lamina medullaris dorsalis; LPO, lobus parolfactorius; PALP, palaeostriatum primitivum; NI, neostriatum intermedialis; PM, nucleus preopticus medialis; SM, tractus septomesencephalicus; TOM, tractus opticus marginalis; V, ventricle. (Scales represent 1 mm.) b: retention of radioactivity in the whole homogenate and nuclear fraction 1 h after an intravenous injection of [^{3}H]testosterone in the ring dove. The anatomical landmarks used in the dissection are given in Stern (1972). The sample labelled 'hypothalamus' included the preoptic area. Data are given in c.p.m. ^{3}H/μg protein ± S.E.M. (Reproduced by permission of Zigmond et al., 1972a).

moving all the cell organelles by centrifugation. When such cytosol fractions from different regions of the rat brain were incubated with a low concentration of [³H]oestradiol, more binding per unit weight of tissue was found in the preoptic—hypothalamic and amygdaloid samples than in the sample from the cerebral cortex (Eisenfeld, 1970; Zigmond, 1971). It should be noted that most of the behavioural and physiological effects of implanting oestradiol in the rat brain have been found from implants in the preoptic—hypothalamic and amygdaloid regions (see Zigmond, 1975 for references).

By keeping the amount of tritiated oestradiol in the incubation constant and adding increasing amounts of non-radioactive oestradiol one can demonstrate the saturability of the binding (Eisenfeld, 1970). If, instead of adding unlabelled oestradiol, other compounds are added, the specificity of the process can be studied. For instance neither progesterone nor testosterone affect the binding of oestradiol even at fairly high concentrations (Eisenfeld, 1970). This finding is important since both these hormones circulate in the plasma of female rats at concentrations higher than that of oestradiol.

Proteolytic enzymes inhibit oestradiol binding when added to incubations of brain cytosols, while enzymes which degrade nucleic acids have no effect. These results suggest that the binding macromolecule is a protein, or at least that a protein component is essential for binding to occur.

Oestradiol has also been found to bind to macromolecules in the cytosol *in vivo* (Zigmond, 1971). However, with increasing time after an injection more of the labelled hormone is found bound in the cell nucleus (Zigmond and McEwen, 1970; Mowles *et al.*, 1971). As with cytosol binding, nuclear binding of oestradiol is highest in the preoptic—hypothalamic and amygdaloid regions and is highly stereoselective (Zigmond and McEwen, 1970). Oestradiol-17α, a stereoisomer of oestradiol-17β (the naturally occurring hormone in most species), does not interfere with [³H]oestradiol-17β retention unless injected in extremely high doses (Zigmond and McEwen, 1970). The oestradiol found in hypothalamic cell nuclei appears to be bound to a macromolecule (Mowles *et al.*, 1971).

Nuclear retention of oestradiol can also be demonstrated *in vitro* by incubating pieces of hypothalamic tissue with labelled hormone (Chader and Villee, 1970; Clark *et al.*, 1972). However, purified hypothalamic nuclei will not retain the hormone unless they are incubated in the presence of cytosol (Chader and Villee, 1970). These results, particularly if taken in the context of what is known about oestradiol binding in peripheral target cells, favour a scheme in which oestradiol first binds to a macromolecule in the cytosol and that this steroid—macromolecule complex then moves into the nucleus. At present we do not known what happens to the hormone after it enters the nucleus. However, the data of Anderson *et al.* (1973) suggest that 12 h after a subcutaneous injection of oestradiol the steroid—macromolecule complex is no longer present in the nucleus.

Autoradiographic techniques especially designed for use with diffusible substances such as steroids have allowed the precise mapping of oestradiol-concentrating cell bodies in the rat brain (Stumpf, 1970; Pfaff and Keiner, 1973). The highest concentration of labelled cells was in the preoptic area, the medial basal hypothalamus and the amygdala; however, labelled cells were also found elsewhere in the brain. Many questions remain to be answered concerning the 'oestradiol-concentrating cells'. For instance, in some cases it is not clear whether the cells are neurones or glia (Pfaff and Keiner, 1973). In the case of labelled neurones it would be useful to determine where the cells project, what neurones synapse on them and what transmitters are involved in these different connections.

TESTOSTERONE AND PROGESTERONE AS 'PREHORMONES'

The main difference between the interaction of oestradiol with its target tissues and that of testosterone or progesterone concerns the metabolism of the hormones. The vast majority (>85%) of the tritium found in the hypothalamus, uterus or anterior pituitary gland 1—2 h after an injection of [^{3}H] oestradiol is still present as [^{3}H] oestradiol, whereas [^{3}H] testosterone and [^{3}H] progesterone are more extensively metabolised. After an [^{3}H] testosterone injection [^{3}H] dihydrotestosterone, [^{3}H] androstenedione and [^{3}H] 3α-androstanediol were found in the brain (Stern and Eisenfeld, 1971; Sholiton et al., 1972) (Fig. 2). Two findings suggest that these metabolites are at least partially formed in the brain itself. First of all they are found in the brains of rats whose livers — the main peripheral site of steroid metabolism — have been removed (Sholiton et al., 1972) and secondly the brain contains all the enzymes necessary for the formation of these metabolites — 5α-steroid reductase, 3α-steroid dehydrogenase and 17β-steroid dehydrogenase (Rommerts and Van der Molen, 1971) (Fig. 2). After an injection of [^{3}H] progesterone in the rat, [^{3}H] pregnanedione and [^{3}H] 3α-hydroxy-5α-pregnan-20-one were found in the brain (Raisinghani et al., 1968; Wade and Feder, 1972). (Fig. 3). The same enzyme, 5α-steroid reductase, seems to be responsible for the conversion of testosterone to dihydrotestosterone and progesterone to pregnanedione. The two enzymatic activities have a similar regional and subcellular distribution and progesterone interferes with the conversion of testosterone in vitro, probably by competitive inhibition (Rommerts and Van der Molen, 1971; Snipes and Shore, 1972; Denef et al., 1973). 5α-steroid reductase activity in the brain is highest in the brain stem and diencephalon.

Recently a great deal of interest has been directed at another route of androgen metabolism, i.e. to oestrogens. Brain tissue from neonatal and adult rats of both sexes have been shown to convert androstenedione to oestrone in vitro (Naftolin et al., 1972; Reddy et al., 1974). (Fig. 4.) This conversion was most prominent in the anterior hypothalamus and in a sample containing the amygdala and hippocampus.

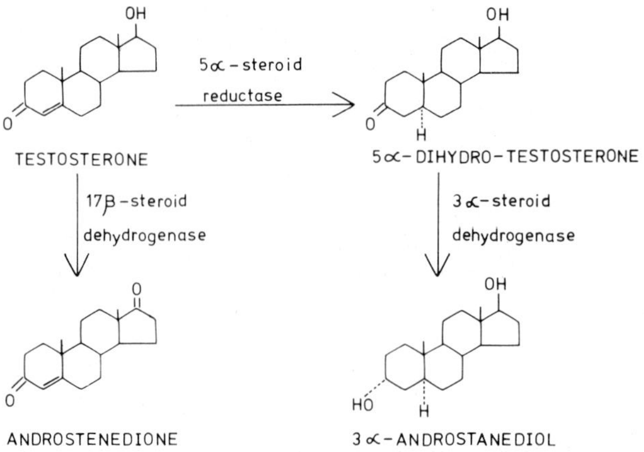

116

Fig. 2. Androgen metabolism in the rat brain. After testosterone is injected peripherally, testosterone, 5α-dihydrotestosterone, androstenedione, and 3α-androstanediol can be found in the brain. These metabolites of testosterone may be formed in the brain since the enzymes 5α-steroid reductase, 3α-steroid dehydrogenase and 17β-steroid dehydrogenase are present in this tissue. The figure shows the chemical structures of the four androgens and possible routes of formation of the metabolites from testosterone.

The functional significance of the metabolism of testosterone and progesterone in the brain is not yet clear. In the case of certain of the peripheral effects of testosterone it seems likely that dihydrotestosterone is the active androgen and that testosterone can be considered a 'prehormone' (see Zigmond, 1975 for references). Dihydrotestosterone may also act in the brain both in the control of pituitary gonadotrophin release and in the regulation of male sexual behaviour, although in the latter case there seem to be species differences (Feder, 1971; Phoenix, 1973; Hutchison, Chapter 7 in this volume). Oestradiol has also been shown to mimic some of the behavioural effects of testosterone, raising the question of whether aromatisa-

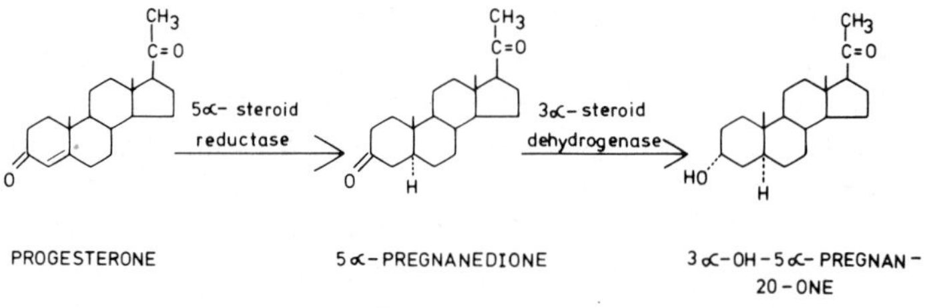

Fig. 3. Progestin metabolism in the rat brain. After progesterone is injected peripherally, progesterone, 5α-pregnanedione and 3α-hydroxy-5α-pregnan-20-one were found in the brain. The enzymes 5α-steroid reductase and 3α-steroid dehydrogenase may catalyse the formation of these metabolites of progesterone in the brain.

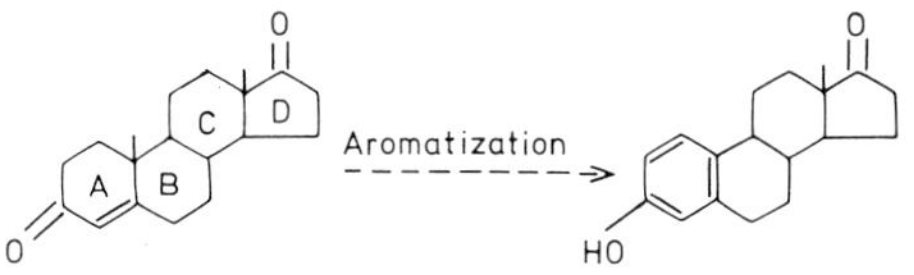

Fig. 4. Aromatisation of androgens in the rat brain. Aromatisation here refers to the formation of an aromatic A ring. The process is illustrated by the conversion of androstenedione to oestrone.

tion is involved in the control of these behaviours. The evidence on this point is still inconclusive (see Zigmond, 1975).

ANDROGEN RETENTION IN BIRDS

Our knowledge concerning the localisation of steroid hormones in the avian brain is primarily restricted to androgens. In ring doves half-an-hour after an intravenous injection, the labelled hormone was two to three times more concentrated in the preoptic—hypothalamic area than in the cerebrum or plasma (Stern, 1972). A metabolite tentatively identified as dihydrotestosterone was also concentrated in the preoptic—hypothalamic area. These tissue differences in the retention of labelled androgens were abolished by pretreating animals with a large dose of unlabelled testosterone (Stern, 1972). The specificity of the retention process, however, needs further study, since the retention of 'dihydrotestosterone' by the hypothalamus was also reduced by pretreatment with progesterone and corticosterone (Stern, 1972). Nevertheless, these steroids might be interfering with androgen metabolism rather than androgen binding.

Subcellular fractionation studies showed a concentration of labelled hormone within cell nuclei in the preoptic—hypothalamic area but not in the cerebrum (Zigmond et al., 1972a). When the concentration of radioactivity was compared in nuclei isolated from the two brain regions, the concentration was 15 times higher in those isolated from the preoptic—hypothalamic areas. (Fig. 1b). A similar magnification in the regional differences in hormone retention was found in the rat for oestradiol and corticosterone when studies at the whole tissue level were extended to the subcellular level (McEwen et al., 1970b; Zigmond and McEwen, 1970).

The anatomical distribution of 'androgen-concentrating cells' has been examined autoradiographically in the adult chaffinch and in the chick. One hour after an intramuscular injection of [3H]testosterone into castrated chaffinches, heavily labelled cells were found in the medial preoptic area and medial hypothalamus (Zigmond et al., 1972b and 1973). At this time more than 90% of the radioactivity in nuclei isolated from the whole chaffinch brain could be accounted for by [3H]testosterone and [3H]dihydrotestosterone. As was found for a number of hormones in the rat, labelled cells were not restricted to the diencephalon. While an investigation of the entire chaffinch brain has not been completed, labelled cells have been found in the lateral septum and in the midbrain in the nucleus intercollicularis, the periventricular zone above the optic ventricles and the central grey (Zigmond *et al.*, 1972b and 1973). The number of binding sites per labelled cell or the affinity of the sites for the hormone seems to differ in different parts of the brain. Thus the cells in the medial hypothalamus were, in general, more heavily labelled than those in the lateral septum (Zigmond, 1975).

After an injection of [^{3}H] testosterone the most heavily labelled cells in the chick brain were found in the nucleus preopticus paraventricularis magnocellularis (Meyer, 1973). Cells in this nucleus and in the nucleus hypothalamicus anterior medialis showed no decrease in the extent of labelling from 0.5—3 h after hormone administration, suggesting a long-term retention of the hormone. Three hours after an injection heavily labelled cells were also found in the nucleus supraopticus, nucleus preopticus medialis, nucleus paraventricularis magnocellularis, nucleus hypothalamicus posterior medialis and the nucleus rotundus.

In contrast to these results with [^{3}H] testosterone, it is interesting to note that few labelled cells were found in the hypothalamus of the duck after an injection of [^{3}H] corticosterone (Rhees *et al.*, 1972). As in the rat, most of the 'glucocorticoid-concentrating cells' were found in the hippocampus and in the septum (Gerlach and McEwen, 1972; Rhees *et al.*, 1972; Warembourg, 1974). Unlike the rat (see Sar and Stumpf, 1973), the distribution of labelled cells was reported to be similar in the duck after [^{3}H] corticosterone or [^{3}H] progesterone (Rhees *et al.*, 1972).

WHO CARES WHERE THE HORMONE GOES?

The study of hormone binding in the brain could be relevant to the physiology of behaviour for two reasons. First, this approach may provide a means for identifying target cells for different hormones throughout the brain. Second, the binding of the hormone to a specific macromolecule may represent the initial step in producing the hormone's effect.

While hormone implantation is an essential technique for establishing that an area of the brain is responsive to a particular hormone and for establishing the behavioural and/or physiological consequences of the hormone's interaction with cells in the region, the technique has two disadvantages. Because the hormone diffuses out from the site of the implant the anatomical resolution that can be achieved is somewhat limited. Also, because the entire procedure of implantation, behavioural testing, and histological determination of the implant sites is time-consuming, it is impractical to study the responsiveness of cells throughout the brain.

On the other hand, a comprehensive autoradiographic examination of cells in the brain which concentrate a particular hormone is feasible. In fact quite complete atlases describing the binding of both oestradiol and corticosterone in the rat have already been published (Stumpf, 1970; Pfaff and Keiner, 1973; Warembourg, 1974). The problem is that although the correlation between sites of hormone binding and sites of hormone action is good (McEwen *et al.*, 1972; Zigmond, 1975), it is not clear whether every cell which concentrates a particular hormone actually responds to that hormone. Furthermore, even were we to make that assumption we could not predict in which of the multiple behavioural and physiological effects produced by the hormone this cell participates. Thus the two techniques should be used together — autoradiography to locate potential target sites and hormone implantation to suggest the role of that site in the hormone's actions.

The question of whether steroid binding to a specific macromolecule represents the interaction of the hormone with its receptor also remains unanswered for the moment. In the best studied case, *i.e.* oestradiol, the chemical specificity of the binding correlates well with what is known about the chemical specificity of the hormone's action. In other words potent oestrogens such as oestradiol-17β and diethylstilboestrol bind to the same macromolecule, while compounds with weak oestrogenic activity such as oestradiol-17α, testosterone, and progesterone bind only weakly, if at all. Nevertheless, we must await further knowledge on the biochemical and neurophysiological changes produced by different steroids before we can fully evaluate the role of this binding in steroid hormone action.

REFERENCES

Anderson, J.N., Peck, E.J., Jr. and Clark, J.H. (1973) Nuclear receptor complex: accumulation, retention, and localisation in the hypothalamus and pituitary. *Endocrinology*, 93, 711—717.
Berthold, A.A. (1849) Transplantation der Hoden. *Arch. Anat. Physiol.*, 16, 42—46.
Chader, G.J. and Villee, C.A. (1970) Uptake of oestradiol by the rabbit hypothalamus. Specificity of binding by nuclei *in vitro*. *Biochem. J.*, 118, 93—97.
Clark, J.H., Campbell, P.S. and Peck, E.J. (1972) Receptor estrogen complex in the nuclear fraction of the pituitary and hypothalamus of male and female immature rats. *Neuroendocrinology*, 77, 218—228.
Denef, C., Magnus, C. and McEwen, B.S. (1973) Sex differences and hormonal control of testosterone metabolism in rat pituitary and brain. *J. Endocr.*, 59, 605—621.
Eisenfeld, A.J. (1967) Computer analysis of the distribution of [^{3}H]estradiol. *Biochim. biophys. Acta (Amst.)*, 136, 498—507.
Eisenfeld, A.J. (1970) ^{3}H-estradiol: *in vivo* binding to macromolecules from the rat hypothalamus, anterior pituitary and uterus. *Endocrinology*, 86, 1313—1318.
Feder, H.H. (1971) The comparative actions of testosterone propionate and 5α-androstan-17β-ol-3-one propionate on the reproductive behaviour, physiology and morphology of male rats, *J. Endocr.*, 51, 241—252.

120

Gerlach, J.L. and McEwen, B.S. (1972) Rat brain binds adrenal steroid hormone: radioautography of hippocampus with corticosterone. *Science*, 175, 1133—1136.

Hart, B.L. and Haugen, C.M. (1968) Activation of sexual reflexes in male rats by spinal implantation of testosterone. *Physiol. Behav.*, 3, 735—738.

Harris, G.W. (1955) *Neural Control of the Pituitary Gland*. Arnold, London.

Hutchison, J.B. (1971) Effects of hypothalamic implants of gonadal steroids on courtship behaviour in Barbary doves (*Streptopelia risoria*). *J. Endocr.*, 50, 97—113.

Hutchison, J.B. (1975) Hypothalamic mechanisms of sexual behaviour, with special reference to birds. In J.S. Rosenblatt, R.A. Hinde, E. Shaw and C. Beer (Eds.), *Advances in the Study of Behaviour, Vol. 6*. Academic Press, New York, in press.

King, R.J.B. and Mainwaring, W.I.P. (1974) *Steroid—Cell Interactions*. Butterworths, London.

Komisaruk, B.R., Adler, N.T. and Hutchison, J. (1972) Genital sensory field: enlargement by estrogen treatment in female rats. *Science*, 178, 1295—1298.

Kow, L.M. and Pfaff, D.W. (1974) Effects of estrogen treatment on the size of receptive field and response threshold of pudendal nerve in the female rat. *Neuroendocrinology*, 13, 299—313.

Laumas, K.R. and Farooq, A. (1966) The uptake *in vivo* of (1,2-^{3}H)progesterone by the brain and genital tract of the rat. *J. Endocr.*, 36, 95—96.

McEwen, B.S. and Pfaff, D.W. (1970) Factors influencing sex hormone uptake by rat brain regions. I. Effects of neonatal treatment, hypophysectomy, and competing steroid on estradiol uptake, *Brain Res.*, 21, 1—16.

McEwen, B.S. and Weiss, J.M. (1970) The uptake and action of corticosterone: regional and subcellular studies on rat brain. In D. De Wied and J.A.W.M. Weijnen (Eds.), *Pituitary, Adrenal and the Brain, Progr. Brain Res., Vol. 32*. Elsevier, Amsterdam, pp. 200—212.

McEwen, B.S., Pfaff, D.W. and Zigmond, R.E. (1970a) Factors influencing sex hormone uptake by rat brain regions. II. Effects of neonatal treatment and hypophysectomy on testosterone uptake. *Brain Res.*, 21, 17—28.

McEwen, B.S., Weiss, J.M. and Schwartz, L.S. (1970b) Retention of corticosterone by cell nuclei from brain regions of adrenalectomized rats. *Brain Res.*, 17, 471—482.

McEwen, B.S., Zigmond, R.E. and Gerlach, J.L. (1972) Sites of steroid binding and action in the brain. In G.H. Bourne (Ed.), *Structure and Function of the Nervous System*, Academic Press, New York, pp. 205—291.

Meyer, C.C. (1973) Testosterone concentration in the male chick brain: an autoradiographic survey. *Science*, 180, 1381—1383.

Mowles, T.F., Ashkanazy, G., Mix, E. and Sheppard, M. (1971) Hypothalamic and hypophyseal estradiol binding complexes. *Endocrinology*, 89, 484—491.

Naftolin, F., Ryan, K.J. and Petro, Z. (1972) Aromatization of androstenedione by the anterior hypothalamus of adult male and female rats. *Endocrinology*, 90, 295—298.

Pfaff, D.W. and Keiner, M. (1973) Atlas of estradiol concentrating cells in the central nervous system of the female rat. *J. comp. Neurol.*, 151, 121—158.

Phoenix, C.H. (1973) The role of testosterone in the sexual behavior of laboratory male Rhesus. *Symp. IVth Int. Congr. Primal., Vol. 2, Primate Reproductive Behaviour*. Karger, Basel, pp. 99—122.

Raisinghani, K.H., Dorfman, R.I., Forchielli, L.L., Gyernnek, L. and Geutler, G. (1968) Uptake of intravenously administered progesterone, pregnanedione and pregnanolone by the rat brain. *Acta Endocrinol. (Kbh)*, 57, 395—404.

Reddy, V.V.R., Naftolin, F. and Ryan, K.J. (1974) Conversion of androstenedione to estrone by neural tissues from fetal and neonatal rats. *Endocrinology*, 94, 117—121.

Rhees, R.W., Abel, J.H., Jr. and Haack, D.W. (1972) Uptake of tritiated steroids in the brain of the duck (*Anas platyrhynchos*). An autoradiographic study. *Gen. comp. Endocr.*, 18, 292—300.

Rommerts, F.F.G. and Van der Molen, H.J. (1971) Occurrence and localization of 5α-steroid reductase, 3α- and 17β-hydroxysteroid dehydrogenases in hypothalamus and other brain tissues of the male rat. *Biochim. biophys. Acta (Amst.)*, 248, 489—502.

Sar, M. and Stumpf, W.E. (1973) Neurons of the hypothalamus concentrate (^{3}H) progesterone or its metabolites. *Science*, 182, 1266—1268.

Sholiton, L.J., Jones, C.E. and Werk, E.E. (1972) The uptake and metabolism of (1,2-^{3}H)-testosterone by the brain of functionally hepatectomized and totally eviscerated male rats. *Steroids*, 20, 399—415.

Snipes, C.A. and Shore, L.S. (1972) Metabolism of progesterone *in vitro* by neural and uterine tissues. *Fed. Proc.*, 31, 236. (Abstract.)

Stern, J.M. (1972) Androgen accumulation in hypothalamus and anterior pituitary of male ring doves; influence of steroid hormones. *Gen. comp. Endocr.*, 18, 439—449.

Stern, J.M. and Eisenfeld, A. (1971) Distribution and metabolism of ^{3}H-testosterone in castrated male rats; effects of cyproterone, progesterone and unlabelled testosterone. *Endocrinology*, 88, 1117—1125.

Stumpf, W.E. (1970) Estrogen-neurons and estrogen-neuron systems in the periventricular brain. *Amer. J. Anat.*, 129, 207—218.

Wade, G.N. and Feder, H.H. (1972) Effects of several pregnane and pregnene steroids on estrous behaviour in ovariectomized estrogen-primed guinea pigs. *Physiol. Behav.*, 9, 773—775.

Warembourg, M. (1974) Etude radioautographique des retroactions centrales des corticosteroides ^{3}H chez le rat et le cobaye. In P. Dell (Ed.), *Neuroendocrinologie de l'Axe Corticotrope*. Colloques de l'Inserm, Paris, in press.

Zigmond, R.E. (1971) *Anatomical and Chemical Specificity of Gonadal Hormone Retention in the Rat Brain*. Ph.D. Thesis, Rockefeller University, New York.

Zigmond, R.E. (1975) Binding, metabolism, and action of steroid hormones in the central nervous system. In L.L. Iversen, S.D. Iversen and S.H. Snyder (Eds.), *Handbook of Psychopharmacology*. Plenum Press, New York, in press.

Zigmond, R.E. and McEwen, B.S. (1970) Selective retention of estradiol by brain cell nuclei in specific regions of the ovariectomized rat. *J. Neurochem.*, 17, 889—899.

Zigmond, R.E., Stern, J.H. and McEwen, B.S. (1972a) Retention of radioactivity by brain cell nuclei in the ring dove after injection of ^{3}H-testosterone. *Gen. comp. Endocr.*, 18, 450—453.

Zigmond, R.E., Nottebohm, F. and Pfaff, D.W. (1972b) Distribution of androgen-concentrating cells in the brain of the chaffinch. In *Proceedings of the Fourth International Congress of Endocrinology, Washington, D.C., 1972*. Excerpta Medica, Amsterdam (Abstract) p. 136.

Zigmond, R.E., Nottebohm, F. and Pfaff, D.W. (1973) Androgen-concentrating cells in the midbrain of a songbird. *Science*, 179, 1005—1007.

7

TARGET CELLS FOR GONADAL STEROIDS IN THE BRAIN: STUDIES
ON STEROID-SENSITIVE MECHANISMS OF BEHAVIOUR

J.B. HUTCHISON

*M.R.C. Unit on the Development and Integration of Behaviour, Madingley, Cambridge
(Great Britain)*

INTRODUCTION

With the recent development of biochemical methods for the identification
of sex hormones in peripheral blood plasma, new information is becoming
available on both long-term and short-term fluctuations in plasma levels of
gonadotrophic and steroidal hormones in animals undergoing normal reproductive cycles. In the case of the major androgenic steroid, testosterone,
it has been possible to correlate peripheral levels with changes in the reproductive behaviour. For example, plasma testosterone titres in starlings
(*Sturnus vulgaris*) are elevated during the period of the reproductive cycle
when courtship and territorial aggressive behaviour are at their peak, and
decline after breeding has terminated (Temple, 1974). Short-term elevations
of plasma luteinizing hormone and testosterone can be induced by sexual
stimulation in the bull (Katongole *et al.*, 1971) and the male rat (Purvis and
Haynes, 1974), and by the act of copulation in male rabbits (Saginor and
Horton, 1968; Haltmeyer and Eik-Nes, 1969). Similarly, plasma testosterone
levels are higher immediately before and during orgasm in the human male
than at control sampling periods (Fox *et al.*, 1972).

Although it is possible to draw correlations between peripheral plasma
levels of sex hormones and behaviour, very little is known at present about
the way in which fluctuations in these hormones at the periphery relate to
the functioning of brain mechanisms of behaviour. However, it is becoming
clear that the brain can be regarded as a steroid target, because certain areas,
notably the hypothalamus, contain high-affinity, steroid-binding macromolecules (McEwen *et al.*, 1972; see also Zigmond, Chapter 6 in this volume).
A number of studies (reviewed by Davidson, 1972; Hutchison, 1975) have also
drawn attention to the hypothalamus by demonstrating that the preoptic—
anterior hypothalamic complex is both steroid-sensitive and closely associated
with male and female sexual behaviour. Since these areas can be localized in

terms of concentrations of cells containing 'receptors' for steroid hormones, there is a possibility that some of the functional properties of mechanisms of sexual behaviour can be attributed directly to these steroid-sensitive cell systems which may be influenced by long- or short-term changes in peripheral hormone level.

The purpose of this chapter will be to describe some work which suggests first, that the hypothalamus plays a part in determining the structure of pre-copulatory courtship behaviour in the male Barbary dove (*Streptopelia risoria*), by means of a system which is differentially sensitive to androgen concentrations within the hypothalamus, and second, that this system is unstable and organized so that the sensitivity of the hypothalamus to androgen in relation to courtship behaviour is variable and depends on the endocrine state of the animal.

DIRECT ACTION OF ANDROGEN ON BRAIN MECHANISMS OF MALE COURTSHIP

Birds are particularly useful for the study of hormone-sensitive brain mechanisms of behaviour in view of their elaborate, visual courtship displays which are stereotyped and easily quantified. The courtship display of the male Barbary dove consists initially of a rapid alternation of aggressive displays (termed 'chasing' and 'bowing') which cause the female to retreat, and a nest-orientated display (termed 'nest-soliciting') where the male selects a potential site for the nest, causing the female to approach (Hutchison, 1970a; Lovari and Hutchison, 1975). (Fig. 1.) These courtship displays of the male decline and disappear rapidly after castration and are reinstated by intramuscular injections of testosterone propionate, indicating that male courtship is androgen-dependent and that testosterone may be the effective steroid (Hutchison, 1970a). Further evidence for this is that testosterone has been identified in peripheral blood plasma of sexually active males (Hutchison and Katongole, 1975).

Studies with intracranial testosterone propionate implants in castrated male doves have indicated that qualitatively normal male courtship can be obtained only from implants in the preoptic and anterior hypothalamic areas (Fig. 1; Hutchison, 1967 and 1971; Barfield, 1971). Fragmentary courtship is sometimes obtained from implants in areas of the brain which are adjacent to the preoptic and anterior hypothalamic areas (see Zigmond, Chapter 6 in this volume, Fig. 1). But the effectiveness of implants in restoring courtship can be related to their proximity to the preoptic—anterior hypothalamic area, suggesting that these fragmentary displays are due to low concentrations of hormone diffusing to this area (Hutchison, 1971). In general, the area from which the most complete behavioural response can be induced corresponds fairly well with the area that contains nuclear binding sites for testosterone, and the area delimited by autoradiography in the male chaffinch brain which contains the greatest number of labelled cells following [3H]testosterone injection (Zigmond, 1975).

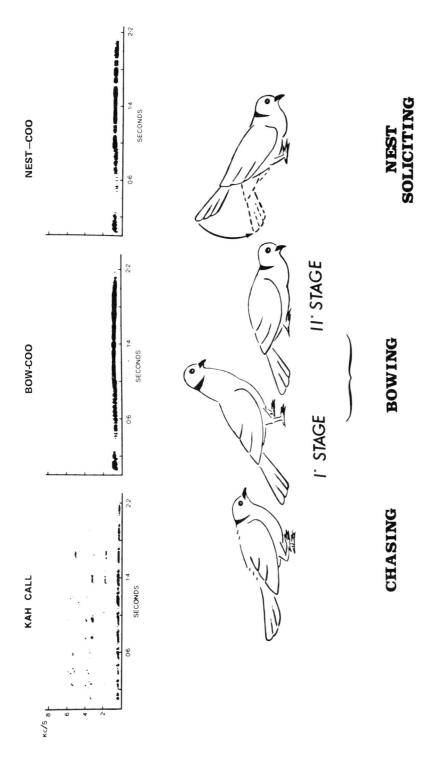

Fig. 1. Aggressive (chasing and bowing) and nest-orientated components (nest-soliciting) of male courtship typically shown by a sexually active male Barbary dove during a short (<2 min) interaction with a female. Sound spectrographs of vocalizations accompanying each behaviour pattern are included. For a detailed explanation see Lovari and Hutchison, 1975.

Further support for the view that testosterone acts directly on hypo-thalamic mechanisms of male sexual behaviour in the avian brain has come from studies of the effects of intracranial implants of testosterone propionate on copulatory behaviour. Barfield (1969) working with the domestic fowl has shown that the androgen-sensitive area associated with male copulatory behaviour is restricted to the preoptic area, whilst in castrated male doves, androgen sensitivity associated with copulatory behaviour is localized in both the preoptic and anterior hypothalamic areas (Barfield, 1971). In male chicks, precocious copulation has been induced by testosterone propionate implants in the anterior hypothalamus (Gardner and Fisher, 1968) and inhibited by progesterone implants in the preoptic area (Meyer, 1972); progesterone is known to antagonize the behavioural effects of testosterone on the avian brain (Komisaruk, 1967).

HYPOTHALAMIC CONCENTRATION OF ANDROGEN AND COURTSHIP

Circumstantial evidence suggests that the type of courtship displayed by the male dove is related to the level of circulating androgen. Thus, in castrates, the aggressive components of courtship, particularly bowing, decline and disappear more rapidly (within 1—3 days) than the nest-orientated com-ponents (15—20 days) (Hutchison, 1970a). Since it can be assumed that endogenous androgens are metabolized and disappear from peripheral plasma within minutes after castration, the aggressive courtship displays may depend more on relatively high concentrations of androgen in the hypothalamus than does nest-orientated behaviour. This hypothesis can be tested by mani-pulating hypothalamic androgen level directly. Testosterone can be elevated to different levels in the hypothalamus by the use of solid implants of dif-fering surface areas, or the effects of testosterone on the hypothalamus can be reduced by means of an antagonist to the action of testosterone.

Using the first of these experimental approaches, the effects of three types of testosterone propionate implants with differing surface areas have been compared (Fig. 2a). The low-diffusion implants resulted in courtship be-haviour in which the nest-orientated components were restored to precastra-tion levels, but the aggressive components were almost absent. By contrast, both the high- and medium-diffusion implants restored more complete courtship displays, but a larger proportion of males with high-diffusion implants displayed aggressive courtship for longer durations than males with medium-diffusion implants (Fig. 2a).

In the second of these experimental approaches, the antagonistic effects of progesterone on the action of testosterone in inducing male courtship (Erickson et al., 1967) were used to reduce the effectiveness of testosterone propionate implants acting on the hypothalamus. These antagonistic effects are presumably mediated at the level of the hypothalamus, because the be-

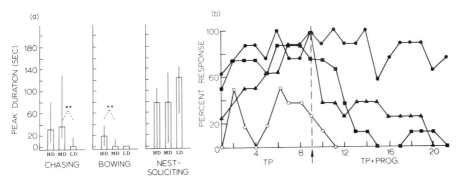

Fig. 2. a: effects of testosterone propionate implants with different surface areas implanted into the anterior hypothalamus and preoptic regions of doves. HD, high-diffusion implants (55—85 μg); MD, medium-diffusion implants (25—55 μg); LD, low-diffusion implants (hormone contained in bore of 27 g tubing). The data are expressed as medians (bars) and ranges (vertical lines). The peak duration is the longest of the daily durations of display of a pattern; the daily duration is the sum of the durations of bouts of display of the pattern within the 3-min test with a female. (Data derived from Hutchison, 1970b). ** $P < 0.02$, Mann—Whitney U-test, two-tailed. b: effects of progesterone on androgen-induced courtship activity in doves. Castrates were treated intramuscularly with 300 μg testosterone propionate (TP) for 8 successive days. On the ninth day each castrate received 300 μg TP and a similar dose of progesterone (PROG). Treatments were continued for 20 days. ■, chasing; ○, bowing; ▲, nest-soliciting; and ●, strutting.

havioural effects of intrahypothalamic implants of testosterone propionate in castrated male doves are suppressed by systemic progesterone (Hutchison, in preparation). The prediction was made that as progesterone concentration in the hypothalamus increased, selectively blocking the effects of testosterone, the aggressive components would decline relative to the nest-orientated components. This was found to be the case in males treated with progesterone and testosterone propionate (300 μg of each hormone/day; Hutchison, 1975). The effects on courtship were similar to those obtained by castration. Thus aggressive behaviour declined rapidly and disappeared within 3—6 days of the start of the combined treatment with testosterone propionate and progesterone. Nest-orientated behaviour continued to be displayed by the majority of males until 10—20 days after the initial treatment and then disappeared (Fig. 2b). Control birds, treated with testosterone propionate alone, continued to display full courtship until injections ceased. Strutting which normally occurs between bouts of aggressive and nest-orientated courtship was not significantly affected by the combined testosterone and progesterone therapy, suggesting that differences in rate of decline between courtship components was a selective effect of progesterone, and not due to the adverse effects of this hormone on the general metabolism or activity of progesterone-treated birds.

Taken together, the results of the two methods of manipulation of hypothalamic testosterone level suggest that when hypothalamic concentration of testosterone is high, both aggressive and nest-orientated behaviour will be displayed, whereas when testosterone concentrations are lower, aggressive behaviour will be absent. It can be suggested that the anterior hypothalamus distinguishes between testosterone concentrations by means of a system organized so that mechanisms in the brain associated with aggressive behaviour have a higher sensitivity threshold to testosterone than those associated with nest-orientated behaviour (Hutchison, 1974a). Conceivably, the focus for this threshold system could lie within the anterior hypothalamus and might involve cells which are differentially sensitive to testosterone. Thus certain cell populations associated with nest-orientated behaviour may have a greater affinity (see Zigmond, Chapter 6 in this volume) for binding testosterone than other testosterone-sensitive cell populations associated with aggressive behaviour.

HYPOTHALAMIC SENSITIVITY TO ANDROGEN AND COURTSHIP

Effects of prolonged androgen deficit

It has been known for some time that the type of sexual behaviour induced by androgen therapy in castrated males may be similar quantitatively to behaviour shown before castration. In male guinea pigs (Riss and Young, 1954) and in rats (Larsson, 1966) there is a positive correlation between precastration and posttreatment copulatory behaviour induced by testosterone propionate. Similarly, in male doves, there is a strong probability that the courtship evoked by androgen therapy will be similar structurally to precastration courtship: males that displayed the aggressive components of courtship in the absence of nest-orientated courtship before castration behaved similarly afterwards if treated systemically or implanted intrahypothalamically with testosterone propionate (Hutchison, 1970b and 1971). A question which has not received a great deal of attention so far is whether sexual behaviour can be restored in quantitative terms to precastration levels. Studies with male doves (Hutchison, 1970b) have shown that courtship, particularly the aggressive component, bowing, is seldom restored to precastration levels in castrates even with dosages as large as 300 μg of testosterone propionate per day. Moreover, a proportion of castrated male doves (approximately 29%) show no courtship response to intramuscular testosterone propionate injected daily, although the majority respond to therapy within 1—3 days (Hutchison, 1975). When the daily dosage is doubled (600 μg/day), all of the castrates that previously failed to respond begin to display courtship. These observations suggest that a decline in behavioural responsiveness to intramuscular androgen occurs after castration; this decline appears to occur more rapidly in some individuals than others. Given this decline in effectiveness of systemic testosterone, and bearing in mind the known action of androgen on hypothalamic areas associated with male courtship behaviour, a question can be phrased concerning the functioning of the steroid-sensitive system. Does the sensitivity of the hypothalamus to androgen in relation to courtship behaviour remain stable throughout long-term changes in circulatory androgens?

This question can be answered experimentally by studying the effects of exogenous testosterone in conditions where a prolonged endogenous androgen deficit is imposed experimentally. In such conditions, deficits in the behavioural response of castrates to systemic treatments may, of course, be due to peripheral metabolic changes which alter the rate of transport of hormone to the brain. Therefore, the hypothesis that a change occurs specifically in brain mechanisms underlying the behaviour can only be tested by measuring the effectiveness of intrahypothalamic implants of testosterone propionate in restoring courtship at different periods after castration. When this is carried out at 15, 30 and 90 days after castration, there are distinct differences between the behavioural responses of the groups of castrates. Whereas implants of testosterone propionate are highly effective in restoring courtship in 15-day castrates (Fig. 3), their effectiveness is lower in 30-day castrates, particularly with respect to bowing behaviour, and implants are almost completely ineffective in initiating the display of courtship in 90-day castrates (Hutchison, 1974b). The behavioural effects of testosterone propionate are therefore diminished after castration, suggesting that the threshold of sensitivity of the preoptic—anterior hypothalamic area to testosterone rises after castration.

Because the effects of testosterone on courtship appear to be mediated by cell populations in the anterior hypothalamic—preoptic area, changes in the steroid-retention properties of these cells may be responsible for behavioural deficits in long-term castrates implanted with testosterone propionate. Is there any evidence that the steroid-binding properties of anterior hypothalamic cells change after castration? On the basis of studies in rats of the inverse relationship between peak uptake of [^{3}H] oestradiol and the period between ovariectomy and intravenous injections of hormone (McGuire and Lisk, 1969), Lisk (1971) has suggested that oestradiol receptor molecules in the hypothalamus become inactivated in the prolonged absence of circulating oestrogen. This hypothesis was not established conclusively because labelled oestradiol was injected intravenously, and delays in peak uptake of oestradiol may have been due to peripheral factors rather than the steroid-binding properties of hypothalamic cells. However, it can be suggested tentatively that one factor responsible for the ineffectiveness of testosterone implants in long-term castrates may have been increased degradation of hypothalamic 'receptor' macromolecules that bind testosterone.

Apart from the possible effects on testosterone binding in the hypothalamus, the metabolism of testosterone may be affected by prolonged androgen deficit. Increased 5α-steroid reductase activity has been shown to

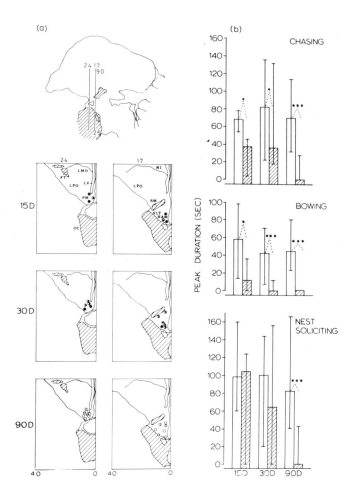

Fig. 3. Comparison between courtship shown before castration and following implantation of testosterone propionate in the preoptic—anterior hypothalamic area 15, 30 and 90 days after castration (a), each symbol (circle) denotes the tip of an implant; shading within the symbol indicates the behavioural response: filled circle, chasing, bowing and nest-soliciting; right-half shading, chasing and nest-soliciting; bottom-half shading, chasing and bowing; vertical line, chasing; horizontal line, nest-soliciting and open circle, no response. Peak durations of courtship shown before castration are indicated by open bars; postimplantation peak durations by cross-hatched bars. The data are expressed as medians and ranges, and are derived from Hutchison, 1974b. Precastration and postimplantation peak durations are compared statistically *$P < 0.05$, ***$P < 0.001$ (Wilcoxon matched pairs test, one-tailed). FT, tractus frontothalamicus; HAM, nucleus hypothalamicus anterior medialis; LMD, lamina medullaris; LPO, lobus parolfactorius; LV, lateral ventricle; NI, neostriatum intermediale; OC, optic chiasma; PM, nucleus preopticus medialis; SM, tractus septomesencephalicus; VLT, nucleus ventrolateralis.

occur in the hypothalamus of male rats following castration (Denef *et al.*, 1973), suggesting that there may be increased conversion of testosterone to dihydrotestosterone after castration (see also Zigmond, previous chapter). Similar changes in enzyme activity have not yet been demonstrated in doves, but increased conversion of testosterone to dihydrotestosterone could conceivably reduce the effective levels of testosterone available to act on androgen-sensitive brain mechanisms underlying courtship behaviour. Intrahypothalamic implants of both dihydrotestosterone and an esterified form, dihydrotestosterone acetate, in castrated doves were found to be relatively ineffective in inducing courtship (Hutchison, 1975). This is consistent with findings in castrated male rats where 5α-dihydrotestosterone propionate has been found to be far less effective in inducing male copulatory behaviour than testosterone propionate, whether injected systemically (Feder, 1971) or implanted into the anterior hypothalamus (Johnston and Davidson, 1972). Therefore, increased conversion of testosterone to dihydrotestosterone might have the effect of eliminating levels of testosterone in the hypothalamus that would normally be effective for activation of male courtship.

The main conclusion from this study is that the responsiveness of brain mechanisms of male courtship behaviour to testosterone declines after the elimination of gonadal androgen; the focus for this decline may be in the preoptic—anterior hypothalamic complex. Therefore, in addition to activating courtship behaviour in males receiving sensory stimulation, androgen may maintain the hormonal sensitivity of hypothalamic mechanisms associated with male courtship behaviour.

Effects of environmental factors

Photoperiodic regulation of testicular activity is characteristic of many species of birds (reviewed by Lofts *et al.*, 1970). An example of the degree to which photoperiod can influence testicular activity is provided by the quail (*Coturnix coturnix*). Long photoperiods imposed on males that have been maintained on a short photoperiod stimulate a surge of luteinizing hormone that induces testicular growth within one day of the onset of the long photoperiod (Follett and Farner, 1966). Although the effects of a prolonged androgen deficit in endogenous androgen on the steroid-binding properties of hypothalamic cells may be one factor in the decline in sensitivity of the hypothalamus to implanted testosterone, environmental factors such as photoperiod or social stimuli may also play a significant role. These may well act via extrahypothalamic mechanisms to block the response of the anterior hypothalamus to implanted testosterone propionate, or to inhibit the response of the extrahypothalamic mechanisms associated with courtship behaviour which normally respond to androgen action on the hypothalamus.

Of these environmental factors, it seems likely that photoperiod might influence the hypothalamic mechanisms of male courtship behaviour in doves, because implants of testosterone propionate are found to be more effective in initiating courtship in long-term (90-day) castrates maintained on a long photoperiod (13 h/day) than in castrates maintained on a short photoperiod (8.5 h/day) (Hutchison, 1974c). This difference could be due to at least three influences of photoperiod: (a) direct effects of higher gonadotrophin levels on the brain in the 13 h/day group, in view of the increased aggressiveness of starlings (Davis, 1957) induced by systemic treatment with gonadotrophin preparations; (b) the effects of low gonadotrophin level in the 8.5 h/day group on testosterone uptake in the hypothalamus, in view of the decrement in testosterone uptake by the preoptic area of the rat caused by hypophysectomy (McEwen et al., 1970) or (c) the direct effects of long photoperiod on brain mechanisms mediating aggressive behaviour, in view of the observation that photoperiod is critical for the initiation of nest-building in female canaries by systemic oestradiol treatment, an effect thought to be independent of gonadotrophin level (Steel and Hinde, 1972).

These variables are difficult to separate experimentally. However, as an initial step in assessing the role of photoperiod in the activation of male courtship by testosterone in doves, an experiment was carried out in which photoperiod was used to determine hormonal condition prior to castration and implantation. The rationale behind the experiment was to lower gonadotrophin level, and consequently testicular activity, by maintaining males on a short daylength for a period before castration, and then to compare the effects of implants in these castrates with the behavioural responses obtained from implanted castrates maintained on a long photoperiod. An assumption implicit in this experiment was that the elevation of gonadotrophin level normally induced by castration, would be lower in males maintained on a short photoperiod (Hinde et al., 1974). As yet, there is no direct evidence from measurement of gonadotrophin levels in doves for this assumption. The groups involved in the experiment were maintained on either a 14 h/day or a 6 h/day photoperiod (Hutchison, 1975) for a month prior to castration and during the experimental period. The short photoperiod group showed a pronounced decline in testicular weight and courtship behaviour, suggestive of a decline in gonadotrophin levels. Thirty days after castration, males of both groups received intrahypothalamic implants of testosterone propionate. There were no significant differences between the groups in the duration of courtship responses elicited by the implants, but the short photoperiod group displayed aggressive courtship during significantly fewer daily tests (Hutchison, 1975). If it can be assumed that the level of gonadotrophin differed immediately prior to castration and subsequently between these groups, the overall similarity in behavioural response to implants suggests that gonadotrophin level and photoperiod have little influence on the level of display of courtship induced by testosterone. However, there may be a direct influence of photoperiod or gonadotrophin level to prolong testosterone action on anterior hypothalamus. This effect has still to be confirmed.

The conclusions discussed so far have been derived from studies of male doves which have had an androgen deficit imposed by either manipulation of photoperiod or elimination of gonadal androgen. The question is whether these conclusions are relevant to males undergoing a seasonal fluctuation in endogenous androgen level. To determine whether a seasonal decline in the responsiveness of brain mechanisms of male courtship behaviour occurs, males that had undergone some testicular atrophy were brought into the laboratory during late autumn, tested for courtship, castrated and 30 days later implanted intrahypothalamically with testosterone propionate. No courtship was shown before castration. The response of these castrates to implants consisted almost entirely of low levels of chasing (Hutchison, 1975), and was not comparable with the response to implants of males that had been sexually active prior to castration. This preliminary result suggests that there may well be a seasonal decline in responsiveness to androgen of hypothalamic mechanisms of male courtship behaviour.

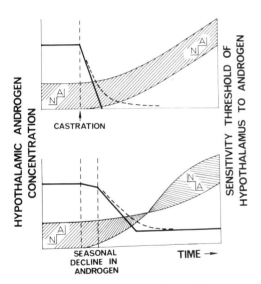

Fig. 4. Model of changes in hypothalamic sensitivity to androgen which may be related to male courtship behaviour in doves. Following elimination of testicular androgen by castration, hypothalamic thresholds of sensitivity to androgen are elevated and interrelationships between thresholds of aggressive (A) and nest-orientated (N) components are maintained. Following seasonal decline in endogenous androgen, thresholds of sensitivity are elevated, but interrelationships between thresholds of aggressive and nest-orientated components are inverted. Heavy line, hypothalamic androgen concentration; heavy dotted line, residual effects of androgen on hypothalamic cells; hatched area; threshold of sensitivity of the hypothalamus to androgen; stepped line, relationships between sensitivity thresholds of mechanisms underlying aggressive courtship and nest-orientated courtship. Figure derived partly from Hutchison, 1974a.

In view of the differences in threshold of sensitivity to androgen between mechanisms of aggressive and nest-orientated behaviour described in the preceding section, a further question is whether the interrelationships between the sensitivity thresholds remain stable during a long-term decline in circulating androgen imposed either by castration or occurring during the seasonal cycle. These relationships would appear to be stable in long-term castrates implanted with testosterone propionate because nest-orientated courtship, which has the lower sensitivity threshold to testosterone, was elicitable from some long-term castrates, and aggressive courtship, particularly bowing, was almost absent. However, of the sexually inactive males implanted with testosterone propionate during late autumn, when endogenous androgen levels were presumably at their lowest, the majority displayed the aggressive components of courtship, and there were no nest-orientated displays. This result would suggest that the thresholds of responsiveness of the aggressive components of courtship to androgen were lower than those for the nest-orientated components in this group. Therefore, the threshold relationships of the sexually inactive males would appear to be the reverse of those in the long-term castrates that had been sexually active prior to castration (Fig. 4). Supplementary evidence for this was provided by determination of plasma testosterone levels in a sample of intact males undergoing normal seasonal

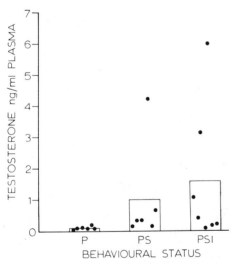

Fig. 5. Individual plasma testosterone levels assayed by a competitive protein-binding technique (Katongole, 1971). A comparison is made between a group which displayed aggressive courtship without the nest-orientated component during 3-min tests for courtship (P), and a group which displayed full aggressive and nest-orientated courtship (PS). A further group of incubating males are included (PSI). The histograms represent the means. (Data derived from Hutchison and Katongole, 1975.)

development that consistently showed the aggressive components of courtship unaccompanied by nest-orientated behaviour (Hutchison and Lovari, in preparation). These males had significantly lower plasma testosterone levels than males that displayed both aggressive and nest-orientated components (Fig. 5; Hutchison and Katongole, 1975), suggesting that the sensitivity threshold relationships of the aggressive males are the reverse of the groups that displayed full courtship.

CONCLUDING COMMENTS

There has been a rapid expansion in information on the actual levels of hormonal steroids in peripheral blood plasma. However, the relationships between fluctuations in plasma hormone level and reproductive behaviour have not been studied to any extent, particularly with respect to the way in which changing hormonal conditions may influence the functioning of brain mechanisms of behaviour. For an understanding of these relationships, it will be necessary to explore further the physiological processes underlying the action of hormones on brain mechanisms of behaviour.

There is increasing evidence that steroid sex hormones influence reproductive behaviour by direct action on a discrete area of the brain, the anterior hypothalamic—preoptic complex. This paper describes a series of experiments which throw light on two possible modes of action of androgen on steroid-sensitive mechanisms underlying male courtship behaviour in the Barbary dove. These are first, that the level of testosterone in the hypothalamus can determine the structure of male courtship behaviour and second, that the sensitivity of the androgen-sensitive area to implanted testosterone can be manipulated by varying the level of circulating androgen. In more general terms it can be suggested that the activating effects of a sex hormone on mechanisms of sexual behaviour depend both on effective hormone concentration in the hypothalamus and the response threshold of the steroid-sensitive system to androgen.

There is now some understanding of the way in which steroid hormones influence brain cells of birds and mammals, particularly with regard to the binding of steroids to intracellular macromolecules. Certain as yet unspecified cells in the preoptic—anterior hypothalamic area contain 'receptors' which bind androgenic steroids. As this brain area also appears to be the focus for the action of androgen on brain mechanisms of male sexual behaviour, it is very likely that binding of androgen to steroid-specific target cells in the hypothalamus forms the initial stage of the process of activation of male sexual behaviour. If this is so, and there is no direct evidence linking steroid-binding with mechanisms of behaviour, then some of the functional characteristics of the androgen-sensitive areas in the hypothalamus mentioned above may be related to changing steroid-binding properties of cellular receptors within the hypothalamus.

136

Acknowledgements

I am very grateful to Dr. Rosemary Hutchison for comments on the manuscript and to Mr. Les Barden for technical assistance.

REFERENCES

Barfield, R.J. (1969) Activation of copulatory behaviour by androgen implanted into the preoptic area of the male fowl. *Horm. Behav.*, 1, 37—52.
Barfield, R.J. (1971) Activation of sexual and aggressive behaviour by androgen implanted into the male ring dove brain. *Endocrinology*, 89, 1470—1476.
Davidson, J.M. (1972) Hormones and reproductive behavior. In H. Balin and S. Glasser (Eds.), *Reproductive Biology*. Excerpta Medica, Amsterdam, pp. 877—918.
Davis, D.E. (1957) Aggressive behavior in starlings. *Science*, 126, 253.
Denef, C., Magnus, C. and McEwen, B.S. (1973) Sex differences and hormonal control of testosterone metabolism in rat pituitary and brain. *J. Endocr.*, 59, 605—621.
Erickson, C.J., Bruder, R.H., Komisaruk, B.R. and Lehrman, D.S. (1967) Selective inhibition by progesterone of androgen-induced behaviour in male ring doves (*Streptopelia risoria*). *Endocrinology*, 81, 39—45.
Feder, H.H. (1971) The comparative actions of testosterone propionate and 5α-androstan-17β-ol-3-one propionate on the reproductive behaviour, physiology and morphology of male rats. *J. Endocr.*, 51, 241—252.
Follett, B.K. and Farner, D.S. (1966) The effect of the daily photoperiod on gonadal growth, neuro-hypophysial hormone content and neurosecretion in the hypothalamo-hypophysial system of the Japanese quail (*Coturnix coturnix japonica*). *Gen. comp. Endocr.*, 7, 111—124.
Fox, C.A., Ismail, A.A.A., Love, D.N., Kirkham, K.E. and Loraine, J.A. (1972) Studies on the relationship between plasma testosterone levels and human sexual activity. *J. Endocr.*, 52, 51—58.
Gardner, J.E. and Fisher, A.E. (1968) Induction of mating in male chicks following preoptic implantation of androgen. *Physiol. Behav.*, 3, 709—713.
Haltmeyer, G.C. and Eik-Nes, K.B. (1969) Plasma levels of testosterone in male rabbits following copulation. *J. Reprod. Fert.*, 19, 273—277.
Hinde, R.A., Steel, E. and Follett, B.K. (1974) Effect of photoperiod on oestrogen-induced nest building in ovariectomized or refractory female canaries (*Serinus canarius*). *J. Reprod. Fert.*, 40, 383—399.
Hutchison, J.B. (1967) Initiation of courtship by hypothalamic implants of testosterone propionate in castrated doves (*Streptopelia risoria*). *Nature (Lond.)*, 216, 591—592.
Hutchison, J.B. (1970a) Differential effects of testosterone and oestradiol on male courtship in Barbary doves (*Streptopelia risoria*). *Anim. Behav.*, 18, 41—51.
Hutchison, J.B. (1970b) Influence of gonadal hormones on the hypothalamic integration of courtship behaviour in the Barbary dove. *J. Reprod. Fert.*, Suppl. 11, 15—41.
Hutchison, J.B. (1971) Effects of hypothalamic implants of gonadal steroids on courtship behaviour in Barbary doves (*Streptopelia risoria*). *J. Endocr.*, 50, 97—113.
Hutchison, J.B. (1974a) Differential hypothalamic sensitivity to androgen in the activation of reproductive behavior. In O. Schmitt and F.G. Worden (Eds.), *The Neurosciences, 3rd Study Volume*. MIT Press, Cambridge, Mass., pp. 593—599.
Hutchison, J.B. (1974b) Post-castration decline in behavioural responsiveness to intrahypothalamic androgen in doves. *Brain Res.*, 80, 1—10.

Hutchison, J.B. (1974c) Effect of photoperiod on the decline in behavioural responsiveness to intra-hypothalamic androgen in doves (*Streptopelia risoria*). *J. Endocr.*, 63, 583—584.

Hutchison, J.B. (1975) Hypothalamic mechanisms of sexual behavior, with special reference to birds. In J.S. Rosenblatt, R.A. Hinde, E. Shaw and C. Beer (Eds.), *Advances in the Study of Behavior, Vol. 6*. Academic Press, New York, in press.

Hutchison, J.B. and Katongole, C.B. (1975) Plasma testosterone and aggressive courtship in the male Barbary dove (*Streptopelia risoria*). *J. Endocr.*, 65, in press.

Johnston, P. and Davidson, J.M. (1972) Intracerebral androgens and sexual behavior in the male rat. *Horm. Behav.*, 3, 345—357.

Katongole, C.B. (1971) A competitive protein-binding assay for testosterone in the plasma of the bull and the ram. *J. Endocr.*, 51, 303—312.

Katongole, C.B., Naftolin, F. and Short, R.V. (1971) Relationship between blood levels of luteinizing hormone and testosterone in bulls, and the effects of sexual stimulation. *J. Endocr.*, 50, 457—466.

Komisaruk, B.R. (1967) Effects of local brain implants of progesterone on reproductive behavior of ring doves. *J. comp. physiol. Psychol.*, 51, 32—36.

Larsson, K. (1966) Individual differences in reactivity to androgen in male rats. *Physiol. Behav.*, 1, 255—258.

Lisk, R.D. (1971) The physiology of hormone receptors. *Amer. Zoologist*, 11, 755—767.

Lofts, B., Follett, B.K. and Murton, R.K. (1970) Temporal changes in the pituitary—gonadal axis. *Mem. Soc. Endocr.*, 18, 545—575.

Lovari, S. and Hutchison, J.B. (1975) Behavioural transitions in the reproductive cycle of Barbary doves (*Streptopelia risoria*, L.). *Behaviour*, in press.

McEwen, B.S., Pfaff, D.W. and Zigmond, R.E. (1970) Factors influencing sex hormone uptake by rat brain regions. II. Effects of neonatal treatment and hypophysectomy on testosterone uptake. *Brain Res.*, 21, 17—28.

McEwen, B.S., Zigmond, R.E. and Gerlach, J.L. (1972) Sites of steroid binding and action in the brain. In G.H. Bourne (Ed.), *Structure and Function of the Nervous System, Vol. 5*. Academic Press, New York, pp. 205—291.

McGuire, J.L. and Lisk, R.D. (1969) Localization of estrogen receptors in the rat hypothalamus. *Neuroendocrinology*, 4, 289—295.

Meyer, C.C. (1972) Inhibition of precocial copulation in the domestic chick by progesterone brain implants. *J. comp. physiol. Psychol.*, 79, 8—12.

Purvis, J. and Haynes, N.B. (1974) Short-term effects of copulation, human chorionic gonadotrophin injection and non-tactile association with a female on testosterone levels in the male rat. *J. Endocr.*, 60, 429—439.

Riss, W. and Young, W.C. (1954) The failure of large quantities of testosterone propionate to activate low drive male guinea pigs. *Endocrinology*, 54, 232—235.

Saginor, M. and Horton, R. (1968) Reflex release of gonadotrophin and increased plasma testosterone concentration in male rabbits during copulation. *Endocrinology*, 82, 627—630.

Steel, E. and Hinde, R.A. (1972) The influence of photoperiod on oestrogenic induction of nest building in canaries. *J. Endocr.*, 55, 265—278.

Temple, S.A. (1974) Plasma testosterone titers during the annual reproductive cycle of starlings (*Sturnus vulgaris*). *Gen. Comp. Endocr.*, 22 470—479.

Zigmond, R.E. (1975) Binding, metabolism and action of steroid hormones in the central nervous system. In L.L. Iversen, S.D. Iversen and S.H. Snyder (Eds.), *Handbook of Psychopharmacology, Vol. 1*. Plenum Press, New York, in press.

8

THE ROLE OF THE AVIAN HYPERSTRIATAL COMPLEX IN LEARNING

EUAN M. MACPHAIL

Laboratory of Experimental Psychology, University of Sussex (Great Britain)

INTRODUCTION

This chapter will discuss the effects of lesions of the avian hyperstriatal complex on learning, and attempt to relate the findings to comparable mammalian work. The subject matter is being approached from the standpoint of experimental psychology, a bias that will show itself, first, through the assumption that the long-term goal of the research is a better understanding of the nature of intelligence, and second, through a concern (that may occasionally seem excessive) with procedural details of behavioural situations. By way of introduction it seems appropriate to make explicit the reasons for (a) the use of lesions to investigate learning processes, and (b) using birds as experimental subjects.

Brain lesions and learning

Historically, brain lesions were seen initially as a technique to help establish localisation of 'capacities' such as intellect, sight, emotion, and so on, these capacities being seen as discrete functions whose independent existence was taken to be self-evident. During the nineteenth and early twentieth centuries, a fairly good account of localisation of primary cortical sensory areas was indeed achieved, primarily by the use of this technique. As is well known, efforts to localise the intellect did not fare as well. The point of interest here, however, is not just that the attempts failed, but, rather, the question whether their success would have amounted to any real advance in our understanding of the nature of intellectual functioning. For it does not seem that the discovery that man reasoned with, say, his frontal lobes, would in itself answer any of the questions that primarily interest psychologists; most psychologists, for example, now believe that man reasons (in some sense) with his nervous system, or parts of it, rather than with his ventricles, as was the predominant mediaeval view (Clarke and Dewhurst, 1972). But this belief does not help explain *how* man reasons — the physiological counterpart of the question merely changes from 'how do the ventricles work?' to 'how does the brain work?'. If, therefore, the localisation of intellectual function does not help explain that function, and if psychologists are concerned with explanation of intellectual function, what interest can lesion experiments hold for psychologists? An answer to this question requires a closer look at the notion of intellectual function; for as long as 'the intellect' is regarded as a discrete and indivisible faculty, then lesions may indeed serve only to localise that faculty. But the attempts that psychologists make to explain intellectual function indicate that they do not believe any such thing about the intellect; learning theories, for example, are theories about what logical devices are minimally necessary to produce the observed intelligent behaviour of animals and man, and about the interaction of such devices. Psychologists with a particular interest in learning, that is, ascribe a certain logical structure to intelligence, and once the notion of structure is accepted, then another role for the lesion technique, besides that of localisation, emerges. Lesions may now be used to help expose structure through behavioural breakdown. If intelligent behaviour is brought about by the interaction of a number of discrete logical devices, then it may be possible, by the use of lesions, to disturb the activity of some (possibly only one) of these devices while leaving others relatively unaffected. The resulting change in behaviour may help make clear the role played in intact animals by the damaged mechanism or mechanisms. Lesions at different sites might disrupt other mechanisms, and the series of breakdowns of behaviour so obtained might indicate the nature of at least some of the mechanisms involved in normal intelligent behaviour.

So far, a case has been made for the possibility that lesions might produce revealing breakdowns of normal behaviour. But it is clear that the technique need not necessarily succeed; if, for example, the logically discrete mechanisms are all closely intermingled anatomically, then it would not be possible to selectively disrupt some mechanisms rather than others, and the changes in behaviour that were obtained would, presumably, represent only a general decline in performance over the entire range of intelligent behaviour. But of course we do not know the anatomical organisation of the mechanisms of intelligence, just as we do not know how many such mechanisms there are, or what role each performs. The usefulness or otherwise of lesions may, it appears, be determined only by experiment — do the behavioural breakdowns obtained imply selective disruption, or general deterioration?

There are by now sufficient examples from mammalian research to show that breakdowns can be obtained that appear to be of interest to psychologists; perhaps the most striking is the effect on memory storage found in humans with bilateral hippocampal damage (Scoville and Milner, 1957). Such patients have access to memories laid down prior to operation, but are in general unable to store information received subsequent to operation for longer than a few seconds. This particular behavioural breakdown is widely used by psychologists as evidence for a distinction between a short-term and a long-term memory store (*e.g.* Baddeley, 1972). Other well-known breakdowns include the selective effect of frontal lesions on delayed response performance in monkeys (Mishkin and Pribram, 1956), the opposing effects of septal lesions in cats on active and passive avoidance learning (McLeary, 1961), and the disruption by hippocampal lesions of reversal (but not aquisition) of simultaneous discriminations (Kimble and Kimble, 1965). Such effects all raise questions about the organisation of normal behaviour in those situations: what device, for example, is required for delayed response that is not required for, say, simultaneous discriminations with long intertrial intervals? The various possible answers to such questions, derived essentially from theoretical accounts of learning, may be subsequently tested by examining what further deficits occur in the lesioned animals. Where the pattern of deficits agrees with that predicted by a proposed mechanism, then one more piece of evidence for the existence of such a device is available.

Role of comparative studies in physiological psychology

Not unnaturally, much of the rationale for the use of birds by purely behavioural psychologists applies equally to their use by physiological psychologists. Birds, pigeons in particular, have been widely used in the laboratory, partly as a result of the development, by Skinner and his colleagues, of an automated apparatus in which pigeons perform in a stable manner. Like rats, pigeons are cheap, easy to obtain, and breed well in captivity; an advantage over rats is their superior vision, which makes them particularly suitable subjects from which to obtain visual generalisation gradients. Although these factors have clearly weighed heavily with behavioural psychologists, the fact remains that birds have been little investigated by physiological psychologists; it is not altogether easy to see why this neglect has arisen. It may be that the rat brain is regarded as being basically similar to the human brain, so that accounts of human brain function may reasonably be expected to benefit more from an understanding of the rat as opposed to any avian brain. Another reason for the neglect may, however, be that psychologists have used various species solely for the purpose of convenience, and not because the study of more than one species has any virtue in itself; that is, psychologists may have lost sight of the theoretical value of comparative research. Paradoxical as it may seem, at least one justification for research into the avian brain is precisely that it may well aid understanding of mammalian brain function.

One relevant argument relies upon the notion of behavioural breakdown introduced in the preceding section. Suppose that a particular lesion in mammalian brain causes a characteristic behavioural syndrome, changes in behaviour, that is, in a variety of situations. Such a syndrome might be the

142

result of the disruption of a single mechanism that plays a role in all of these situations, or of several mechanisms, some of which are involved in some situations, some in others. The resolution of this question is clearly of considerable importance in obtaining a convincing account of the syndrome and its theoretical significance. It may be that smaller lesions in the same species show that the syndrome can be broken down, but negative results of such attempts leave open the possibility that there are, nevertheless, diverse mechanisms involved and that these mechanisms are so anatomically intermingled that discrete lesioning is simply not possible. In case this is so, it may be beneficial to investigate species whose neuroanatomical organisation appears very different from that of mammals, but which do perform well in learning situations. Two obvious candidates are birds and fish.

It may be as well to point out here that, although Skinner and his colleagues have emphasised the similarities in behaviour amongst many species used in operant experiments, other workers have carried out behavioural experiments for reasons very similar to those proposed here. For example, Sutherland and Mackintosh (1971) have argued that if a number of disparate behavioural phenomena, found in members of one species, are explained as being due to the operation of common mechanisms, then, other things being equal, either all or none of these phenomena should be obtained in the members of other species.

The normal pattern of behaviour of a species may, then, be broken down by either physiological or comparative techniques, and similar rationales underly each approach. Neither approach has received much attention from learning theorists in the past. This lack of interest appears to have stemmed from two somewhat peculiar beliefs: the first, that intelligent behaviour was produced by a few simple mechanisms, and the second, related to the first, that such mechanisms were common to all animals that could learn, differences in 'intelligence' mainly reflecting differences in the number of bonds that might be formed, or, perhaps, their rate of formation. Perhaps few psychologists by now find either of those beliefs tenable — at any rate, physiological psychology has grown rapidly in recent years, and the time now seems overdue for the comparative study of brain function to enjoy increased interest.

EXPERIMENTAL STUDIES

Behavioural effects of hyperstriatal lesions

The rationale proposed here for the use of lesions gives little or no indication of the brain sites that might profitably be explored; the most sensible course to follow would seem to be the systematic lesioning of one brain area after another, using a variety of test situations, each situation designed so that it might reasonably by expected to tap a different set of mechanisms from the others. This is in effect the course that has been followed in the investigation of mammalian brain, and although so far relatively few studies have been carried out on the avian brain, enough information has already emerged to show that at least one area, the hyperstriatal complex, appears to yield interesting results.

The first extensive study of the effects of brain lesions on learning in mammals was carried out by Lashley (1929), who used lesions of varying extents in different parts of rat neocortex. As is well known, Lashley failed to localise any mechanisms specific for learning within the cortex, one of his conclusions being that, for complex tasks such as maze-learning, the deterioration in learning ability following cortical injury depended on the extent of cortical damage, but was independent of its locus. An early experiment on the avian brain came to a similar conclusion: Layman (1936) tested chickens on a discrimination problem in which the birds had to choose between either a circle or a triangle on each trial, and compared lesioned birds to unoperated controls. His lesions invaded a variety of hemispheric areas, including the dorsolateral corticoid area, the hyperstriatal complex, the archistriatum, palaeostriatum and neostriatum. He found no single area that appeared critical for learning the discrimination, but a modest positive correlation (+0.39) between the total amount of hemispheric damage and errors.

Tuge and Shima (1959) examined the formation of conditioned reflexes in two groups of pigeons: the first group received damage to the corticoid surface and to dorsal hyperstriatal areas, and the second group received extensive damage, involving the hyperstriatal complex, neostriatum, and ectostriatum, along with some invasion of palaeostriatal and archistriatal areas. The unconditioned stimulus was an electric shock to the leg, and the conditioned stimulus was either a light or a metronome; responses measured were changes in heart and respiration rates. Birds in the first group showed essentially normal formation of conditioned reflexes, whereas birds in the second group showed no sign of learning. One somewhat more interesting observation was that two birds from the first group were incapable of forming a discrimination between a red and a white light as conditioned stimulus.

The first convincing demonstration of involvement of hyperstriatal areas in a learning situation was provided by Zeigler (1963a), who ran three groups of birds, one with extensive lesions of the hyperstriatal complex, a second with lesions primarily of the neostriatum, and an unoperated group. The birds were run on either the acquisition or retention of both a pattern and a brightness discrimination, using a successive go—no-go procedure, in which only one response key was available in the experimental chamber, and the subject's task was to learn to respond when the key showed the positive stimulus, and not to respond when the key showed the negative stimulus. Hyperstriatal birds showed deficits relative to both control groups in both situations, being more severely affected in the pattern than the brightness discriminations. Pritz *et al.* (1970) carried out a similar experiment on

144

retention of successive brightness, pattern and colour discriminations by birds with lesions of the Wulst (that is, the dorsal and accessory hyperstriatum), and found deficits in brightness and pattern discriminations, but not on the colour discrimination.

An experiment which, more than any other, served to establish the hyperstriatal complex as an area of interest to psychologists was carried out by Stettner and Schultz (1967). They used quail as subjects, and tested normal birds and birds with extensive lesions of the Wulst and surrounding areas on the acquisition and serial reversals of a simultaneous pattern discrimination. Their procedure was as follows: on each trial, two response keys were illuminated, one with horizontal and the other with vertical stripes (the sides on which the two patterns appeared being varied irregularly), and responses to only one of the patterns were reinforced. Once a bird had acquired the discrimination to a criterion of 90% correct choices for three consecutive sessions, the reinforcement values of the two stimuli were reversed, and reversal training began, with the former positive stimulus now being the negative stimulus. There were 25 reversals in all, and each reversal was taken to a criterion of 16 correct out of the last 20 trials of a daily session. Stettner and Schultz found first, that acquisition performance was unaffected by the lesions, and second, that reversal performance was severely disrupted, the disruption taking the form of perseverative responding to the former positive stimulus at the outset of each reversal.

The fact that acquisition performance was unaffected indicates that the lesions did not cause any gross sensory, motor or motivational deficits, and suggests that at least some of the mechanisms involved in learning are intact. The most plausible account of the disruption of reversal is that reversal learning involves some process that is not essential for acquisition, and that

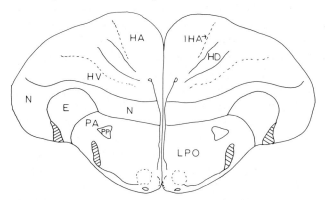

Fig. 1. Drawing of a transverse section through the pigeon forebrain, based on the Karten and Hodos (1967) atlas. Abbreviations: E, ectostriatum; HA, hyperstriatum accessorium; HD, hyperstriatum dorsale; HV, hyperstriatum ventrale; IHA, nucleus intercalatus hyperstriati accessorii; LPO, lobus parolfactorius; N, neostriatum; PA, palaeostriatum augmentatum and PP, palaeostriatum primitivum.

this process has been selectively affected by the lesion. It can be seen that this is precisely the kind of behavioural breakdown that may prove revealing to psychologists; a further ground for interest is that a similar dissociation between acquisition and reversal performance is obtained by lesions in certain parts of the mammalian limbic system, so that the way is open for a systematic comparison of the pattern of deficits found in hyperstriatal birds with that found in mammals with limbic lesions.

Anatomical and electrophysiological investigations of the hyperstriatum

It is appropriate at this stage to give some consideration to what is known of hyperstriatal function, besides its possible role in learning. Fig. 1 illustrates the location of those forebrain structures that are of primary interest here; it can be seen that the avian forebrain differs markedly from the familiar mammalian pattern, the most striking difference being the absence of any overlying layered cortical mantle. Our knowledge of forebrain organisation has recently been considerably enlarged by the work of Karten and his colleagues, and a brief account of their principal conclusions, particularly as regards the hyperstriatum, will be given here.

A main proposition is that the avian homologues of the mammalian corpora striata (the caudate nucleus, putamen, and globus pallidus) are to be found in the palaeostriatal complex and the parolfactory lobe of the avian hemispheres (Nauta and Karten, 1968; Zeier and Karten, 1971; Karten and Dubbeldam, 1973). The posterior and medial archistriatum may be the homologue of the mammalian amygdala, but the anterior two-thirds of the archistriatum, neostriatum, and hyperstriatum, have no homologous structures in the mammalian brain. However, these striatal structures may well contain neurones that are functionally homologous to neurones in mammalian neocortex; for example, neurones in Field L of the caudal neostriatum receive direct afferents from the nucleus ovoidalis of the thalamus, and that nucleus in turn receives afferents from the nucleus mesencephalicus lateralis pars dorsalis, which is the avian homologue of the inferior colliculus (Karten, 1968). Their proposal is, then, not that Field L corresponds to the mammalian auditory neocortex, but that certain neurones in Field L may be homologous with certain neurones in the auditory cortex.

The avian hyperstriatal complex consists of the Wulst and the ventral hyperstriatum; the Wulst, in turn, consists of the dorsal hyperstriatum, the acessory hyperstriatum, and the intercalated nucleus of the acessory hyperstriatum (IHA). Recent work has shown that neurones in the IHA may be homologous with striate cortex neurones in that they receive afferents from thalamic nuclei to which retinal ganglion cells project directly, and have small receptive fields with a columnar arrangement (Revzin, 1969; Karten *et al.*, 1973).

There are three other relevant, but less well documented pieces of information concerning the Wulst. First, there is electrophysiological evidence for both auditory and cutaneous sensory projections to the area (Adamo and King, 1967; Delius and Bennetto, 1972); second, there is a suggestion, based on the existence of direct efferents from the anterior Wulst to pontine and spinal regions (these efferents appearing comparable to a component of the mammalian pyramidal tract) that the anterior Wulst may be involved in the control of movement (Adamo, 1967; Karten et al., 1973). A final observation concerns the ventromedial Wulst, which receives afferent input from the dorsomedial anterior thalamic complex (Karten et al., 1973); the mammalian anterior thalamic complex projects to the cingulate gyrus, and forms part of the 'Papez circuit', and Karten et al. (1973) therefore suggest that the ventromedial Wulst may be homologous with part of the mammalian limbic system.

This condensed review is sufficient to show that the Wulst is a complex area, and one that is most unlikely to be functionally homogeneous; it should, moreover, be noted that most of the experiments reviewed in the preceding section used extensive lesions, which invaded not only the Wulst, but parts of the ventral hyperstriatum and the hippocampal complex, two areas of whose possible function virtually nothing is known.

THEORETICAL INTERPRETATIONS

Perceptual deficit hypothesis

One interpretation of hyperstriatal lesions is that they result in a perceptual deficit which manifests itself in disruption of visual discrimination performance. Much of the work described previously can be accommodated by this notion; Tuge and Shima (1959) found that two hyperstriatal pigeons, which would form normal defensive conditioned reflexes, could not, nevertheless, form a discrimination between a white light and a red light as conditioned stimuli. Both Zeigler (1963a) and Pritz et al., (1970), showed disruption by hyperstriatal lesions of successive discriminations based on pattern or brightness. Finally, the work of Karten et al. (1973) and Revzin (1969) provides good anatomical and electrophysiological grounds for supposing some role of the Wulst in visual perception.

No attempt has so far been made to specify the precise nature of the proposed perceptual deficit, and it should be emphasised (as it is by Pritz et al.) that there is no reason to suppose that hyperstriatal birds suffer from any gross sensory impairment. Zeigler, for example, writes: "There were no obvious signs of visual deficits in the general behaviour of those subjects which showed deficits in visual discrimination learning . . . the performance deficits . . . are not attributable simply to gross visual disturbances" (Zeigler, 1963a; p. 173). Pritz et al. made observations that entirely bear out those of Zeigler, and in addition emphasised that their birds showed no sign of any deficit in motor control or feeding that might account for their impaired performance.

There are, however, experimental results which appear to pose serious difficulties for virtually any version of the perceptual deficit hypothesis. Stettner and Schultz (1967), for example, in the experiment described earlier, found no effect of hyperstriatal lesions on acquisition of a simultaneous pattern discrimination; Hodos *et al.* (1973) similarly found no effect of lesions in the Wulst on the retention of simultaneous discriminations of either brightness or pattern.

In summary, whereas successive discriminations of pattern or brightness are disrupted by hyperstriatal lesions, simultaneous discriminations using similar stimuli are immune. One possible resolution of this difficulty relies on the fact that hyperstriatal lesions appear, in successive discriminations, to have more effect the more difficult the discrimination. For example, Zeigler (1963a) found a greater effect of lesions on pattern as opposed to brightness discrimination, and the brightness discrimination was the easier of the two; similarly, Pritz *et al.* (1970) found no effect of their lesions on retention of a colour discrimination, and in their experiment also, the original acquisition of the colour discrimination was an easier task than that of either the pattern or the brightness discrimination. Hyperstriatal lesions, that is, appear to have an effect in proportion to the sensory demand made on the subjects, and it could be argued that simultaneous discriminations make less sensory demand than do successive discriminations; in a simultaneous discrimination, both stimuli may be compared on each trial, whereas in successive discriminations, the stimulus on the key must be compared to some sort of engram or trace of the alternative stimulus.

The basic objection to this account is that it fails to provide any explanation of the deficit in reversal learning reported by Stettner and Schultz (1967); it is difficult to see any sense in which the reversal of a simultaneous discrimination makes any more sensory demand than its original acquisition. It may, of course, be that the reversal deficit is not related to the successive discrimination deficit — it is, as has been pointed out, hardly unlikely that the Wulst has diverse functions. However, parsimony demands that a careful search be made for a single hypothesis that can accommodate all the data satisfactorily.

The first stage in such a search should, perhaps, focus on the procedural differences between simultaneous and successive discriminations; the only proposal made so far is that the latter may be more difficult from the sensory point of view. However, it should be noted that successive discriminations make a very different demand on response output than do simultaneous discriminations; whereas in simultaneous discriminations, subjects must learn which of two stimuli to approach, in successive go—no-go discriminations subjects must learn *not* to approach a stimulus that has previously, as a consequence of pretraining procedures, acquired approach strength. The task that is, is essentially one of learning to inhibit a response. Now there is evidence (*e.g.* Kimble, 1968) to show that lesions of various parts of the mammalian limbic system appear to affect a mechanism whose task in intact animals is to inhibit responses; if a corresponding mechanism exists in birds, then damage to it would, naturally, result in poor performance in go—no-go discriminations of the type described, but have little effect on visual discriminations that do not require withholding of response, such as, for example, those involving a choice between two simultaneously presented stimuli.

The next section will examine in detail the notion that hyperstriatal lesions cause deficits in response inhibition. Before that discussion, one important general point arising from this analysis of the perceptual deficit hypothesis should be made: this is, that no one learning situation can be used to analyse any deficit satisfactorily. Performance in any behavioural situation makes demands on sensory capacity, response capacity, motivational state, and, of course, learning mechanisms. Where a disruption in behaviour does occur, new situations must of necessity be used before any of the possible sources of deficit may be ruled out. One way in which most of the alternatives to the perceptual deficit interpretation of the successive discrimination impairment might be ruled out would be an experiment to show that successive discriminations of comparable difficulty, but based in another modality (*e.g.* audition) are not disrupted by hyperstriatal lesions; regrettably, no such experiment has yet been conducted although its outcome would clearly be of considerable interest.

Response disinhibition hypothesis

Tuge and Shima (1959) argued that their failure to establish a discrimination in two lesioned birds indicated that the hyperstriatum was involved in 'internal inhibition', although, of course, an alternative account is that a visual deficit was responsible. Reynolds and Limpo (1965) were the first to discriminate between the two hypotheses, and to provide tentative support for a response disinhibition account. These authors found that dorsal hyperstriatal lesions affected performance in fixed-interval but not fixed-ratio components in a multiple schedule. Here again, it is important to clarify the precise demands of the task. There was a single response key; when it was red, the first response that the pigeon made after 4 min had elapsed obtained food (fixed-interval component); when the key was green, the fifty-fifth response made by the pigeon obtained food (fixed-ratio component). Responses made prior to the end of the 4-min interval are without effect in the fixed-interval component, so that intact birds, after some training, pause at the beginning of each component, their probability of responding gradually increasing throughout the interval. In fixed-ratio components, the more rapidly a bird responds, the sooner food is obtained, and trained birds respond accordingly with high rates of response. Their birds were trained preoperatively, and, on return to the test situation, both hyperstriatal and control birds responded rapidly on the fixed-ratio component but in fixed-interval components, hyperstriatal birds failed to show the initial pause characteristic of normal performance on such schedules. The birds did, on the other hand, show a drop in response rate when the key light changed from green to red, marking a change from the fixed-ratio to a fixed-interval component, so that no sensory deficit was apparent. Reynolds and Limpo also observed that lesioned birds, although showing no initial pause, frequently broke off responding abruptly in the middle of intervals, and so concluded that their effect could not be *simply* a disinhibition of responding.

The proposal that hyperstriatal lesions cause difficulty in response inhibition appears to give a satisfactory account of the deficits reported so far. Acquisition and retention of simultaneous discriminations should be immune to lesions (*e.g.* Stettner and Schultz, 1967; Hodos *et al.*, 1973), as subjects are required to approach a stimulus on each trial. Reversal of simultaneous discriminations should be disrupted since the first stage of reversal performance for normal (as for lesioned) birds consists of learning not to approach the former positive stimulus — clearly, the alternative stimulus cannot be approached without such an inhibition being achieved. The pattern of reversal disruption reported by Stettner and Schultz, *i.e.* perseverative responding to the former positive stimulus, reinforces this account. Finally, performance in successive go—no-go discriminations should be affected since the basic requirement for solution of such discriminations is precisely inhibition of responding in negative trials. It might be argued that the response disinhibition hypothesis would predict disruption of all successive discriminations, and not depend on the stimuli used, whereas in fact Pritz *et al.* (1970) found that hyperstriatal birds were affected on discrimination of pattern and brightness, but not of colour; but it has already been pointed out that the occurrence or otherwise of disruption may depend on the difficulty of the task, and a reasonable proposal is that the ease of a successive discrimination indicates precisely that little inhibition need be generated for its solution, so that indeed lesions might have little apparent effect.

Response inhibition is a loosely-defined notion, pointing only to the fact that disruption appears to occur when a task requires withholding of response. Failure to inhibit responses might be a symptom of any of a number of underlying disturbances. It could, for example, be due to an essentially motor deficit, although it should be pointed out, as it was by Douglas (1967) in his discussion of hippocampally ablated rats that "ablated animals do not continue to walk until they bump into a wall (as do animals with caudate lesions)" (Douglas, 1967; p. 434). Alternatively, failure to inhibit responses could be due to a more central deficit like, for example, a failure to register negative reinforcement efficiently, or to a reduced motivational effect of aversive events. Before attempting to specify any of these alternatives as the basic deficit, however, the first step is clearly to see how general the disinhibition effects in fact are.

Macphail (1969) carried out an experiment that was designed to answer two questions: first, did the lesions of, for example, Stettner and Schultz (1967), tap a general system of response inhibition, and second, is the critical locus of those lesions in the Wulst or some other structure adjoining the Wulst, such as, for example, the hippocampal complex? The relevance of this second question is underlined by the fact that hippocampal lesions in mammals cause a pattern of deficits that has been interpreted as being due to a loss of inhibition. The pattern includes, for example, disruption of (a) simultaneous discrimination reversal performance; (b) passive avoidance (*i.e.* learning to withhold a response which is punished); (c) simple extinction and (d) successive go—no-go discriminations (Kimble, 1968; Swanson and Isaacson, 1967).

In the Macphail (1969) experiment, lesions were placed in the medial hyperstriatum accessorium of five pigeons, and subjects were tested on (a) acquisition and reversal of a simultaneous brightness discrimination; (b) extinction in a simple runway and (c) a passive avoidance task. Most unexpectedly the birds showed no deficit in reversal performance and took fewer trials than controls to reach criterion in both the extinction and the passive avoidance situations. A sixth bird, having a more extensive lesion, showed the opposite pattern of results, *i.e.* poor reversal performance and an increased number of trials in both the extinction and passive avoidance tasks. Each pattern of deficits does, however, suggest that general systems of response inhibition and facilitation may have been tapped. The five birds having the small lesions showed an abnormal excess of inhibition in both the extinction and passive avoidance tasks; in the discrimination situation, although showing no change in the number of errors, the birds did in fact refuse to choose either stimulus within the maximum time allowed on each trial (2 min) significantly more often than the controls. Increased inhibition is seen in cats with cingulate gyrus lesions (McLeary, 1961), and this raises the intriguing possibility that the medial Wulst does indeed contain an area with functions similar to those of the mammalian cingulate cortex, in accordance with the suggestion of Karten *et al.*, (1973). At present, however, such a speculation must remain tentative; one problem, for example, is that Stettner and Schultz' lesions clearly invaded the medial Wulst, and yet their birds showed no sign of excessive inhibition — in fact, quite the reverse. It may, of course, be that the effects of medial Wulst damage are overridden by the effects of lesions elsewhere in the Wulst, but it should be noted that in cats simultaneous lesions of the cingulate cortex and the septal area lead to excessive inhibition, not to a loss, of response inhibition (Lubar, 1964); that is to say, septal lesions in cats, which by themselves lead to a loss of inhibition, are overridden by the effects of cingulate cortex lesions, and not *vice versa*.

The sixth bird showed signs of a loss of response inhibition in all three situations, and so was comparable to hippocampectomised rats. As this bird's lesion (unlike those of the other five subjects) invaded the hippocampal complex, the possibility that hippocampal damage is responsible for the loss of inhibition remained. A second investigation (Macphail, 1971), used birds with lesions of lateral regions of the hyperstriatal complex; these lesions caused extensive damage of both the ventral and dorsal hyperstriatum, but left the hippocampal region intact. Two experiments were carried out, each relevant to the notion that the lesions would obtain deficits in response inhibition. The first experiment was a successive red—green go—no-go discrimination, using free operant techniques. The birds were operated on before training began, and the first stage of the experiment consisted in obtaining stable baseline rates of response to a single key which could be either red or green. In both red and green components, responses were reinforced on a variable interval schedule having a mean interval of one minute. In variable interval schedules, rewards are dependent on responses, but are delivered at unpredictable moments in time. Accordingly, trained birds respond at a relatively uniform rate in components of such schedules. Once a bird achieved a stable baseline performance, a discrimination was introduced by omitting all reinforcements in red components. In intact birds, the introduction of a discrimination has two effects, one obvious and the other less so: first, the rate of response in the negative components declines and second, the rate of response in the positive components increases. This latter phenomenon is known as behavioural contrast (Reynolds, 1961), and is of considerable theoretical interest, a widely-held view being that it is a consequence of emotional responses (*e.g.* frustration) generated by unexpected non-reward in the negative components (Scull *et al.*, 1970). Two questions may therefore be asked concerning the results of this experiment: first, do lesioned birds show higher rates of response in negative components (so indicating a deficit in ability to withhold responses), and second, do they show any increase or decrease in behavioural contrast? The results showed that as a consequence of the lesions there was a significant deficit in negative component performance, *i.e.* lesioned birds responded more rapidly — but this was not a large effect presumably, again, because the discrimination was an easy one. There was no significant difference between the groups in positive component performance, lesioned birds on average showing a somewhat larger behavioural contrast effect than controls.

The absence of any effect of the lesions on behavioural contrast carries two implications relevant to a more detailed analysis of the response inhibition deficit. First, normal behavioural contrast implies that the lesioned animals must have registered the omission of reinforcement as efficiently as controls; that is, the failure to inhibit responses adequately does not seem due to any failure to detect alterations in reinforcement contingencies. Second, normal contrast provides some indirect evidence against the possibility that lesioned birds inhibit responding less successfully because they do not find non-reinforced responses as aversive as controls. For, if aversiveness can be assessed by its emotional consequences, and if contrast is an index of those consequences, this result shows that at least as much aversion occurs in lesioned as in control animals.

Following completion of the successive discrimination, birds were tested on acquisition and six reversals of a position discrimination; the overall design was similar to that used by Stettner and Schultz (1967), except that the positive stimulus was the right-hand key in acquisition, the left-hand key in the first reversal, and so on, both keys being green. The results of this experiment were clear-cut: lesioned birds showed no acquisition deficit (making slightly, but not significantly, fewer errors on average than controls) but made significantly more errors throughout the reversal series.

The two parts of the Macphail (1971) experiment therefore provide evidence indicative of a loss of response inhibition following lesions confined to the hyperstriatal complex. Stettner and Schultz, in their report (1967) of reversal deficits, used lesions that were essentially dorsal to the ventral hyperstriatum, and the lesions of Reynolds and Limpo (1965), which disrupted fixed-interval pauses, appear to have spared the ventral hyperstriatum altogether. The evidence therefore points to the critical focus for these effects being in the Wulst itself. Such an interpretation also seems applicable to the results recently reported for chicks by Benowitz (1972; Benowitz and Lee-Teng, 1973) who found a deficit in chicks with extensive lesions involving both hyperstriatal and hippocampal structures in retention of a one-trial passive avoidance situation, and Benowitz and Lee-Teng found disruption by similar lesions of reversal of a shape discrimination. Benowitz and Lee-Teng propose that their results are due to hippocampal invasion, but a more parsimonious interpretation would seem to be that damage to the Wulst is responsible. Of course, until the effects of lesions restricted to the hippocampal complex are investigated, such a preference must be based on grounds of parsimony alone — that is, there is as yet no compelling evidence for or against the view that the avian hippocampus is involved in response inhibition.

If the Benowitz and Benowitz and Lee-Teng results are included, there is now a fairly impressive amount of evidence for the generality of the hyperstriatal effects. Deficits suggestive of loss of response inhibition have been shown in three species of birds — in pigeons (*e.g.* Reynolds and Limpo, 1965; Macphail, 1971), quail (Stettner and Schultz, 1967) and chicks (Benowitz and Lee-Teng, 1973), and in a variety of situations, including reversals of both position (Macphail, 1971) and pattern discriminations (Stettner and Schultz, 1967), successive go—no-go dicriminations of pattern (Zeigler, 1963a), brightness (Pritz *et al.*, 1970), colour (Macphail, 1971), and passive avoidance tasks (Benowitz, 1972; Macphail, 1969 — one bird only); these deficits also arise when different types of reinforcement have been used— although most experiments have used food as the reinforcement for appetitive tasks, Benowitz and Lee-Teng used warm air as a reward, and whereas Macphail (1969) used shock as a punishment, the Benowitz (1972) situation involved avoidance of an aversive taste solution.

Although the loss of response inhibition hypothesis provides a generally successful account of the deficit pattern described for hyperstriatal birds, there is the difficulty, pointed out by Reynolds and Limpo (1965) that their birds occasionally had long pauses in the middle of fixed-interval components, suggesting that the deficit was not *simply* a loss of response inhibition. A similar informal (and irritating) observation in this laboratory has been that hyperstriatal birds performing reversals frequently stop responding altogether after long runs of perseverative errors.

A recent experiment (Macphail, 1975) studied this 'halting' phenomenon quantitatively. Two groups of birds having hyperstriatal lesions, one with anterior, the other with posterior lesions, a control operated group, with neostriatal lesions, and an unoperated control group, were tested on the acquisition and reversal of a simultaneous position discrimination. On each occasion on which a bird failed to complete a trial within 5 min of its commencement, food was automatically made available in the hopper for 8 sec, and this was usually successful in restoring responding in the birds. Both groups of hyperstriatal birds were severely impaired relative to both control groups over the reversal series (and also, in this experiment, in acquisition); the hyperstriatal birds showed the characteristic runs of perseverative errors, and did, as anticipated from the early observations, halt significantly more frequently than the controls.

As the lesioned animals consistently made many more errors than controls, it cannot be concluded, of course, that hyperstriatals are more likely than controls to halt after a given number of errors; the point of this result is that hyperstriatals did in fact stop responding rather than change keys, albeit after a large number of errors. *In parenthesi* it should be noted that a deficit in simultaneous discrimination acquisition in some experiments, but not in others, is found not only in hyperstriatal pigeons, but also in hippocampal rats (compare, for example, Kimble, 1963 and Olton, 1972). A notion currently being investigated in this laboratory is that acquisition deficits occur when extensive pretraining has been given. How such a notion, if verified, would bear on the hypotheses being reviewed here is not altogether clear and the phenomenon will not be discussed further in this paper.

The Macphail (1975) results, like the Reynolds and Limpo observations, pose a problem for the response inhibition hypothesis, which would seem to predict runs of errors, but no tendency to stop responding altogether. The difficulty of the hyperstriatal birds appeared to lie in an unwillingness to peck the alternative key despite having abandoned responding on the previously positive key.

Response shift hypothesis

The response inhibition account of mammalian hippocampal function appears to face a difficulty similar to that described for hyperstriatal pigeons at the end of the preceding section. Olton (1972) tested hippocampal rats on the acquisition of a simultaneous brightness discrimination. He found that both normal and hippocampal rats soon adopted consistent position habits in his apparatus, but slowly began to discriminate, in terms of latencies, between the positive and the negative stimuli; at this stage, while still consistently choosing one side, a rat would respond more rapidly if the positive stimulus was on that side than if the negative stimulus was there (the Mahut Effect). However, whereas normal rats soon broke their position habits and solved the task, hippocampals persisted in position responding, although by now responding very much more slowly to the negative than to the positive stimulus. Here again it should be emphasised that although hippocampal rats were in effect performing a successive discrimination on their preferred side, it is clear from Olton's data that they were slower than controls in acquiring the discrimination; the point of interest is that, even having acquired the discrimination, hippocampal rats did not select the alternative side on trials on which the negative stimulus appeared on their preferred side. Olton interpreted his result as showing that hippocampal lesions disrupt a mechanism for response shift rather than response suppression; hippocampal rats, according to this view, can suppress responding in response to non-reward, but do not, as normals do, shift responding as a consequence of non-reward.

To see whether such an account might also apply to pigeons with hyperstriatal lesions, a second experiment was carried out with the same four groups of pigeons (Macphail, 1975). The task was a successive go—no-go discrimination, with discrete trials. On positive trials, two green keys were illuminated, and as soon as a bird pecked the keys five times in all, food was delivered; on negative trials, the same two keys were illuminated with red light, and stayed lit for 4 sec, independent of whether the birds responded or not. In both positive and negative trials, the number of responses to each key was recorded. It should be emphasised that in both types of trial, the outcome was independent of which key was chosen; in positive trials, a bird could make all five responses to one key, or three to one key and two to the other, and in negative trials, of course, no responses had any effect. Training continued until there occurred 20 consecutive trials with five responses on all positive trials, and none on all the negative trials.

The main results of this experiment were as follows: first, neither group of hyperstriatal birds showed a significant increase in either trials or errors to criterion (although the trend was in the direction of an increase), and this supports the view that on simple go—no-go discriminations, hyperstriatal birds show little or no impairment. Second, the percentage of total negative responses to the preferred key was significantly higher for the hyperstriatal, as opposed to the control groups (the preferred key simply being that on which more than half a given bird's negative responses occurred). Whereas the control birds made on average approximately 64% of their negative responses on one key, hyperstriatal birds made approximately 76% of their negative responses on the same key. In other words, control birds showed a significantly greater tendency to shift their responses in response to non-reward than did hyperstriatals.

The Macphail (1975) and Olton (1972) results provide evidence that, at least in some situations, both hyperstriatal lesions in birds and hippocampal lesions in rats cause difficulty not in response inhibition, but in response shift. The question that now arises is whether a response shift account can accommodate the entire pattern of deficits.

An immediate difficulty facing such an interpretation concerns those situations in which only one stimulus, and so only one apparent response, is available; such situations include, for example, successive go—no-go discriminations, simple extinction and passive avoidance. Hippocampal rats show deficits in all three situations (Swanson and Isaacson, 1967; Kimble, 1968) and pigeons with hyperstriatal lesions show deficits in difficult go—no-go discriminations (Zeigler, 1963a; Pritz et al., 1970), passive avoidance (Benowitz and Lee-Teng, 1973) and possibly in simple extinction (Macphail, 1969 — one subject).

One effort to provide a unitary account may be made by the proposal that the three situations (successive discriminations, extinction and passive avoidance) do involve a shift of response in that subjects must learn to shift from approaching a stimulus to avoiding that stimulus. But this account abolishes the distinction between response suppression and response shift that is at the heart of Olton's hypothesis; if inhibiting of response is a shift of response, then hippocampal rats, since they did succeed in jumping more slowly to the negative stimulus, did not suffer from any simple deficit in response shift.

Attention shift hypothesis

It appears from the preceding sections that neither the response inhibition nor the response shift hypothesis can accommodate the pattern of deficits. A third account proposes that the syndrome is the result of changes in attention in lesioned animals. Such an account has already been proposed elsewhere for hippocampal rats (Douglas, 1967), as well as for hyperstriatal pigeons (Hodos et al., 1973). It will be recalled that Karten et al. (1973) described a system of direct visual projections from the retina to the thalamus, and from the thalamus to the IHA of the Wulst; however, Hodos et al. (1973) have shown that lesions of this system at either the thalamic or the hyperstriatal level do not have any marked effect on the retention of simultaneous discriminations of either brightness or pattern. A second forebrain area in birds that is concerned in vision is the ectostriatum, which receives afferents from the thalamic nucleus rotundus, which in turn receives projections from the tectum. Lesions of the ectostriatum do cause severe deficits in both brightness and pattern discriminations, and a current proposal is that the thalamofugal pathway, of which the IHA is a part, may modulate the visual input to the tectofugal system, of which the ectostriatum is a part (Karten et al., 1973). Hodos et al. (1973) have suggested, therefore, that the thalamofugal system may be involved in selective attention within the visual domain.

There is substantial evidence for abnormalities of attention in hippocampal rats, one typical study being that of Crowne and Riddell (1969). These authors presented a novel stimulus (paired presentation of a houselight and a loud tone) to hippocampal and control rats under two conditions; in one condition, the rats were pressing a panel for water reward while thirsty, and in the other, rats were in the same box, but satiated. When the novel stimulus occurred during panel-pressing, hippocampal rats showed·virtually no sign of orienting to it; they also showed no deceleration in heart rate (a component of the normal orienting response pattern of intact rats), and virtually no decrease in the rate of panel-pressing. When, however, the same novel stimulus was presented in the satiated condition, the hippocampal rats showed a quite normal orienting response, as measured by the heart rate deceleration. That is, hippocampal rats do not shift their attention to novel stimuli when actively attending to other stimuli, but do attend normally when no alternative attention-capturing stimulus is present.

Although there are a number of reports indicating some role for the hippocampus in attention, the evidence available on hyperstriatal pigeons is very thin as yet. Zeigler (1963b) reported that hyperstriatal pigeons, placed for one hour daily in an activity cage, showed no decline in level of activity over the hour, in contrast to controls, which showed considerably less activity in the third 20-min as compared to the first 20-min. This may be evidence of slower habituation in hyperstriatals, although it should be pointed out that the cage was not novel. Adamo and Bennett (1967) found a tendency in chickens with Wulst lesions to orient more frequently to a sound stimulus repeated every 3 min, but only when the stimulus was of short (0.2 sec) rather than long (2 sec) duration. Lesioned animals also made many more incorrect orientations than controls — that is, they tended to look in a direction other than that from which the sound originated. Each of these experiments, then, provides some evidence that hyperstriatal birds are less likely to shift attention from a stimulus despite repetition of that stimulus. A further general indication that the hyperstriatum may be involved in the directing of attention is provided by the observation of Cohen and Pitts (1967) that stimulation of discrete sites in the Wulst of unanaesthetised pigeons obtained orientation to some point in the visual field; where birds were semi-restrained in a harness, the head was moved, and "stimulation of the same site in the unrestrained animal produced a coordinated pattern of head and body movement having the same directionality seen in the semi-restrained case" (Cohen and Pitts, 1967; p. 330).

Douglas and Pribram (1966) proposed that the mammalian hippocampus is involved in 'gating out' stimuli associated with non-reward, and, given that attention is a continuous process in conscious animals, such a proposal is essentially identical to the notion that the hippocampus is involved in shifting attention as a consequence of non-reward. The detailed application of this hypothesis to the results described previously might be as follows: whereas normal rats in extinction, passive avoidance, and in negative components of successive discriminations, shift attention from the negative or non-reinforced stimulus, hippocampals persist in attending to, and so in responding to, the stimulus. It should be emphasised that this account, in common with any explanation of lesion-induced behavioural changes, carries implications concerning the behaviour of normal animals; specifically, it is proposed that normals do not stop responding in, say, an extinction situation, because the positive stimulus has become neutral — normals, according to this account, stop responding *before* the positive stimulus has lost its approach strength, and it is the residual strength which induces lesioned animals to continue responding.

The failure of hippocampal rats in Olton's (1972) experiment similarly reflects their failure to shift their attention from the negative stimulus in what has essentially become a successive discrimination on one side of the apparatus. The hypothesis does have difficulty with the Crowne and Riddell (1969) experiment, since normals in this experiment shifted attention from the rewarded panel to a novel stimulus, and this shift was not, of course, a consequence of non-reward; this experiment, therefore, suggests that hippocampal rats may suffer from a general deficit in ability to shift attention under all circumstances in which shifts occur in normals.

The notion that hyperstriatal pigeons suffer from a general deficit in ability to shift attention is particularly plausible in the light of the anatomical data provided by Karten and his colleagues. However, one problem facing such an interpretation is the fact that hyperstriatal lesions appear to have little, if any, effect on easy go—no-go discriminations; presumably the stimuli in an easy discrimination are especially salient, so that it is not clear why attention is so rapidly shifted from them in lesioned birds. But this difficulty loses its force when the interpretation of the behaviour of normal birds is considered; either normals in easy discriminations cease responding to the negative stimulus because it has lost approach strength (*i.e.* they do not shift attention from it) — in which case no deficit should be expected in lesioned birds, or normals find it, in fact, particularly easy to shift attention from the negative stimulus when the difference between it and the positive stimulus is particularly marked — in which case, little demand is made on the ability to shift responding, so that lesions might indeed have little effect.

The concept of attention shift that has been introduced here is quite as loosely defined as was that of response inhibition. This generality in turn reflects the fact that no current theory of attention in animals appears to give a convincing account of the deficit pattern obtained. One objection to the Douglas and Pribram (1966) theory has already been advanced; another, more fundamental objection, concerns their notion that negative stimuli are 'gated out' at a relatively early stage of perceptual processing. If intact rats are presented with a stimulus that has been associated with foot shock, they show a marked inhibition of ongoing appetitive behaviour during that stimulus (the conditioned emotional response); hippocampal rats show a markedly reduced conditioned emotional response (Brady, 1958). In this situation, then, normals apparently attend *more* to an aversive stimulus than do the lesioned animals, a finding that is clearly very difficult for the Douglas and Pribram model. It is, of course, satisfactorily accommodated by both the response disinhibition hypothesis and the notion that hippocampals are less easily distracted than normals.

A prominent theory of selective attention in intact animals is that developed by Sutherland and Mackintosh (1971); according to this theory, animals performing a visual discrimination must learn first, to switch in an analyser for the dimension relevant to the discrimination (*e.g.* brightness), and, second, to attach appropriate response strengths to stimuli at different points along that dimension (*e.g.* approach dim, avoid bright). Within this theory, switches of attention involve switches from one dimension to another; but the difficulty facing hyperstriatal pigeons performing reversals is not one of switching dimensions, since the relevant dimension does not change. The problem is one of switching attention from one stimulus to another at a different point along the same dimension.

The position adopted here, then, is that hyperstriatal lesions may damage a mechanism that is involved in shifts of attention between particular stimuli, rather than between one dimension and another. A closer specification of such a mechanism would be an essay in learning theory which is not yet warranted by the available data. It must be emphasised that, at this stage, although it is difficult to see how existing theories of attention might account for the breakdowns observed, it is clearly premature to claim that these data show any important inadequacy in those theories. It is possible that further experiments may show that only a minor modification or extension of an existing theory of attention is required; but it is also possible that the experiments will indicate the necessity for a major change in theory, and it is this possibility that should stimulate research.

CONCLUDING COMMENTS

Further speculation concerning avian hyperstriatal function must await new experimental results, and although it would, of course, be possible to discuss possible hippocampal functions in mammals in considerably more detail, that would not be relevant here. Parallels have been drawn frequently in this review between the hyperstriatum and the hippocampus, and some remarks about the utility of this exercise are appropriate. It should be emphasised that it is not proposed that the two areas are homologous; anatomically, this is clearly not so — if there is an avian homologue of the mammalian hippocampus, it is almost certainly the avian hippocampal complex (Ariëns Kappers *et al.*, 1936). There is, however, a family resemblance between deficits brought about by the two types of lesion, and it may well be that the use of situations that have proved sensitive to hippocampal lesions will aid the analysis of hyperstriatal lesions. This, then, is an almost atheoretical hypothesis — merely the proposal that tasks that have been successfully used with hippocampal mammals should be used with hyperstriatal birds. The pursuit of the parallel may in the end show that the resemblance is illusory — that the reasons for which hippocampals fail at reversal learning are quite different from those for which hyperstriatals fail; on the other hand, the resemblance may be so close that it may be concluded that some core deficit is common to both types of lesion; this in turn would suggest a similar behavioural mechanism in both cases. To establish the existence of such a mechanism in two such distantly related classes of animal would argue its considerable importance for learning, and warrant further close analysis of its properties.

The value of comparisons with mammals as a means of generating hypotheses for experiments using birds may be further illustrated by summarising the course of Zeier's research on archistriatal function in the pigeon. In an early experiment Zeier (1968) tested the effect of archistriatal lesions on a simple (variable interval) free operant schedule of key-pecking for food. His rationale consisted of two points: first, that there was a widely-accepted anatomical homology between the mammalian amygdala and the avian archistriatum, and, second, that the mammalian amygdala was involved in the motivational control of learning in mammals. Zeier tentatively interpreted the results of this experiment (an increase in response rate in archistriatal subjects) as suggesting that his archistriatal subjects were relatively insensitive to reinforcement density, this notion in turn being derived from the Douglas and Pribram (1966) model of amygdalar function in mammals. In a subsequent experiment (Zeier, 1969) the schedule used, which differentially reinforced low rates of response (DRL), was chosen specifically to test a prediction drawn from the application of the Douglas—Pribram model to archistriatal subjects; the prediction was upheld *i.e.* performance on the schedule, which is particularly error-sensitive, was improved by the lesions. However, Zeier's subsequent work (Zeier and Karten, 1971; Zeier, 1971), showed, first, that there are, on anatomical grounds, four subdivisions within the avian archistriatum, and second, that the only region of the archistriatum from which lesion-induced improvements in DRL performance may be obtained (the anterior two-thirds of the structure) is anatomically *not* homologous with the mammalian amygdala, being in fact more comparable to the neocortex. Thus in Zeier's work, an extremely interesting finding, the improvement in DRL performance following lesions of the anterior archistriatum, emerged as a consequence of drawing parallels between avian and mammalian areas, despite the fact that the supposed anatomical homology between the areas was not in fact applicable to the critical site involved. It will not be surprising if many of the parallels that we currently draw subsequently turn out to be erroneous; what Zeier's work shows is that even erroneous parallels may be fruitful.

The very lack of information on the behavioural effects of hyperstriatal lesions makes it at once less likely that speculations made now concerning hyperstriatal function will be correct and more important that such speculation should be attempted. In the present state of knowledge, the most pressing need is for revealing data; whether the results of a given experiment will be revealing or not depends on whether the experiment taps the appropriate behavioural mechanisms. Experiments are therefore designed in the light of the most plausible speculation currently available, and an experimenter's attention may be focussed on certain possible functions of the relevant brain area. If the position taken here has any validity then it will be more instructive to design experiments aimed at assessing the ability of hyperstriatal birds in situations requiring shifts of attention than experiments assessing, for example, the effects of lesions on pattern discriminations.

Exactly which behavioural situations require shifts of attention is, of course, a question for learning theorists, and one to which there is no complete answer as yet. This, in the end, is precisely the point: if a convincing account of hyperstriatal function does emerge, then lesioned birds may be used to test behavioural notions. For example, if a large body of evidence supports the view that the hyperstriatum is involved primarily in shifts of attention, then the view that extinction situations involve shifts of attention will be strengthened, and behavioural theories may be modified in consequence. Brain lesion analysis and learning theory should enjoy a reciprocal relationship: at present, analysis of hyperstriatal function is heavily dependent on learning theory for hypotheses to be used in the design of sensitive situations, and the hope is that eventually lesion-induced deficits will point the way to new hypotheses for learning theorists.

Acknowledgements

Preparation of this manuscript was aided by a grant from the U.K. Medical Research Council.

I am grateful to N.J. Mackintosh, A. Dickinson, and G. Hall for their comments on an earlier draft of this manuscript.

REFERENCES

Adamo, N.J. (1967) Connections of efferent fibres from hyperstriatal areas in chicken, raven, and African lovebird. *J. comp. Neurol.*, 131, 337—356.
Adamo, N.J. and Bennett, T.L. (1967) The effect of hyperstriatal lesions on head orientation to a sound stimulus in chickens. *Exp. Neurol.*, 19, 166—175.
Adamo, N.J. and King, R.L. (1967) Evoked responses in the chicken telencephalon to auditory, visual and tactile stimuli. *Exp. Neurol.*, 17, 498—504.
Ariëns Kappers, C.U., Huber, G.C. and Crosby, E.C. (1936) *The Comparative Anatomy of the Nervous System of Vertebrates, Including Man.* Macmillan, New York.
Baddeley, A.D. (1972) Human memory. In P.C. Dodwell (Ed.), *New Horizons in Psychology 2.* Penguin, Harmondsworth, pp. 36—61.
Benowitz, L. (1972) Effects of forebrain ablations on avoidance learning in chicks. *Physiol. Behav.*, 9, 601—608.
Benowitz, L. and Lee-Teng, E. (1973) Contrasting effects of three forebrain ablations on discrimination learning and reversal in chicks. *J. comp. physiol. Psychol.*, 84, 391—397.
Brady, J.V. (1958) The paleocortex and behavioural motivation. In H.F. Harlow and C.N. Woolsey (Eds.), *Biological and Biochemical Bases of Behavior.* University of Wisconsin Press, Madison, Wisc., pp. 193—235.
Clarke, E. and Dewhurst, K. (1972) *An Illustrated History of Brain Function.* Sanford Publications, Oxford.
Cohen, D.H. and Pitts, L.H. (1967) The hyperstriatal region of the avian forebrain: somatic and autonomic responses to electrical stimulation. *J. comp. Neurol.*, 131, 323—336.
Crowne, D.P. and Riddell, W.I. (1969) Hippocampal lesions and the cardiac component of the orienting response in the rat. *J. comp. physiol. Psychol.*, 69, 748—755.
Delius, J.D. and Bennetto, K. (1972) Cutaneous sensory projections to the avian forebrain. *Brain Res.*, 37, 205—221.
Douglas, R.J. (1967) The hippocampus and behavior. *Psychol. Bull.*, 67, 416—442.
Douglas, R.J. and Pribram, K.H. (1966) Learning and limbic lesions. *Neuropsychologia*, 4, 197—220.
Hodos, W., Karten, H.J. and Bonbright, J.C. (1973) Visual intensity and pattern discrimination after lesions of the thalamofugal visual pathway in pigeons. *J. comp. Neurol.*, 148, 447—468.
Karten, H.J. (1968) The ascending auditory pathway in the pigeon (*Columba livia*). II. Telencephalic projections of the nucleus ovoidalis thalami. *Brain Res.*, 11, 134—153.
Karten, H.J. and Dubbeldam, J.L. (1973). The organization and projections of the paleostriatal complex in the pigeon (*Columba livia*). *J. comp. Neurol.*, 148, 61—90.
Karten, H.J. and Hodos, W. (1967) *A Stereotaxic Atlas of the Brain of the Pigeon (Columba livia).* Johns Hopkins Press, Baltimore, Md.
Karten, H.J., Hodos, W., Nauta, W.J.H. and Revzin, A.M. (1973) Neural connections of the 'visual Wulst' of the avian telecephalon. Experimental studies in the pigeon (*Columba livia*) and owl (*Speotyto cunicularia*). *J. comp. Neurol.*, 150, 253—278.
Kimble, D.P. (1963) The effects of bilateral hippocampal lesions in rats. *J. comp. physiol. Psychol.*, 56, 273—283.
Kimble, D.P. (1968) Hippocampus and internal inhibition. *Psychol. Bull.*, 70, 285—295.
Kimble, D.P. and Kimble, R.J. (1965) Hippocampectomy and response perseveration in the rat. *J. comp. physiol. Psychol.*, 60, 474—476.
Lashley, K.S. (1929) *Brain Mechanisms and Intelligence.* University of Chicago Press, Chicago, Ill.
Layman, D.H. (1936) The avian visual system. I. Cerebral functions of the domestic fowl in pattern vision. *Comp. Psychol. Monogr.*, 12, 1—58.

162

Lubar, J.F. (1964) Effect of medial cortical lesions on the avoidance behavior of the cat. *J. comp. physiol. Psychol.*, 58, 38—46.

McLeary, R.A. (1961) Response specificty in the behavioral effects of limbic system lesions in the cat. *J. comp. physiol. Psychol.*, 54, 605—613.

Macphail, E.M. (1969) Avian hyperstriatal complex and response facilitation. *Commun. behav. Biol.*, 4, 129—137.

Macphail, E.M. (1971) Hyperstriatal lesions in pigeons: effects on response inhibition, behavioral contrast, and reversal learning. *J. comp. physiol. Psychol.*, 75, 500—507.

Macphail, E.M. (1975) Avian hyperstriatal function: response inhibition or response shift? *J. comp. physiol. Psychol.*, in press.

Mishkin, M. and Pribram, K.H. (1956) Analysis of the effects of frontal lesions in monkeys. II. Variations of delayed response. *J. comp. physiol. Psychol.*, 49, 36—40.

Nauta, W.J.H. and Karten, H.J. (1968) A general profile of the vertebrate brain with sidelights on the ancestry of the cerebral cortex. In F.O. Schmitt (Ed.), *The Neurosciences: Second Study Program*. Rockefeller University Press, New York, pp. 7—26.

Olton, D.S. (1972) Behavioral and neuroanatomical differentiation of response-suppression and response-shift mechanisms in the rat. *J. comp. physiol. Psychol.*, 78, 450—456.

Pritz, M.B., Mead, W.R. and Northcutt, R.G. (1970) The effects of Wulst ablation on color, brightness, and pattern discrimination in pigeons. *J. comp. Neurol.*, 140, 81—100.

Revzin, A.M. (1969) A specific visual projection area in the hyperstriatum of the pigeon (*Columba livia*). *Brain Res.*, 15, 246—249.

Reynolds, G.S. (1961) Behavioral contrast. *J. exp. Anal. Behav.*, 4, 57—71.

Reynolds, G.S. and Limpo, A.J. (1965) Selective resistance of performance on a schedule of reinforcement to disruption by forebrain lesions. *Psychon. Sci.*, 3, 35—36.

Scoville, W.B. and Milner, B. (1957) Loss of recent memory after bilateral hippocampal lesions. *J. Neurol. Neurosurg. Psychiat.*, 20, 11—21.

Scull, J., Davies, K. and Amsel, A. (1970). Behavioral contrast and frustration effect in multiple and mixed fixed-interval schedules in the rat. *J. comp. physiol. Psychol.*, 71, 478—483.

Stettner, L.J. and Schultz, W.J. (1967) Brain lesions in birds: effects on discrimination acquisition and reversal. *Science*, 155, 1689—1692.

Sutherland, N.S. and Mackintosh, N.J. (1971) *Mechanisms of Animal Discrimination Learning*. Academic Press, New York.

Swanson, A.M. and Isaacson, R.L. (1967) Hippocampal ablation and performance during withdrawal of reinforcement. *J. comp. physiol. Psychol.*, 64, 30—35.

Tuge, H. and Shima, I. (1959) Defensive conditioned reflex after destruction of the forebrain in pigeons. *J. comp. Neurol.*, 111, 427—445.

Zeier, H. (1968) Changes in operant behavior of pigeons following bilateral forebrain lesions. *J. comp. physiol. Psychol.*, 66, 198—203.

Zeier, H. (1969) DRL-performance and timing behavior of pigeons with archistriatal lesions. *Physiol. Behav.*, 4, 189—193.

Zeier, H. (1971) Archistriatal lesions and response inhibition in the pigeon. *Brain Res.*, 31, 327—339.

Zeier, H. and Karten, H.J. (1971) The archistriatum of the pigeon: organization of afferent and efferent connections. *Brain Res.*, 31, 313—326.

Zeigler, H.P. (1963a) Effects of endbrain lesions upon visual discrimination learning in pigeons. *J. comp. Neurol.*, 120, 161—181.

Zeigler, H.P. (1963b) Effects of forebrain lesions upon activity in pigeons. *J. comp. Neurol.*, 120, 183—194.

9

INTERHEMISPHERIC INTERACTIONS

HANS ZEIER

Swiss Federal Institute of Technology, Department of Behavioural Sciences, Zurich (Switzerland)

The evolution of the vertebrate nervous system is believed to proceed from bilateral symmetry to the formation of specialised asymmetric brain functions in man. In organisms with bilateral symmetry one hemisphere is the mirror image of the other. In order to organise behaviour, the hemispheres interact and cooperate with each other. Bilateral correlation is provided by commissural and decussational connections within the brain and can be studied with various physiological and behavioural methods. The visual system of birds seems to be particularly suitable for such investigations. Visual information is extremely important for birds and their visual system is accordingly highly developed. The main advantage of using birds is that, in contrast to mammals, their visual pathways completely cross in the optic chiasm. The direct visual input can, therefore, be restricted to one hemisphere by stimulating the opposite eye exclusively. Unlike mammals, it is not necessary to cut the optic chiasm in order to lateralise one eye's information. The present paper reviews some of the attempts made in birds to investigate hemispheric interactions with behavioural methods.

INTERHEMISPHERIC TRANSFER*

Formation of bilateral engrams during monocular training

A pigeon monocularly trained to perform a brightness, pattern or colour discrimination task in a Skinner-box, is also able to perform this task with the untrained eye (Catania, 1965; Ogawa and Ohinata, 1966), even if a previously acquired discrimination response is reversed (Catania, 1965). Such interhemispheric transfer of colour or pattern discrimination performance

---

* In the terminology of transfer the terms 'interhemispheric', 'interocular' and 'eye-to-eye' transfer are all used for describing the same general phenomenon. The first refers more to the central loci involved, while the latter two indicate the channel used.

can be prevented by cutting the supraoptic decussation (Meier, 1971). Successful interhemispheric transfer has also been reported for various other behaviours, *e.g.* the following response in an imprinting situation (Moltz and Stettner, 1962) or the suppression of the innate pecking response of newly hatched chicks caused by an aversive stimulus (Cherkin, 1970). Birds with eyes positioned on the sides of their heads have only a restricted binocular visual field and, therefore, may accomplish learning with only one eye and its corresponding hemisphere. It seems to be biologically significant that these experiences are shared by both hemispheres. This could be achieved either through the formation of a unilateral engram which is available to the untrained visual pathway via the commissural systems, or, alternatively, through the laying down of engrams for the monocularly learned response in both hemispheres. According to Mello (1968), the latter seems to be the case in birds: pigeons monocularly trained on pattern or colour discrimination tasks were also able to perform the learned discrimination habits with the untrained eye, even if the optic tectum receiving the direct input during original training was lesioned before the transfer test took place. This finding suggests that monocular training induces bilateral memory traces. However, in mammals this issue is rather controversial. Results obtained in the cat suggest the existence of two contributions of the trained to the untrained hemisphere in a transfer test situation (Myers, 1961 and 1965). The first would occur at the time of initial training and establish a secondary, less well-defined memory trace in the untrained hemisphere. The second contribution would take place at the time of actual testing in that the untrained hemisphere gets access to the firm trace of the trained hemisphere over interhemispheric connections. The existence of the latter mechanism is further supported by experiments using the spreading depression technique (Leão, 1944; Bureš *et al.*, 1974). With this method reversible decortication can be achieved by applying KCl to the cortex. If a rat learns to perform a sensory discrimination or a complex motor response while one hemisphere is blocked by spreading depression, it acts like a naive animal when tested while the other hemisphere undergoes spreading depression. However, if between training and testing a few trials are allowed without depression of electroencephalographic activity, interhemispheric transfer occurs (Bureš, 1959; Russell and Ochs, 1961; Burešová and Bureš, 1969). It seems likely that during the interdepression trials the memory trace stored in the trained hemisphere is replicated in the untrained hemisphere. Apparently, interhemispheric exchange of information is much more efficient if both hemispheres receive the same information simultaneously.

Conditions for interhemispheric transfer

Certain types of monocular experiences lead to a unilateral engram which is not available through the untrained eye. The occurrence of interhemispheric transfer depends on several factors, *e.g.* the behavioural testing situation and amount of training used, the difficulty of the problem to be learned, the position of the relevant stimulus in the visual field, the motivation, the type of reinforcement, the innate response tendencies and previous experience and activity of the opposite hemisphere at the time the experience should be transferred. These factors interact and can even cancel out each other.

Beritoff and Chichinadze (1936) monocularly trained pigeons to approach a food box or to stay away from it, depending on a discriminative stimulus (colour or pattern) projected onto the wall. When they covered the trained eye and exposed the naive eye, the pigeon approached the food box as soon as the stimulus light occurred, but no longer discriminated between go and no-go signals. A similar lack of transfer was reported for the pigeon by Levine (1945a and b). In two test situations he found no interhemispheric transfer when the discriminative stimuli were presented vertically in front of the bird, and successful transfer when they were presented in a horizontal position directly below the beak. The first test was a pecking response for grains of corn cut in the shape of triangles and rectangles; the first type being loose and the latter glued onto the floor. The second test was a modified Lashley jumping stand where the animals had to learn to discriminate between figures or differences in brightness. As there were no restrictions for the visual field of the eye which had not been blindfolded, it is difficult to determine which parts of the retina were actually used for performing the discrimination response. In Levine's experiment, there was also no intraocular transfer; here the shift in the position of the stimuli from subrostral to anterostral resulted in a loss of discrimination even if the bird had both eyes open (Levine, 1952). Catania (1965), on the other hand, using an operant conditioning situation, found complete interhemispheric transfer for pattern and form discrimination irrespective of the position of the stimuli. These results were later confirmed by several other studies (Ogawa and Ohinata, 1966; Cuénod and Zeier, 1967; Meier, 1971). The differences between Levine's and Catania's findings may be due to differences in the testing procedures and to different amounts of discrimination training. Levine's tests involved a few hundred responses, Catania's a few thousand. In the latter experiment the pigeon may, therefore, have been considerably overtrained, which probably favoured complete interhemispheric transfer.

Lack of interhemispheric transfer is generally observed if the task to be learned and transferred is rather difficult to perform. This is true for the pigeon with sequential learning tasks (Zeier, 1966). In an instrumental conditioning chamber with two pecking keys, pigeons were monocularly trained to make one pecking response to the left and then one to the right key. This response sequence was rewarded with 5 sec free access to grains presented in a food hopper, followed by 5 sec time-out in complete darkness. The other two response possibilities allowed (*i.e.* two pecks to the left or starting with

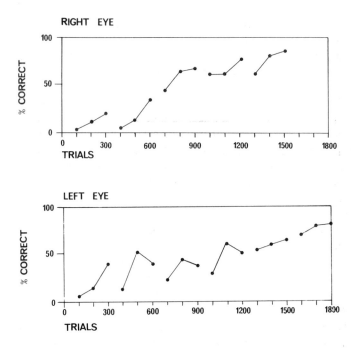

Fig. 1. Typical performance curve of a pigeon trained to perform a sequential task (pecking first the left and then the right key) with one eye and then tested for transfer with the other. Note that there are no savings with the second eye.

a peck to the right key) terminated the trial with a time-out of 10 sec and then the next trial was initiated. Each daily training session consisted of 300 discrete trials. The pigeons were wearing the goggles shown in Fig. 2, which restricted their view to the frontal visual field. One goggle was covered with a black cap while the other was left open.

A typical performance curve from this task is presented in Fig. 1. In the beginning the score remained rather low, because alternation between the two keys occurs with a much smaller probability than perseveration on either key. However, within a few sessions the criterion of 75% correct responses within two successive blocks of 100 trials was reached. When tested with the untrained eye, the animals needed about the same amount of training as with the first eye. Similar results were obtained if the reversal of the first task, namely pecking first the right and then the left key, had to be learned with the second eye. Therefore one can assume that neither positive nor negative transfer exists on this task. Interestingly, under these monocular learning conditions the birds needed about three times as many trials for reaching the 75% correct criterion than when under binocular conditions. With one eye occluded the pigeons showed a strong preference for pecking the key located on the same side as the eye which was open. The restriction of the visual field due to the goggle probably generated additional difficulties on this complex sensorimotor task.

Interhemispheric transfer can be further affected (a) by early experience and (b) by innate response predispositions.

(a) Siegel (1953) reared ring doves under conditions which allowed their retinae to be stimulated only by light intensity differences, but prevented perception of visual form. They were then trained on a Lashley-type jumping apparatus to discriminate between circles and triangles. In contrast to normally reared birds, visually deprived birds showed only partial transfer of the monocularly learned discrimination habit from the trained to the un-trained eye and more, but still not complete, transfer from a single eye to both eyes functioning simultaneously.

(b) Response modifications which run counter to an innate response tendency can fail to show interhemispheric transfer. The innate depth avoidance response of newly hatched chicks on a visual cliff disappears when chicks are reared on the glass over the deep side. If chicks are exposed to this situation with only one eye open, this response modification does not transfer to the other eye (Zeier, 1970). This perceptual experience is apparently not strong enough to overrule the innate predisposition. However, it can be achieved with reinforcers which apparently generate stronger memory traces.

If newly hatched chicks are presented with a small metallic bead, nearly all animals peck it within a few seconds. This response can be suppressed by a single presentation of a target coated with an aversive liquid such as methyl anthranilate (Lee-Teng and Sherman, 1966). If this one-trial learning situation is experienced with one eye, the chicks will then avoid pecking the target with the other eye (Cherkin, 1970). However, the extinction of this suppression behaviour, achieved by prolonged presentation of an uncoated target after the original avoidance learning, does not transfer to the eye which has not directly experienced the extinction procedure (Benowitz, 1973). Apparently, the aversive experience generates a strong engram in both hemispheres which can only be extinguished by direct input to each hemisphere and not by interhemispheric cross-talk.

Mirror image reversal

An interesting phenomenon is the so-called mirror image reversal (Mello, 1965). Pigeons monocularly trained to peck a key when a $45°$ oblique line is displayed on this key and not to peck when its mirror image, a $135°$ tilted line is there, choose the mirror image of the trained stimulus when tested with the naive eye. Such reversals occur for left—right, but not for up—down mirror images. Discrimination of left—right mirror images appears rather difficult for bilaterally symmetrical animals. Corballis and Beale (1970) showed that pigeons may learn to respond differentially to mirror image stimuli by developing asymmetrical observing responses and so encode the left—right difference into something easier to discriminate. A pigeon viewing with the left eye may learn to peck the key when a line appears on the lower part of the key and not to peck when it is on the upper part. Such responding would then be based on an up—down and not a left—right difference and lead to the observed mirror image reversal phenomena. Such a masking hypothesis was also postulated for chiasm-sectioned monkeys who show the same mirror image reversal effect (Hamilton and Tieman, 1973).

HEMISPHERIC DOMINANCE AND LATERALISATION

In mammals, the information flow over the interhemispheric connections is usually sufficient to allow the formation of bilateral engrams. Therefore, no hemispheric dominance or lateralisation occur. However, certain types of trained behaviour can become lateralised. If, for example, rats are trained to grasp small food pellets out of a narrow glass tube, they prefer to do this persistently with the left or right forepaw. Burešová et al. (1958) established such a forelimb preference in rats and then taught them to run through a multiple T-maze. While the grasping response was inhibited by blocking the hemisphere contralateral to the preferred forepaw with cortical spreading depression, the retention of the maze habit was the same irrespective of which hemisphere was suppressed. These findings suggest that hemispheric dominance in rats is restricted to motor control of the forelimb, whereas general motor reactions and complex habits are usually performed with symmetrical participation of both hemispheres. Monkeys, however, show some concentration of learning in one hemisphere (Gazzaniga, 1963). This trend finally ends in the extreme asymmetry of language representation in humans, probably due to the incapability of the callosal connections to handle the information flow associated with the highest brain functions (Bogen and Bogen, 1969).

Lateralised control of vocal behaviour seems not to be a uniquely human phenomenon. Nottebohm (1971 and 1972) showed that in the chaffinch, a European song bird, calls, songs and subsongs are under dominant control of the left hypoglossal nerve. If he sectioned this nerve in adult birds, drastic changes occurred in that most of the song components were lost and could not be regained with practice. Section of the right hypoglossal nerve resulted in few, if any, losses. If the same operations were performed in a bird which was still in the plastic song stage, or had not yet started to develop its song, normal song developed under the control of the intact side. Nottebohm suggested that during the ontogeny of the chaffinch the left hypoglossal nerve innervates its syringeal musculature before the right one does. This primacy of innervation would then lead to greater vocal control from the be-

Fig. 2. Behavioural test situation. The front doors of the outer and inner boxes are left open for demonstration. The pigeon is wearing goggles on which colour filters can be mounted in order to provide different visual input to the two eyes. The behaviour can be observed by closed circuit television. The stimuli are displayed with a projector.

ginning of vocal ontogeny and become confirmed and further developed as a result of vocal learning.

Hemispheric dominance can also become developed in learning situations where different information is simultaneously presented to each half of the brain, so that the animal's capacity to handle this information is taxed (Palmers and Zeier, 1974). In the bird different information can be provided to each hemisphere by mounting goggles over their eyes and covering one with a red and the other with a green filter (Mello, 1967). Such a procedure is depicted in Fig. 2. In our test situation the pigeon wearing the goggles is restrained within a small box in front of the pecking keys. He can reach out of it, peck the key and eat the reinforcement which is delivered in a food magazine placed below the pecking key. This restraining device keeps the pigeon in a defined position in order to always have both eyes stimulated simultaneously. The pecking keys and the goggles both have the same red and green filter, respectively. The filters used absorb practically all white light if put together. There are no lights within the testing chamber except the red and green key lights and, during reinforcements, the white light of

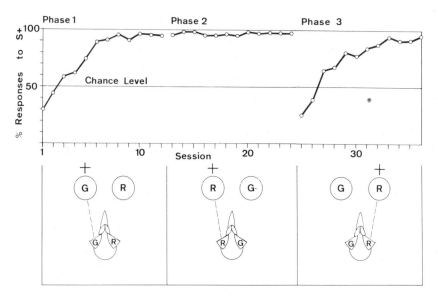

Fig. 3. Combined learning curves of four pigeons trained to perform a two choice discrimination task. The lower part indicates filter position and reinforcement conditions. S+ was reinforced with a probability of 80%, S— with 20%. Within one phase the position of the goggle filters remained the same, while the position of the stimuli alternated between trials in a random sequence. (Reproduced by permission of Palmers and Zeier, 1974.) R and G indicate red and green filters, respectively.

the food magazine. Therefore, in this test situation the pigeon can, for example, approach the red key only with the aid of the eye having the red filter, since the other eye having a green filter cannot see red lights and therefore is not able to detect the location of the red key. Hence, each choice response made is triggered by the hemisphere receiving the corresponding input. In our experiment (Palmers and Zeier, 1974) four naive pigeons were trained in a Skinner-box with two pecking keys. They had goggles with filters corresponding to the key lights and were restrained as shown in Fig. 2. They had to learn a red—green discrimination in daily sessions of 100 discrete trials. A trial was started by presenting a red and a green stimulus. A response switched them off and during reinforced trials provided free access to food for 4 sec. Following a time-out of 10 sec in complete darkness, the next trial was initiated. Unreinforced trials were terminated by a 14 sec time-out. The position of the red and green key lights alternated in a random sequence, in order to avoid position habits.

During the first 12 sessions (phase 1) the pigeons had a green filter in front of their left eye and a red filter in front of their right eye. A response to the green key (S+) led to reinforcement with a probability of 0.8, whereas a response to the red key had a reinforcement probability of only 0.2.

Fig. 3. shows the learning curves. The data are pooled together as the individual differences were small. In the beginning the animals had no preference for either key and pecked them randomly. Since each eye could only detect the position of one key, both hemispheres must have been active. But within a few sessions, all animals learned to peck the green key almost exclusively. For the next phase the reinforcement conditions and the filter position on the goggles were reversed. If, in the previous phase, the pigeons had learned a red—green discrimination, they would now peck the green key using the opposite eye and thereby gain fewer reinforcements. However, they now pecked the red key and so continued to use the eye and hemisphere trained during phase 1. Apparently the pigeons learned to suppress stimuli from the opposite eye irrespective of the colour information. That the observed behaviour is not caused by rapid learning was proved by phase 3. If only the filters, but not the reinforcement conditions were switched, the performance dropped dramatically. The other eye has first to learn to guide the behaviour in order to reach the same level of reinforcement, which takes at least as many trials as the original acquisition.

When each hemisphere receives a different signal at the same time, the animal's capacity seems to be overtaxed, as it cannot perform two different reactions at the same time. Under this condition one hemisphere becomes dominant and suppresses the other. It makes sense to pay attention to the more relevant information and to neglect the less important one. In the present experiment the animals react to the more important aspect, namely a light signalling food, and neglect the colour dimension introduced by the other stimulus. These results concur with physiological findings (Robert and Cuénod, 1969): the slow potential evoked in the optic tectum of the pigeon

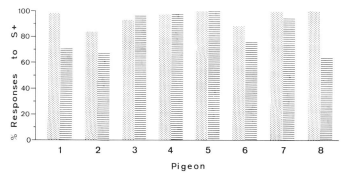

Fig. 4. Pecking preferences made following acquisition of a successive discrimination between S+ (reinforced according to a VI-1 min schedule) and S— (never associated with reinforcements). The columns indicate the percentage of total responses made on S+ during ten alternate 2-min presentations of S+ and S—. The two test sessions were run without any reinforcements, with goggle filters positioned as during training (dotted columns) or reversed (hatched columns). (Reproduced by permission of Palmers and Zeier, 1974.)

by stimulation of the contralateral eye is strongly diminished when a conditioning train of pulses is applied to the opposite tectum. This inhibiting effect lasts from 10 to 80 msec, is maximal around 25 msec and can be abolished by sectioning the tectal and posterior commissures.

If the same colour discrimination is learned with successive presentation of the stimuli, only one hemisphere is receiving information at a given time. As there is no interference with conflicting information, the hemispheres can exchange the information and complete interhemispheric transfer occurs (Palmers and Zeier, 1974). We trained pigeons in a box with only one pecking key to peck this key when a green light was projected on it and not to peck when a red light was there. When this go—no-go discrimination was stable we tested it under extinction conditions. Fig. 4 shows that the animals had a strong preference for the colour reinforced during training, irrespective of the position of the goggle filters. During acquisition only one hemisphere was active at a given time since the other was not receiving any information. Hence there was no interference from the opposite visual channel preventing the formation of bilateral engrams by transhemispheric communication. This may explain why the hemisphere trained to peck did not become dominant, as in the simultaneous discrimination task, but acquired a colour discrimination and was able to transfer it to the opposite hemisphere.

CONFLICT REACTIONS

As lateralisation of engrams does occur under certain conditions, it should be possible to establish conflicting rules of responding for the two hemispheres.

Levine (1945a) and Catania (1965), for example, were able to train pigeons to perform a discrimination response with one eye and the opposite discrimination response with the other. In Catania's experiment the pigeons were trained to peck a key when a vertical line was displayed on it and not to peck when a horizontal line was presented. With the other eye they learned to peck the horizontal line and not to peck the vertical line. If this procedure was repeated daily, alternating the sessions for each eye, the birds finally reacted differently to this discrimination task according to the eye stimulated. Konermann (1966), working with geese, could not establish such a conflict discrimination when the first eye was overtrained, but succeeded when he alternated training of either eye from the beginning. This finding again supports the assumption that monocular training generates an engram in both hemispheres. However, the direct input generates a stronger trace than the transhemispheric input. Therefore it is possible to establish conflict reactions for the two hemispheres by regularly suppressing the weaker indirect input with the stronger direct input.

Comparable results were also obtained in mammals. Myers (1959) trained chiasm-sectioned cats to perform a similar conflict discrimination. As a result,

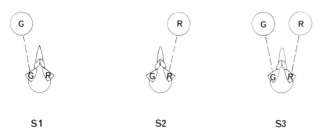

Fig. 5. Three different sets of stimuli presented to birds wearing a red (R) and a green (G) filter, respectively, in front of their eyes. A response made to the lighted key on S1 and S2 was reinforced with a probability of 0.5, while a response on S3 was never reinforced in order to have an even distribution of reinforced responses on the two stimuli.

the animals regularly responded according to the eye being used and neglected the experience gained with the opposite eye, thus proving again that primary sensory input dominates over interhemispheric transmission of information.

Conflicting reactions between the hemispheres can further be established by using the method with the colour filters shown in Fig. 2. In a series of experiments (Palmers, 1972) pigeons having a green filter in front of one eye and a red filter in front of the other were trained to peck three different sets (Fig. 5) of stimuli in a Skinner-box with two pecking keys. In the first set

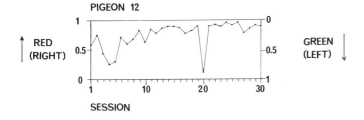

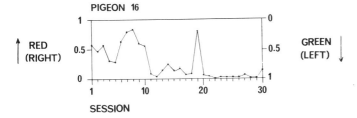

Fig. 6. Typical performance curves of two pigeons in a conflict situation. The ordinate shows for each session the proportion of the pecking responses made to the red and green key when stimulus set S3 was presented.

(S1) the left key was lighted with a green light while the right key remained dark. In the second set (S2) the right key was lighted with a red light and the left remained dark. In the third set (S3) both keys were lighted with the colours positioned as in the two preceding sets. Within a session the three sets were presented in a preprogrammed random order, each set appearing 70 times. In sets S1 and S2 a peck to the lighted key resulted, with a probability of 0.5, in free access to food for 5 sec, followed by 5-sec time-out in complete darkness. Then the next trial was initiated. Pecks to the unlighted key were never reinforced. Unreinforced trials were followed by 10-sec time-out. Trials with S3 were never reinforced in order to have an even distribution of reinforcement to the two pecking keys.

On sets S1 and S2 the pigeons almost exclusively pecked the lighted key. However, on S3 they were in a conflict situation: one eye told them to peck the right key and the other to peck the left. At the beginning, they selected either key with about the same probability, as shown in a typical performance curve in Fig. 6. However, with increased training a preference for one of the keys developed. Apparently, one hemisphere took the lead, became dominant and suppressed the other. Interestingly, this suppression can suddenly become reversed, a phenomenon also observed in experiments on binocular rivalry in humans (Hochberg, 1972).

The probability of pecking the same key on S3 as on a preceding trial with stimulus set S1 or S2 was, on average, just about 0.5 after reinforced as well as after unreinforced trials (Fig. 7). Hence, the reinforcement condition of the stimulus situation with one key light on did not strongly affect the choice made in the conflict situation.

The reaction time in S3 was much higher than in S1 and S2, or during a final session run without colour filters on the goggles so that either eye could see both keys simultaneously (Table I). This finding proves that both eyes received the appropriate stimulation in the experimental setting used. The

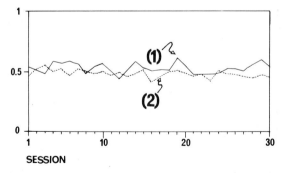

Fig. 7. Probability for pecking the same key on S3 as on a preceding trial with stimulus set S1 or S2 when this preceding trial was reinforced (1) and unreinforced (2), respectively. The curves indicate the group means of 11 pigeons for each session.

TABLE I

Mean reaction time of 11 pigeons on three different sets of stimuli

With filters (session 30)			Without filters (session 31)		
S1	S2	S3	S1	S2	S3
1.4 ± 0.3 sec	1.2 ± 0.3 sec	4.6 ± 1.2 sec	1.4 ± 0.3 sec	1.4 ± 0.3 sec	1.5 ± 0.3 sec

behavioural observations made suggest that the pigeons were really in a conflict and displayed vicarious trial-and-error behaviour when stimulus set S3 was presented: they tried to approach one of the pecking keys but did not hit it, oscillated between the two keys, withdrew and approached them again, often repeating this behaviour several times until they finally pecked a key. Such behaviour was never observed during S1 and S2, or in the situation without the colour filters.

Unresolvable conflicts of two memory traces in mammals also lead to increased reaction times, deterioration of behaviour, and partial dominance of either memory trace. Hatta (1974), for example, trained rats under unilateral cortical spreading depression to perform two competing solutions — one with each hemisphere — of a black—white discrimination. When tested with both hemispheres intact, the animals were in a conflict situation and regressed to innate types of behaviour such as exploration and position responding. Bureš and Burešová (1970) showed that, if in such situations the conflicting tendencies are not well balanced, the behaviour will be determined by the more powerful memory trace.

However, if the interhemispheric conflict can be resolved by some neutral intermediate solution, the two memories synthesise into a behavioural act consisting of a combination of the two responses, which was never before performed by the subject.

HEMISPHERIC INTEGRATION

The experimental results presented above suggest that it must be difficult, if not impossible, for a pigeon to use both hemispheres simultaneously. In order to test this hypothesis we trained pigeons to perform a discrimination task which they could only solve if they integrated and compared the information provided to each eye separately.

In this experiment (Zeier and Graf, in preparation), pigeons having a red filter in front of their left eye and a green filter in front of their right eye were trained on a successive go—no-go type discrimination task in an operant conditioning chamber with one pecking key. They had to peck this key when a positive stimulus was displayed and not to peck it when a negative stimulus

176

was there. The stimuli consisted of two bars, the left one being red and the right one green. Each of them was about 12 mm long and 4 mm wide and was oriented either vertically or horizontally. They were projected from behind onto the pecking key with a Kodak Carousel projector which was controlled by a computer. If the orientation of the two bars was opposite to each other, the stimuli were defined as positive. If the two bars were both displayed either vertically or horizontally, they were regarded as negative. The positive stimuli were always presented for 5 min and reinforced according to a 1-min variable interval schedule. The negative stimuli were never reinforced and remained on until the pigeon did not peck for 1 min. The four stimuli, two positive and two negative, were presented in a random order. Between stimuli there was a time-out of 10 sec in complete darkness. A session lasted until the animal achieved 60 reinforcements. The acquisition training proceeded for 25 sessions. Afterwards, the pecking preferences of the pigeons were tested under extinction conditions. During this test each of the stimuli was presented three times for 2 min in a random order. The pecks made to each stimuli were recorded. Then a second extinction test was made on the following day with goggle filters and the position of the red and green bars on the stimuli reversed. Following two sessions of retraining under the original conditions a final session with a white key light instead of the stimuli was added. This blank test was designed to show whether the pigeons were under the control of the stimuli or of the reinforcement schedule.

A typical performance curve is shown in Fig. 8. With increased training

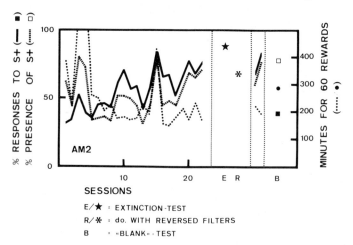

E/★ : EXTINCTION-TEST
R/✳ : do. WITH REVERSED FILTERS
B : «BLANK»-TEST

Fig. 8. Performance curve of pigeon AM2 on a go—no-go type discrimination task requiring interhemispheric integration of the two components of the stimuli displayed. Note how the percentage of the responses and the time of stimulus presentation changes in favour of the positive stimuli with increased training, while the time needed to reach 60 reinforcements decreases. Discrimination is clearly shown on the extinction test, but absent on the blank test run without the stimuli.

pigeon AM2 reacted more and more to the positive stimuli. Their presence increased in relation to the presence of the negative stimuli, and the time needed to reach 60 reinforcements decreased. The preference for the positive stimuli was strongly expressed on the extinction test. Apparently, the bird must have learned to integrate the two components of the stimuli. However, the discrimination was much less efficient, and in some cases even disappeared completely, if the filters were reversed. It seems likely that the pigeons did not learn a general concept such as 'opposite orientations of the bars means positive; same orientations means negative', but rather took the colour dimension into account when encoding the stimuli.

During the final session run without the stimuli (blank test), none of the animals tested showed a significant difference between the number of responses made in the positive phase (VI − 1 min) and the number of responses emitted in the negative phase (DRO). Furthermore, without the stimuli they usually needed much more time to reach the 60 reinforcements than with the stimuli, as was, for example, the case for pigeon AM2 (Fig. 8). These findings strongly support the assumption that pigeons with a performance like AM2 were under the control of the stimuli, and, therefore, able to perform some kind of interhemispheric integration. This discrimination habit was not due to a preference for just one of the two positive stimuli, as is demonstrated for pigeon AM2 in Fig. 9. In this figure, the diagonal running from the lower left to the upper right corner represents the percent-

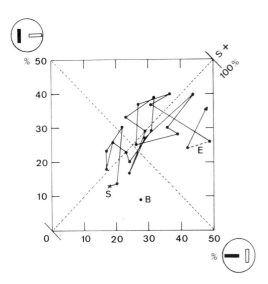

Fig. 9. Distribution of the responses made by pigeon AM2 to the two positive stimuli. Each point indicates one session. Note how the discrimination improves during the training sessions, is rather well balanced between the two positive stimuli and is absent on the final blank test (B) run without the stimuli.

age of responses made to the two positive stimuli compared with the two negative stimuli. A strong preference for one of the two positive stimuli would result in a deviation from this diagonal to the upper left or lower right corner. As shown in Fig. 9, for pigeon AM2 this was certainly not the case.

In our experiment, four out of six pigeons tested showed clear evidence of interhemispheric integration. However, after 25 sessions two pigeons still had not reached a discrimination level above 50%. They remained on this level on the extinction test as well as on the blank test. They were certainly not under the control of the stimuli, but merely learned to adapt to the reinforcement schedule. Apparently, interhemispheric integration is rather difficult for a pigeon to perform. In complicated situations animals tend to regress to strategies easier to perform even if these are less efficient. So the two pigeons failing to perform interhemispheric integration probably just neglected the stimuli and oriented their pecking behaviour to the cues provided by the reinforcement schedule applied.

SUMMARY AND CONCLUSIONS

Birds probably experience their world primarily with one eye and its corresponding hemisphere and then analyse the situation with the other eye. Unilaterally perceived information which cannot directly be confirmed by the other eye, may be transferred to the opposite hemisphere over commissures and decussations. This transfer mechanism does not seem to be an all-or-none process. Very often it may result in some vague engram which probably has to be improved by repeated interactions and comparisons between the hemispheres. However, if there is already something stored in one hemisphere, e.g. an innate response predisposition or an acquired response, this hemisphere neglects conflicting information perceived by the opposite hemisphere, unless it is accompanied by high arousal caused, for example, by a strong reinforcer. In conflict situations one hemisphere tends to become dominant in that it suppresses the other without taking into account what the other is perceiving or telling it what it is doing. Interhemispheric cross-talk is limited in the bird because of the lack of extensive interhemispheric connections, as provided in mammals by the corpus callosum. However, sufficient cross-talk exists to allow a pigeon to perform some simple kinds of interhemispheric integration.

Acknowledgement

This work was supported by the Swiss National Foundation for Scientific Research (Grant 3.2260.74).

REFERENCES

Benowitz, L. (1973) Conditions for the bilateral transfer of monocular learning in chicks. *Brain Res.*, 65, 203—213.
Beritoff, J.S. and Chichinadze, N. (1936) Localisation of visual perception in the pigeon. *Bull. biol. med. Exp. URSS*, 2, 105—107.
Bogen, J.E. and Bogen, G.M. (1969) The other side of the brain: III. The corpus callosum and creativity. *Bull. Los Angeles neurol. Soc.*, 34, 191—220.
Bureš, J. (1959) Reversible decortication and behavior. In M.A.B. Brazier (Ed.), *The Central Nervous System and Behavior.* Josiah Macy Jr. Foundation, New York, pp. 207—248.
Bureš, J. and Burešová, O. (1970) The reunified split brain. In R.E. Whalen, R.F. Thompson, M. Verzeano and N.M. Weinberger (Eds.), *The Neural Control of Behavior.* Academic Press, New York, pp. 211—238.
Bureš, J., Burešová, O. and Křivánek, J. (1974) *The Mechanism and Applications of Leão's Spreading Depression of Electroencephalographic Activity*, Academic Press, New York.
Burešová, O. and Bureš, J. (1969) The effect of prolonged cortical spreading depression on learning and memory in rats. *J. Neurobiol.*, 1, 135—146.
Burešová, O., Bureš, J. and Beran, V. (1958) A contribution to the problem of the 'dominant' hemisphere in rats. *Physiol. bohemoslov.*, 7, 29—37.
Catania, A.C. (1965) Interocular transfer of discriminations in the pigeon. *J. exp. Anal. Behav.*, 8, 147—155.
Cherkin, A. (1970) Eye to eye transfer of an early response modification in chicks. *Nature (Lond.)*, 227, 1153.
Corballis, M.C. and Beale, I.L. (1970) Bilateral symmetry and behavior. *Psychol. Rev.*, 77, 451—464.
Cuénod, M. et Zeier, H. (1967) Transfert visuel interhémispherique et commissurotomie chez le pigeon. *Schweiz. Arch. Neurol. Neurochir. Psychiat.*, 100, 365—380.
Gazzaniga, M.S. (1963) Effect of commissurotomy on preoperatively learned visual discrimination. *Exp. Neurol.*, 8, 14—19.
Hamilton, C.R. and Tieman, S.B. (1973) Interocular transfer of mirror image discriminations by chiasm-sectioned monkeys. *Brain Res.*, 64, 241—255.
Hatta, T. (1974) Interhemispheric competition of antagonistic memories in the rat. *J. comp. physiol. Psychol.*, 86, 481—485.
Hochberg, J. (1972) Perception: II. Space and movement. In J.W. Kling and L.A. Riggs (Eds.), *Woodworth and Schlosberg's Experimental Psychology.* Methuen, London, pp. 475—550.
Konermann, G. (1966) Monokulare Dressur von Hausgänsen, z. T. mit entgegengesetzter Merkmalsbedeutung für beide Augen. *Z. Tierpsychol.*, 23, 555—580.
Leão, A.A.P. (1944) Spreading depression of activity in the cerebral cortex. *J. Neurophysiol.*, 7, 359—390.
Lee-Teng, E. and Sherman, S.M. (1966) Memory consolidation of one-trial learning in chicks. *Proc. nat. Acad. Sci. (Wash.)*, 56, 926—931.
Levine, J. (1945a) Studies in the interrelations of central nervous structures in binocular vision: I. The lack of bilateral transfer of visual discriminative habits acquired monocularly by the pigeon. *J. genet. Psychol.*, 67, 105—129.
Levine, J. (1945b) Studies in the interrelations of central nervous structures in binocular vision: II. The conditions under which interocular transfer of discriminative habits takes place in the pigeon. *J. genet. Psychol.*, 67, 131—142.
Levine, J. (1952) Studies in the interrelations of central nervous structures in vision: III. Localization of the memory trace as evidenced by the lack of inter- and intraocular habit transfer in the pigeon. *J. genet. Psychol.*, 81, 19—27.

180

Meier, R.E. (1971) Interhemisphärischer Transfer visueller Zweifachwahlen bei kommissurotomierten Tauben. *Psychol. Forsch.*, 34, 220—245.

Mello, N.K. (1965) Interhemispheric reversal of mirror-image oblique lines following monocular training in pigeons. *Science*, 148, 252—254.

Mello, N.K. (1967) A method for restricting stimuli to the frontal or lateral visual field of each eye separately in pigeon. *Psychon. Sci.*, 8, 15—16.

Mello, N.K. (1968) The effect of unilateral lesions of the optic tectum on interhemispheric transfer of monocularly trained color and pattern discrimination in pigeon. *Physiol. Behav.*, 3, 725—734.

Moltz, H. and Stettner, L.J. (1962) Interocular mediation of the following response after patterned-light deprivation. *J. comp. physiol. Psychol.*, 55, 626—632.

Myers, R.E. (1959) Interhemispheric communication through corpus callosum: limitations under conditions of conflict. *J. comp. physiol. Psychol.*, 52, 6—9.

Myers, R.E. (1961) Corpus callosum and visual gnosis. In J.F. Delafresnaye (Ed.), *Brain Mechanism and Learning*, Blackwell, Oxford, pp. 481—505.

Myers, R.E. (1965) The neocortical commissures and interhemispheric transmission of information. In G. Ettlinger (Ed.), *Functions of the Corpus Callosum*. Little, Brown, Boston, Mass., pp. 1—17.

Nottebohm, F. (1971) Neural lateralization of vocal control in a passerine bird: I. Song. *J. exp. Zool.*, 177, 229—262.

Nottebohm, F. (1972) Neural lateralization of vocal control in a passerine bird: II. Subsong, calls, and a theory of vocal learning. *J. exp. Zool.*, 179, 35—50.

Ogawa, T. and Ohinata, S. (1966) Interocular transfer of color discriminations in a pigeon. *Ann. Anim. Psychol.*, 16, 1—9.

Palmers, C. (1972) *Interhemisphärische Suppression bei der Taube*. Unpublished Doctoral Thesis, University of Vienna.

Palmers, C. and Zeier, H. (1974) Hemispheric dominance and transfer in the pigeon. *Brain Res.*, 76, 537—541.

Robert, F. and Cuénod, M. (1969) Electrophysiology of the intertectal commissures in the pigeon: II. Inhibitory interaction. *Exp. Brain Res.*, 9, 123—136.

Russell, I.S. and Ochs, S. (1961) One-trial interhemispheric transfer of learning engram. *Science*, 133, 1077—1078.

Siegel, A.I. (1953) Deprivation of visual form definition in the ring dove: II. Perceptual motor transfer. *J. comp. physiol. Psychol.*, 46, 249—252.

Zeier, H. (1966) Ueber sequentielles Lernen bei Tauben, mit spezieller Berücksichtigung des 'Zähl'-Verhaltens. *Z. Tierpsychol.*, 23, 161—189.

Zeier, H. (1970) Lack of eye to eye transfer of an early response modification in birds. *Nature (Lond.)*, 225, 708—709.

Zeier, H. and Graf, M. (1975) Interhemispheric integration in the pigeon. In preparation.

10

HOW BIRDS USE THEIR EYES

MARK B. FRIEDMAN

Department of Psychology, Carnegie-Mellon University, Pittsburgh, Pa. (U.S.A.) *

INTRODUCTION

Although it is obvious both from their behaviour and from their anatomy that birds are highly dependent on vision, quantitative studies have concentrated on visual physiology and psychophysics rather than on visually guided behaviour. Perhaps it is because birds' eyes are proportionately so large compared with the size of their brains that most recent reviews of avian vision have discussed birds' visual capabilities in terms of the constraints imposed by eye structure and function (Sillman, 1973; Pearson, 1972; Kare, 1965; Tansley, 1964; Pumphrey, 1961). Unlike the material covered in these reviews, my concern in this chapter is not with the psychophysical limits of a bird's visual capacity but with what the bird normally does with its eyes. I am interested in how a bird's visual system interacts with the environment during visually guided behaviour. This paper will present the techniques I have developed to measure how birds scan their visual environment and the preliminary results of using these techniques. Although the work described here is behavioural, it has implications for our understanding of the neural organization of avian visual systems.

Since it is technically very difficult to measure where a subject (either animal or human) is looking if the subject's head is permitted to move, there are virtually no quantitative studies on how vertebrates visually scan their environments as they move about. This is despite the extensive work done on how humans with immobilized heads inspect written or pictorial material (Yarbus, 1967). The technical difficulty lies in adequately determining the direction of gaze when the eyes can be moved in the head and simultaneously the head is free to move in space. Most eye movement recording systems cannot distinguish between rotation of the eye (and head) which results in large changes in the orientation of the optic axes and pure translation of the head which does not greatly affect the direction of gaze.

* The work reported here was carried out in the Department of Psychology, University of Edinburgh.

This technical difficulty can be avoided if one studies appropriate species of birds. The eye movements of visual exploration of these species seem to depend on movement of the whole head rather than movement of the eyes in the skull. For these birds it is feasible to measure where objects lie in the visual field by measuring head position and orientation relative to the objects. It is only necessary to confirm that the eyes have not deviated significantly from their rest positions in the skull during visual reorientation. In the behaviour sequences I will be describing, there are no 'eye-in-head' movements larger than the $\pm 2°$ resolution of the total system.

Observations have been made on a variety of species; however, all experimental work has been done with the Barbary dove (*Streptopelia risoria*). Although a range of visually guided behaviours have been studied, including courtship and agonistic interactions between conspecifics, the discussion here is limited to situations of ground feeding by isolated individuals.·

The birds were recorded foraging for food in a one metre diameter cage. They were filmed or videotaped by a remote controlled camera mounted at least 7 m vertically above the centre of the cage. Successive single frames of the recordings were projected onto an electronic digitizing screen where X and Y coordinates were determined for several points on the bird's head and body and for several external points of interest in the cage.

Using these coordinates, a computer analyzed the temporal patterns of the bird's head and body movement. This analysis of the temporal rhythm of movement was in terms of the arbitrary rectilinear coordinate system of the digitizing system and was without reference to objects in the bird's environment.

After deriving the instantaneous height of the bird's head above the ground from the length of shadow it cast in oblique lighting, the computer determined relative positions and orientations of objects in the bird's visual field. Orientation in the visual field was specified in terms of an imaginary sphere of unit radius rigidly attached to the bird's head. The centre of the sphere coincides with the mid-point of a line connecting the centres of the pupils of the bird's eyes. If the sphere is considered as a globe, the poles lie directly over and under the centre of the sphere when the bird's head is in a posture of visual alertness. The equator of the globe is parallel to the ground in this position of alertness. If the perspective projection of an object falls onto the hemisphere above the equator, its angle of elevation (latitude) is considered positive. If it falls below the equator (as do the images of most objects on the ground when the bird's head is not tilted), the angle of elevation is negative. The prime meridian is defined by the vertical plane bisecting the head and beak and is in the direction of the bird's beak. The 90° meridians intersect the equator at the same point as radial lines through the centres of the pupils of the bird's eyes. Positive bearings are to the left side of the bird's head while negative bearings are to the right side.

Although in my study I used spherical coordinates to describe the orientation of objects in the bird's fields of view, in this paper I will use only the bearings in the horizontal plane in order to simplify the discussion. The main points I wish to make are not altered by this simplification of description.

The first of the following sections deals with the temporal pattern of avian eye movements and establishes that the rhythm of movement is related to visual function. The next section presents evidence for specialized use of different areas of the visual field. The third section deals with lateralization of visual function. The final section suggests how these behavioural observations are related to neural aspects of vision in birds.

TEMPORAL PATTERNS OF HEAD MOVEMENT

Pigeons, chickens and other common laboratory birds appear to 'bob' their heads back and forth as they walk. In reality, as such a bird walks its head is thrust forward, then is held nearly stationary while the neck retracts and the body continues to advance under the head. The illusion of 'bobbing' results from the forward and backward movement of the head relative to the body. In the environment, the bird's head is alternately stabilized and thrust forward as Dunlap and Mowrer (1930) demonstrated cinematographically more than forty years ago.

In Fig. 1a changes in head position and bearing are plotted over a typical 3-sec period when a dove first notices, then approaches a seed. The change in head position is the distance in centimetres that a point on the head midway between the eyes moves between successive motion picture frames. Similarly, the head bearing change represents the number of degrees that the head is turned between successive motion picture frames. Since the film was exposed at a constant rate of 16 frames/sec, the figure shows the fluctuation in linear and angular velocity of the head and eyes over time.

The linear and angular changes in position exhibit similar periodicities. One can best see that linear and angular movements occur in the same time

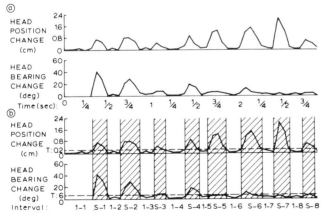

Fig. 1. Changes in a dove's head position and bearing during several seconds.

frame if one considers what happens during the interval starting when either the rate of linear or angular movement exceeds some threshold and finishing when both fall below that threshold of movement. These are the shaded areas in Fig. 1b. The thresholds of 2 mm of head displacement or 6° of head rotation per sixteenth of a second were chosen as about twice the maximum 'noise' observed in the process of photographing and projecting the picture of a bird in the one metre cage.

In man, eye rotation resulting in rapid conjugate change in the direction of gaze is called a saccade and the intervals of relative stability between saccades are called intersaccadic intervals. The same nomenclature for rapid angular changes in bird head position has been retained since, in the absence of substantial eye-in-head movement, these head movements result in corresponding conjugate eye movement. Angular changes in head position are usually accompanied by linear displacements since the line joining the centres of the two eyes does not intersect the axis of head rotation. In fact, the axis about which the head rotates is frequently nearer the caudal region of the neck than the head. There are, however, movements that are primarily linear displacements with little or no accompanying head rotation. There is no human analogue of a rapid conjugate linear eye displacement so I have termed these rapid eye displacements without rotation, saltations from the latin *saltare*, (to leap)*. The stationary phase between saltations is then called an intersaltatory interval. The numbered 'I's' and 'S's' in Fig. 1b are the successive stationary intervals and intervening saccadic or saltatory movements.

I have tabulated the frequency of occurrence of saccades and saltations of various durations and, likewise, the durations of the intervals between successive saccades or successive saltations. Typical histograms for the frequency of occurrence of various durations of a dove's head movements are shown in Fig. 2. The point to note is that the shapes of the distributions and their modal values are very similar for both saltatory and saccadic eye movements. The same is true of their respective intermovement interval durations. This suggests that a common generator mechanism might be responsible for both types of eye movement.

In order for these observations to be important for understanding avian visual mechanisms, two questions must be answered. First, is the observed pattern of head movement directly related to vision, or might it be driven by other sensory inputs or by endogenous rhythms? Second, is the pattern of eye movement limited to doves, pigeons, and other walking birds that appear to 'bob' their heads, or is it found in more diverse species?

With reference to the first question, it is known that vestibular stimulation of a restrained pigeon can result in rhythmical rotations of its head (Huizinga and van der Meulen, 1951). Since the similar periodicities of both

---

* The term was suggested to me by Dr. Colwyn Trevarthen.

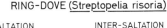

RING-DOVE (Streptopelia risoria)

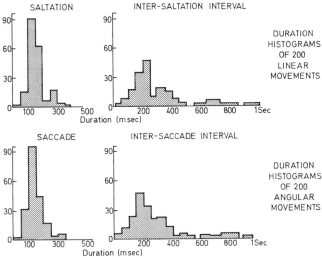

Fig. 2. Frequency histograms of dove head movement durations.

linear and rotational head movements in the dove suggests a common generator mechanism, it is possible that any head movement through inertial space
(*e.g.* movement that stimulates the bird's vestibular system) could elicit
appropriate rhythmic head movement with characteristic slow and fast
phases (Whitteridge, 1960).

An alternative hypothesis follows from the observation that the rhythm of
a dove's head movements during locomotion appears to be synchronized
with its limb movements. Since birds are bipeds whose heads and tails are
often cantilevered out beyond the area covered by their feet, it is possible
that while a bird is walking its head moves reflexively in coordination with
its limbs in order to maintain balance. Daanje (1951) has suggested that, like
the limbs in von Holst's theory, the head might have an endogenously determined preferred rhythm of oscillation. The head's rhythm of movement
would then be loosely coupled to, or entrained by, other bodily rhythms
such as those involved in stepping.

The factors most likely to be involved in the patterning of a dove's head
movements while it is walking are stimuli resulting from movement through
the bird's visual environment, stimuli resulting from movement through
inertial space, and rhythms generated in the mechanics of walking. The
following experiment was performed to try to dissociate these factors.

Successive single frames of recordings of each of six doves walking around
and feeding in a small cylindrical cage (Fig. 3) were analyzed with the com-
puter-based system. Each bird was tested in isolation. Normative histograms
of the dove's head movement patterns were obtained. The velocity of the

186

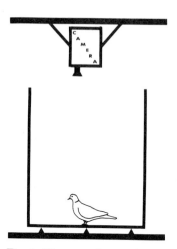

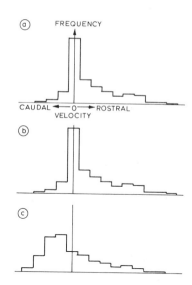

Fig. 3. Diagram of normal test condition.

Fig. 4. Histograms of dove head velocity in three reference frames. a: visual (head velocity relative to cage). b: inertial (head velocity relative to camera support). c: body centred (head velocity relative to bird's back).

horizontal component of the dove's head movement (the caudal to rostral direction being taken as positive) was calculated relative to the cage, to the camera support, and to the bird's body. In the experiment, these measures corresponded to the bird's visual, inertial, and body centred frames of reference. Fig. 4 is a representative set of histograms of the frequency of occurrence, in a typical 1-min period, of head velocities determined at the rate of 16/sec. In this experimental condition, movement through visual space was identical to movement through inertial space with the head alternately holding still and thrusting forward. Since the bird's body continues to walk forward, the first phase of head movement results in the head drifting backward relative to the body while the second phase results in a forward head velocity equal to the head velocity in inertial or visual space less the body velocity. The broadening of the peaks in the last histogram results from the body's frequent slowing-down and speeding-up relative to the cage so that the body velocity is non-uniform over time.

As a control for the effects of the apparatus used in latter parts of the experiment, a small plastic platform was temporarily glued to the feathers on the bird's back. A rod was attached to a pivot on the platform; it was held vertical by a thread which passed over a pulley on the camera to a counterweight (Fig. 5). Analysis of the head movement patterns of the dove walking and feeding with the vertical rod attached showed no significant differences in the velocity histograms when compared with the first condition where the dove was unencumbered. The effect of the rod on feeding behaviours was

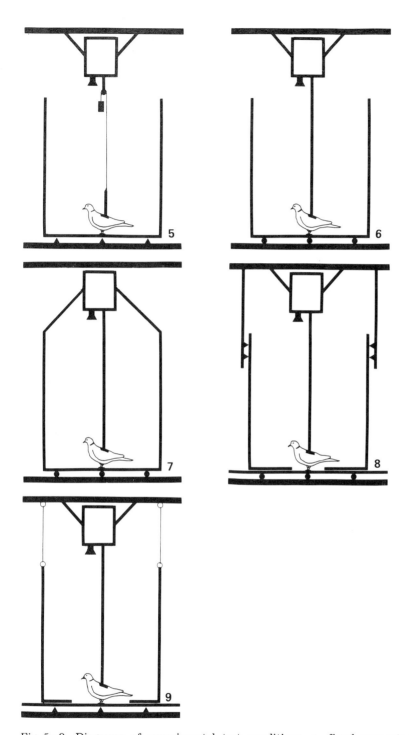

Fig. 5—9. Diagrams of experimental test conditions. ▲, fixed support; ●, ball bearing support.

negligible and the pattern of head movement during locomotion appeared to be normal.

A cage with identical internal dimensions was constructed from 2 mm thick expanded polystyrene with the lower part reinforced with 0.25 mm thick cellulose acetate. The entire cage had approximately the same mass as a dove, about 180 g. The cage was supported over a level glass plate by hundreds of microminiature ball bearings (250 μm polystyrene beads). The dove was placed in this cage and attached by a special joint assembly to a rigid vertical rod which was, in turn, bolted to the camera (Fig. 6). The counterbalanced joint assembly attached the platform on the bird's back to the rigid rod without either supporting the bird or weighing it down. The joint was hinged in such a way as to allow the vertical displacements and pitch and roll movements normal in walking, while it restrained the dove's body from any horizontal rotations or displacements. When the dove took a step, the floor would move under the bird rather than the bird moving across the floor. Similarly, the cage rotated around the bird when the bird tried to turn in the cage. The floor and wall of the cage moved relative to the dove's body in the same as they did in the first two parts of the experiment with the freely moving bird in the fixed cage.

When recordings of the bird walking and feeding in the sliding cage were analyzed, head movements were normal with respect to the visual world (the cage) and the bird's body, but abnormal with respect to the inertial world (the camera supports). Whereas the head continued to be alternately thrust forwards and held still in visual space, with respect to both the body and inertial space it was retracted after the forward thrust. This backward movement in the inertial frame of reference would not normally take place during locomotion. This part of the experiment showed that progression through inertial space is not necessary for periodic head movements.

To test if movement through inertial space could be a sufficient stimulus to induce periodic head movements, the camera and rod assembly was detached from the fixed supports and attached to the sliding cage. The cage, with the dove attached to the rod assembly (Fig. 7) was moved forward or rotated at approximately the dove's normal rate of travel. The dove's head stayed fixed relative to the cage, its visual world, and did not surge in inertial space or oscillate with respect to its body.

To see if head movements were coupled to stepping movements, the camera and rod assembly was remounted on its fixed support and the cage was rigidly attached to the same structure. A 15-cm diameter hole was cut in the centre of the cage floor where the dove stood and a large false floor was placed just beneath the hole for the dove to stand on. This false floor, like the sliding cage, was adjusted to approximate the mass of the dove and was free to slide on miniature polystyrene bearings. After the dove was placed in the cage on the false floor and attached to the joint assembly (Fig. 8), recording was started and continued until the bird spontaneously started walking. In the few instances when spontaneous walking did not occur within 15 min, stepping was induced by gently pulling on the false floor. Each dove was recorded in this set-up on three separate occasions and in each instance the bird either extended its neck gradually forward and ran or it squatted and peered around rapidly. Although it was obvious that the doves were reacting to the absence of visual feedback from their stepping, there was no sign of the head 'bobbing' one sees during normal locomotion. The mechanics of stepping did not induce periodic, coupled head movements.

For a complementary test, the false floor was rigidly attached to the camera support structure. The hole in the cage floor was enlarged to a diameter of 25 cm and the cage was suspended over the false floor by threads strung from the ceiling (Fig. 9). After the dove was attached to the rod from the camera column, the cage was moved by hand and at approximately the same speed as in the sliding cage test conditions. In the sliding cage tests for the sufficiency of vestibular stimulation, moving the cage at these speeds did not induce periodic head movements. In contrast, periodic head movements were induced in the tests with the swinging cage. When the cage was rotated the dove alternately stabilized its head with respect to the cage and saccadically turned its head to a new orientation, showing the classical optomotor response (Huizinga and van der Meulen, 1951). When the cage was displaced at a constant speed in the rostral to caudal direction (up to 20 cm, about one body length), the dove alternately stabilized its head with respect to the cage and thrust it forward with respect to its visual world. Since its visual environment, the swinging cage, was moving, the dove's head alternately oscillated backward and forward in inertial space. The dove showed its normal pattern of head movement with respect to its visual reference frame, but not its inertial frame of reference.

The first two experimental conditions established the range of normal patterns of head movement shown by doves in the cylindrical cage and showed that having a vertical rod attached to a bird's back did not alter the normal pattern of head movement. The condition where the cage moved about the bird rather than the bird about the cage demonstrated that progression through inertial space is not necessary for the normal pattern of head movement. When the whole cage was moved with the bird fixed in the cage, it showed that progression through inertial space is not sufficient, in itself, to induce a normal periodic pattern of head movement. The fifth condition demonstrated that the mechanics of stepping are not sufficient to generate the periodic head movements normally seen. The final condition showed that the mechanics of stepping are not necessary for the normal pattern of head movement but that displacement through visual space is a sufficient condition. These results hold both for linear and angular movements.

Having established that the dove's pattern of eye and head movement is under visual control, the next question is whether this movement pattern is unique to doves, pigeons, and other birds that appear to bob their heads as

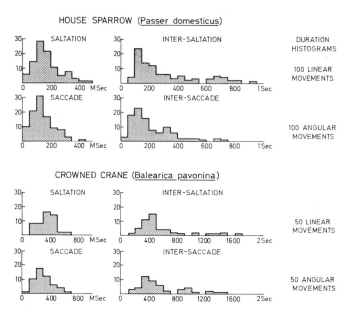

Fig. 10. Frequency histograms of sparrow and crane head movement durations.

they walk. If the pattern of alternately stabilizing the eyes and rapidly reorienting them is unique to these few species then the generality of the findings will be limited accordingly.

Birds that walk and bob their heads tend to be of intermediate size. Small birds with short legs tend to hop rather than walk. If the bird's neck is short, when its body hops the head and eyes, of necessity, must proceed through space in a saltatory fashion. Single frame analysis of films of various hopping passerines shows that their dove-like mode of eye progression is even more pronounced than required by the dynamics of hopping.

If a sparrow is nudged by another bird as it settles at its evening roost, it may make a short hop to an adjacent area. During this hop the bird's head and body move together as a unit. In contrast, the head may precede the body of a sparrow actively foraging for food. As such a bird inspects a food item, it rotates its head saccadically. After deciding to hop closer, the bird's head is thrust forward before the legs start to push the body into the air. Likewise, the head stops and is stabilized in visual space before the body finishes landing from the hop. In the visually alert sparrow, as opposed to the somnolent one, the eyes hop and the body follows.

Sparrows, like doves, have similarly shaped eye movement and interval duration histograms for both saccadic and saltatory eye movements (Fig. 10). Although the saccadic movement and interval duration distributions are skewed slightly toward the shorter durations, their modal values are approximately the same as the saltatory ones. These relationships hold true for a variety of small hopping birds filmed while foraging for food.

Large birds with long legs and necks rarely show head bobbing that includes periodic stabilization of the head in space. Rather, their heads usually accelerate and decelerate without fully stopping. Many of these large birds are more likely to move their eyes in their heads than are small birds. Although I do not have quantitative evidence that they use eye movements to visually stabilize regions of their environment as they walk, that is my impression. I have several film sequences in which walking herons, cranes, and a cassowary show nasal to temporal eye-in-head movements during the slow phases of their heads' surging progression. It is possible that these birds combine head and eye movements to visually explore their worlds in a saltatory fashion. This would be analogous to the combined head and eye movements that primates make during saccadic search (Trevarthen and Tursky, 1969; Bizzi, 1974).

Whether or not long-legged, long-necked birds use saltatory and saccadic eye movements to scan their visual environments as they walk with their heads up, when foraging with their heads near the ground they frequently use these periodic head movements. Frequency *versus* head movement duration histograms from a short section of film of a crowned crane foraging for food in high grass are shown in Fig. 10. In this film segment the bird was standing still and using its long neck to move its head through the grass. Without any body movement, the crane's eyes scan their environment with the same periodic pattern used by the dove and the sparrow.

Clearly most birds do not depend on alternating fixations with rapid reorientations as their sole means of visual scanning. For one thing, it would make flying very difficult. It is equally clear, however, that a great many species of birds of diverse size and shape do use this mode of visual exploration at least some of the time. It is particularly prominent when they forage for food with their heads close to the ground.

I do not mean to imply that most birds move their eyes with the same absolute temporal rhythm. They do not. In general, the longer a species' neck, the higher its head is held above the ground, and the more massive its head, the greater are the durations of its head saccades and saltations. Although the absolute movement durations may vary from species to species, within a given species the angular and linear eye movements tend to have the same temporal distribution. Similarly, the time that the eyes remain stabilized between movements remains the same for both saccades and saltations.

SPATIAL PATTERNS OF HEAD ORIENTATION

It is not a trivial task to determine where a bird is 'looking' since its field of view may be nearly panoramic in the horizontal plane. A man or other primate is usually said to be looking at the place where his visual axes intersect, a visual axis being defined as the line extending from the centre of the fovea outward through the nodal point of the eye. Using the fixation point to determine the location of visual attention is often justified in terms of the fovea's higher resolving power and greater colour discernment. However, even in primates it is a questionable practice to ignore the visual information available from extrafoveal areas of the visual field (Trevarthen, 1968).

At least in primates the visual axes are either parallel or intersect so that a direction of gaze or fixation point can always be determined in terms of the visual axes. In birds it is useless to define where the animal is looking in terms of its fixation point since the bird may have no anatomically definable visual axis or it may have multiple visual axes. In the latter case some of the visual axes may diverge, hence they cannot intersect in the visual field. In any case, since extrafoveal regions of diurnal birds' retinae are relatively thicker, richer in cones, and more highly developed than peripheral areas of primate retinae, it is probable that the foveal—peripheral vision dichotomy is less useful for delineating visual attention in birds than it is in primates (Walls, 1942).

Rather than try to define where a bird is looking from physiological principles, I have tried to use an operational definition of visual attention. Doves were filmed while feeding on seeds scattered on the floor of a cage 1 m in diameter. Successive frames of these films were projected onto the video digitizing apparatus in the reverse temporal sequence from the order in which they were exposed. By projecting the films so that behaviour was analyzed going backward in time, it was possible to determine where a seed that has just been pecked at, had previously been in the bird's visual field. Although this method of analysis is not useful in determining whether a bird has noticed an object which it does not later manipulate, it will determine any regularities in the way different parts of the retinae are used to inspect the visual image of objects that are manipulated.

Fig. 11 is an expanded analysis of the behaviour sequence first presented in Fig. 1. The upper two graphs show the patterns of linear and angular head velocities in an arbitrary frame of reference. The lower three graphs describe the changes in the dove's head position relative to the next seed (P1) that it will eat. The sequence starts when the dove has just finished raising its head after pecking at and eating a seed. The seed that will be eaten next is just over 10 cm from the bird's head and lies in the bird's left visual field at approximately 90° to the direction of the beak. As the bird turns its head saccadically to the left, the image of the seed is stabilized in the visual field at bearings of 14° to the left and then 56° and 69° to the right during successive intersaccadic intervals. There has been little overall change in distance between the bird's head and the seed during this head rotation. The bird's next head movement starts its approach to the seed and simultaneously starts to bring the seed into the frontal part of the right visual field. The bird begins to walk toward the seed with the head moving in a series of saltations

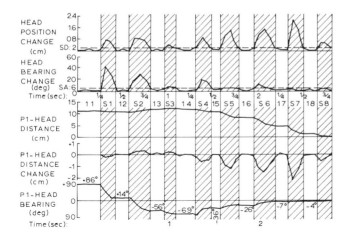

Fig. 11. Temporal and spatial patterns of a dove's head position and motion.

that bring it to a point just above the seed. At the end of 'S8' the eyes converge slightly, and, after the head is momentarily stabilized, the bird pecks at the seed. Following this peck a similar set of graphs would be plotted with head positions and bearings relative to the next seed to be eaten.

One may get a rough idea from Fig. 12 of which areas of a dove's visual field (and, hence, of its retinae) are most frequently involved in guiding the dove to the next seed it will eat. Fig. 12 is a polar histogram in which the angles represent the horizontal bearing of the seed in the dove's visual field and radial distance is proportional to the number of intervals during which the next seed to be eaten was within a particular 10° segment of the visual

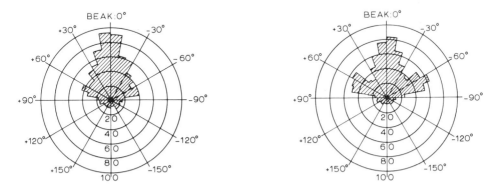

Fig. 12. Frequency histogram of the bearings in a dove's visual field of the next seed to be eaten; interseed spacing less than 2 cm.

Fig. 13. Frequency histogram as in Fig. 13, but interseed spacing greater than 10 cm.

field. The figure summarizes the cumulative distribution of bearings of the seed in the intervals between 1000 head movements during 12 min of feeding on closely spaced seeds (averaging 2 cm between adjacent, randomly scattered seeds). Since the distribution is approximately symmetrical about the 0° direction, left and right visual fields were used equally. The frequency with which the next seed to be eaten was in a particular region of the visual field declined monotonically with increasing magnitude of the seed bearing. In other words, the unsurprising result was that the next seed to be pecked was most likely to be in front of the bird's head.

A more surprising result came from a similar analysis of the frequency distribution of visual bearings when the dove fed on seeds that were much further apart (Fig. 13). The dove made approximately the same number of head movements in these circumstances as it did in the previous condition, but pecked at fewer seeds since the seeds were 10—20 cm apart. Although the data again show a symmetry in the use of the left and right visual fields, the frequency distribution is no longer monotonically related to the absolute magnitude of the visual angle between seed and beak. Instead, the seed is almost as likely to fall within the lateral visual fields between 50° and 75° from the beak as it is to fall within 25° of the beak. It is much less likely to lie in the middle region of a field, from 25° to 50°, than it is to lie in the 25° wide region of the visual field on either side.

What factors are responsible for the increased use of the lateral portions of the visual field when seed spacing is increased from 2 to 20 cm? Referring back to Fig. 11, the seed was in the lateral portions of the dove's visual field during intervals 3 and 4. Although in prior saccades the head had moved slightly toward and away from the seed as shown by the plot 'P1— Head Distance Change', starting with movement 'S4' the head progressively approached the seed and finally, after 'S8', the dove's eyes converged and it pecked at the seed. During the two intersaltatory intervals (I7 and I8) just before the dove converged its eyes and pecked, the seed was located in the frontal visual field within 10° of the midline. Fig. 14 shows part of the same data used in Fig. 13 and supplementary data from additional feeding sessions organized into two separate polar histograms. The prepeck histogram (Fig. 14a) shows the distribution of head—seed bearings when one only considers the two intervals just before the dove's eyes converged and it pecked. As one might expect, the seed is kept in the frontal field of view almost exclusively. The preapproach histogram (Fig. 14b) uses the same data base but only tabulates the seed bearings from the two intervals just before the bird's head starts a steady approach to the next seed to be eaten. It shows that just prior to approaching the seed from about one body length away, the seed is more than twice as likely to be in one of the lateral fields of view than it is to be in the frontal fields.

If a dove were near-sighted for forward directed vision and far-sighted for lateral vision, it would neatly explain the selective use of the frontal and

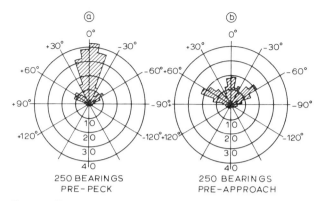

Fig. 14. Frequency histograms of the bearings in the visual field of the next seed to be eaten. a: for the two preapproach intervals. b: for the two prepeck intervals.

lateral parts of the visual field for inspecting near and far objects, respectively. For a closely related species, the pigeon, Catania (1964) has summarized evidence suggesting the view "that the pigeon is near-sighted for stimuli located in front of its beak and far-sighted for stimuli located to either side of its head". Nye (1973) studied the physiological optics of the pigeon eye and concluded that although the bird has the ability to accommodate to near objects in the lateral visual field, at any instant in time it is relatively myopic in the forward direction and hyperopic in the lateral direction.

To test if forward near-sightedness and lateral far-sightedness in the dove might account for the differential use of the forward and lateral visual fields suggested by Fig. 14a and 14b, several seeds were glued to the test cage floor. Doves feeding on scattered seeds were filmed and the bearings of the glued seeds in the visual field on the two head fixations immediately following each of the dove's first three attempts to eat a glued seed were recorded. Since only 1—2% of the seeds were glued to the floor and these were visibly indistinguishable from other seeds, the dove usually ate at least 20 seeds be-

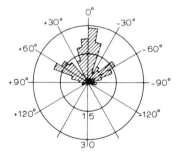

Fig. 15. Frequency histogram of the bearings in the visual field of the seed most recently pecked, during the two intervals after pecking at a glued seed.

fore coming across a seed that wouldn't move when pecked. Fig. 15 summarizes the results for 25 doves. After failing to get the glued seed on the first try, a dove would often fixate a seed with forward vision and then peck a second time. Equally often, the dove would rise to an upright posture and turn to inspect the seed, placing it in a lateral visual field. Since in the latter instances the dove was as close to the seed as when it used frontal vision just before pecking, the selective use of the lateral parts of the visual field cannot be solely due to far-sightedness in that region.

Many species of birds have two areas of increased receptor density in the retinae of each eye, a temporal region used for frontal vision and a central region used for lateral vision (Walls, 1942). The centrally located retinal region usually has the highest receptor cell density in the eye and, if the eye possesses two fovea, it has the deeper, more differentiated fovea. One would suppose that a bird should use the lateral parts of its visual field corresponding to its central region of retinal specialization for tasks requiring high visual acuity. Although casual observation suggests that many birds do frequently turn their heads and use their lateral visual fields to scrutinize objects, laboratory experiments using operant conditioning techniques suggest that the pigeon is very poor at using information from its lateral visual fields for making visual discriminations.

Reynolds (1961) presented pigeons with complex stimuli consisting of colours and shapes on a response key together with lamps of variable intensity to the side of the key in the birds' lateral visual field. He found that the pigeons responded to the side lamps only insofar as these lights affected the appearance of the key (through glare, contrast shift, etc.), implying that pigeons attend to anterior stimuli in preference to lateral ones. While Reynolds was mainly interested in questions of attention, Nye (1973) specifically tested the visual discrimination performance of pigeons for laterally located stimuli. He trained pigeons to peck one of two keys depending on which one of two stimuli was projected on a screen immediately above the keys. The pigeons correctly made red *versus* green and high *versus* low luminance discriminations more than 90% of the time when the stimuli were projected onto a screen in front of the birds' beaks. When a stimulus was presented binocularly in the lateral parts of a pigeon's visual field, the bird was much less accurate in its discrimination. As the test stimuli were progressively moved further into the lateral visual field, the responses became less accurate until at about 75°, when the stimuli were being presented along the eyes' optic axes, the bird responded randomly.

Nye has pointed out that there may have been evolutionary pressure on granivorous birds such as the pigeon to be able to identify objects in front of them while simultaneously being able to detect the movements of potential predators in the lateral visual field. If this were so, then his failure to obtain discriminative responses to stimuli presented in a pigeon's lateral visual fields might be because the lateral fields were 'tuned' to detect different classes of stimuli — *i.e.* motion rather than colour or form. This hypothesis seems unlikely since, in further tests, Nye's pigeons were unable to reliably discriminate between a blank field and a series of moving black and white bars located at about 75° in the lateral fields of view.

In my experiments, doves selectively used the lateral portions of their visual fields to inspect the seeds that they would later eat. The use of these lateral areas was enhanced if there were several types of seed available, including types that were highly preferred relative to the others. This implies that the doves were identifying the seeds, not merely locating them.

Why did my doves regularly use the laterally directed regions of their retinae when Reynold's and Nye's pigeons seemed unable to use similar regions of their eyes to make even elementary visual discriminations? I do not know the answer, but I suspect it involves the nature of the motor response that a bird uses to demonstrate its visual abilities. The doves almost invariably used frontal vision just prior to pecking at a seed but usually inspected the seed with the centrally located, laterally directed areas of their retinae if a locomotor response was necessary to get within striking distance. In the operant tests the pigeons were obliged to demonstrate visual discrimination with pecking responses. It seems implicit in several earlier experiments (Gundlach, 1933) where locomotor responses were utilized in tests of visual acuity that pigeons can make quite fine visual discrimination using lateral vision. A dove or pigeon's nervous system may be structured in such a manner that not only are there constraints on the possible response that a particular visual stimulus may evoke (Hinde and Stevenson-Hinde, 1973) but there are constraints on responses which are related to the region of the eye used to view the particular stimulus.

EYE PREFERENCES

On a long-term basis, there was no evidence that individual doves preferred to guide their feeding movements with one eye rather than the other. As the symmetry of Figs. 12 and 13 demonstrates, the right and left visual fields were used equally often when their use was considered over many minutes and hundreds of head movements.

In contrast, when the head movements of individual doves were sampled on a short-term basis, there were indications that the lateral visual fields were used in 'runs' with first the use of one eye predominating, then the use of the other. This tendency was most apparent when a feeding dove could choose from a variety of seed types, one of which was highly preferred or disliked. The effect was marginal when the bird fed on familiar seed types in familiar surroundings. It was enhanced when the dove was feeding on unfamiliar seeds or was forced, by experimental constraints, to switch from a preferred seed type to a less favoured variety. Although there is no evidence of a systematic differential specialization of either visual field for particular kinds or classes of stimuli, there are indications that relatively novel objects may be recognized when viewed with one eye and not when viewed with the other.

Rather than study this effect with novel varieties of seeds of unknown favour, I chose to use two familiar seeds, hemp and rape, that are similar in size but differ in colour, shape, and texture. I induced the doves to shift their attention from the hemp seeds to the less preferred rape and observed their use of their eyes in the preapproach phase of feeding on the seeds.

Panels of seeds were prepared by coating sheets of hardboard with epoxy and scattering hemp seeds on the surface. After the glue had set, training boards were made by prying about half of the hemp seeds loose from the boards and putting rape seeds in the depressions left by these hemp seeds. Test boards were made by prying off all of the hemp seeds from the boards and randomly filling the resulting depressions with rape and hemp seeds. It was very difficult both for me and for the birds to visually distinguish between the free and the glued seeds since the glue meniscus around each was the same and the depressions under the unglued seeds prevented them from vibrating when the boards were jarred.

Initially, after being placed in an observation cage with several training boards, a dove would mainly approach and peck at hemp seeds. Since these were glued to the boards, the dove could only eat an occasional one that was knocked loose. After the dove had pecked at ten of the approximately 40 seeds on each board, that board was replaced with a fresh one. Following a period of concentrating on the hemp, the dove would start to increase the proportion of its pecks directed toward the rape seeds. When the proportion had risen to 80% (16 pecks directed at rape seeds in 20 successive pecks) the next board to be placed in the cage was a test board on which none of the seeds was glued down.

The orientation in the visual field of the next seed to be eaten was monitored for each of the first ten seeds that the dove pecked. The most interesting results involved determining which eye the dove used to guide itself toward a seed during the transition from preapproach to approach behaviour.

One of the 20 doves tested immediately reverted to concentrating on hemp seeds. Presumably it could visually recognize that they were not glued to the test board. It used its right and left visual fields almost equally often.

The remaining 19 doves all started by eating the rape seeds. They directed at least seven of their first ten pecks toward the rape seeds. Although there was no left—right bias when the whole group was considered, each of these doves tended to use one half of the visual field preferentially just before approaching the next seed it was to eat. In general, the eye that was used just prior to approaching the initial test seed would be used to initiate approaches to at least six of the following nine test seeds. This bias toward continuing to use the same eye to decide which seed to approach was significantly different ($P < 0.01$, Wilcoxon) from the same doves' balanced use of their two eyes prior to training.

Nine doves pecked exclusively at rape seeds on the test board. Whichever eye one of these birds used just prior to approaching its first test seed, it continued to use just before approaching at least seven of the next nine seeds. These birds were fitted with monocular occluders and retested with a fresh test board. After pecking at ten seeds, the occluder would be transferred to the bird's other eye and the bird retested. Five of the nine birds started with the blindfold over the eye that had dominated their approach decisions in the original test situation. The other four birds initially had vision blocked in the eye that was least often used in approaching the rape seeds in the original tests.

Both sequences of monocular test yielded the same result. When the exposed eyes were the ones that the doves had preferentially used in the original binocular tests, the birds pecked at significantly more rape seeds than could be predicted either by random selection or by their own pre-training performance ($P < 0.01$, Wilcoxon). In contrast, these same doves tended to peck at hemp seeds when they could only see with their other eye. Then, their seed selection was more like it had been prior to the binocular training. They pecked at rape seeds significantly less often ($P < 0.01$, Wilcoxon) than they had in either the other monocular test or the original binocular test.

These results suggest that although a dove may use its left and right visual fields equally often in the long-term, there may be short-term preferences. In addition, visual stimuli viewed with one eye may elicit different behaviour than the same stimuli would when viewed with the other eye. When the short-term eye preferences are considered with the differential responses that may be elicited by monocularly viewed stimuli, it implies that normal environmental stimuli may not be processed the same in the left and right visual fields just as they may not be perceived in the same way in the anterior and lateral fields of view. Zeier's chapter in this volume (Chapter 9) provides a more extensive treatment of this subject.

NEURAL IMPLICATIONS

It should be obvious from the preceding sections that avian vision is not simply a passive process in which the eye continuously registers and encodes the surrounding environmental stimuli. Rather, selective parts of the visual field are rhythmically moved to inspect the environment and to investigate particular aspects of it.

Ignoring this, electrophysiological investigations of the pigeon's visual system (Revzin, 1967; Maturana, 1964; Maturana and Frenk, 1963) have tended to adopt the same methodology that proved useful in the analysis of neural coding in the frog's visual system (Lettvin et al., 1959 and 1961; Maturana *et al.,* 1960). In the introduction to their original report, Lettvin and Maturana (1959) explicitly state that they "chose to work on the frog because of the uniformity of his retina, the normal lack of head and eye movements... and the relative simplicity of the connections of his eye to his brain." Since pigeons meet none of these criteria, one must be wary in interpreting the results of neurophysiological experiments in which the bird's head and eyes are immobilized and the visual test stimuli presented arrhythmically.

In experiments with pigeons similar to those with frogs, Maturana (1964) described six classes of retinal ganglion cells that responded maximally to particular configurations of contour, contrast, and movement in their visual receptor fields. It would be interesting to know whether the distribution of these cell types is uniform over the retina or whether certain types might be concentrated in the retinal regions that are used selectively.

Concerning the selective use during feeding of two areas in each eye's visual field, many species of birds have two fovea or at least two highly differentiated areas in each eye (Walls, 1942). It would not be surprising for such birds to selectively use the corresponding parts of their visual fields since these would be areas of higher acuity. The laboratory-reared Barbary doves used in the present experiments do not appear to have two fovea in each eye. Macroscopic and low power microscopic inspection failed to reveal more than one area, and this was a poorly differentiated central one. This implies that the selective use of regions of the dove's visual field is not merely due to increased resolving power at the receptor level, but is a consequence of the organization of the visual system. If the distribution and connectivity of retinal ganglion cell types cannot account for this, then the mechanism will have to be sought in the brain.

Unlike the retinae of doves and other birds that advance their heads in a saltatory fashion, some birds have horizontal bands of increased receptor density in each retina (Duijm, 1958). These birds tend to advance their heads in smooth rather than periodic movements (Whiteside, 1967). If the differentiated retinal band acts like a shift register to neurally stabilize contour relationships, then these birds could use the same central mechanisms for monocular depth perception that saltators use with their optomotor stabilizations. Alternatively, birds with smoothly moving heads could utilize the consequent optic flow fields (Gibson, 1966) to get three-dimensional space perception. In this case, one might expect these birds to have a higher proportion of cells sensitive to moving edges than the pigeon has (Maturana, 1964), since relative distance would be inversely proportional to relative retinal velocity.

In saltating birds free to walk, it would be interesting to know if the retinal ganglion cells would respond to visual stimuli in the same manner during various phases of head and eye movement. Although I have presented evidence that the character of a dove's visual processes may vary across the visual field, I have not been explicit about temporal variations in function. I have no evidence defining the character of the dove's visual processes during periods of head movement and only have indirect evidence about the relative importance the dove attaches to vision during periods of head movement and stabilization.

Actively foraging doves seldom blink or use their nictitating membranes while their heads are stabilized. Although they rarely close their eyes while active, when they do blink it is usually during periods of head motion. Doves frequently draw their nictitating membranes across their eyes. Most of these complete nictitations occur while the head is rapidly moving. If nictitations and eyelid blinks are necessary to maintain and protect the corneal surfaces (Walls, 1942), but unavoidably obstruct pattern vision in the process, it appears that doves prefer to suffer this visual deficit during periods of head movement rather than during periods of head stability.

Many, if not most, saccades and saltations are accompanied by at least a partial nictitation which must block the potential for pattern vision in parts of the visual field. It is possible that rather than being solely related to corneal maintenance, nictitation has a visual function. In man, various visual thresholds are raised during saccadic eye movements (Latour, 1962; Sperling and Speelman, 1965); this suppression of visual information during saccadic reorientation may be important for maintaining the subjective stability of the visual world (Hochberg, 1972). Saccadic suppression in man could be mediated by an inability of the sensory system to respond to rapidly moving patterns (Graham, 1965). Since the dove has a higher flicker fusion threshold than man, its visual system may also have a higher tolerance for rapidly moving patterns. If saccadic suppression were important for some visual functions, then nictitation might provide a necessary mechanical shutter.

I have no direct evidence about the character of the dove's vision during periods of head movement both because these periods are too brief to reliably present controlled stimuli and because vision is often mechanically obscured by the nictitating membranes. Many species of shore birds move their heads in a rhythmical fashion with the duration of the saltations greatly extended. Rather than have head saltations every step as doves might, these foraging shore birds alternate periods of head and body immobility with long smooth head movements during which the body takes several steps. The eyes remain open during these head movements which may last up to several seconds.

I filmed plovers on a beach while catapulting pieces of fish, pieces of bread, and pebbles at them. Upon subsequent analysis of the films, a striking difference in response emerged depending on whether or not the bird's head was moving when my missile was thrown. If the head was stationary, the bird tended to make a directed response: away if the missile was a pebble and toward the end of the object's trajectory if it was edible. Presumably the missiles were identified by their size (pebbles were small so as not to injure the birds), shape, or flight dynamics. If the plover's head was moving when I threw the object, the bird made a non-directed response: coming to a halt and 'freezing' or launching directly into flight. The direction of flight was a compromise between the direction the bird had been walking and the direction away from me. The response did not appear to be related to whether the thrown object was a pebble or food.

Insofar as the plover's slow rhythm of linear head displacement and stabilization is analogous to the dove's alternation of saltation and fixation, it is likely that stimuli detected while the eyes are moving can only elicit undirected startle responses. Object identification and guided movement initiation probably require the eyes to be stabilized in visual space.

The phasic eye movements of doves, pigeons, and other birds whose saltations are relatively brief are probably involved in monocular depth perception (Walls, 1942). These birds usually have very limited regions of binocular overlap and tend to converge their eyes only just before pecking. It is possible that they perceive the solidity of the world with successive fixations in a monocular process analogous to binocular stereopsis. Stereopsis involves fusion of two retinal views and utilization of the relative retinal disparity information. In binocular stereopsis the disparity information is available from the two eyes simultaneously; monocular stereopsis may involve the relative disparity of retinal images on two successive monocular fixations. The neurophysiology of binocular stereopsis may be investigated with the subject's head and eyes stereotaxically stabilized (Hubel and Wiesel, 1970). The mechanism of monocular stereopsis can only be studied if either the bird makes volitional head saltations or the visual stimuli can be rhythmically displaced. In this context, it is interesting to note that stroboscopically evoked responses in the dove's optic tectum last at least 300 msec. This is longer than most saltations and is, in walking doves, about the average duration from the mid-point in one intersaltatory interval to the mid-point of the next stabilization.

Nictitation has been invoked as one means of suppressing distracting visual stimuli during head saltations. Several investigators (Maturana and Frenk, 1965; McGill, 1964; Cowan and Powell, 1963) have found centrifugal fibres running from the isthmooptic nucleus to the retina. These might be involved not only in centrally generated saccadic and saltatory visual suppression, but also in modifying the response characteristics of particular retinal regions during selective use of the visual field. The visual deficits that Rogers and Miles (1972) observed in their isthmooptic lesioned chicks might stem from an inability to use specialized regions of the visual field because of a lack of centrifugal suppressive control.

Whatever the neural mechanisms underlying saccadic and saltatory eye movements in doves, it is apparent that the rhythmical alternation of visual stabilization and rapid reorientation described here is not limited to walking birds with dove-like head movements. As mentioned previously, many long-necked birds rhythmically dart their heads about while inspecting possible food sources and many hopping birds with short necks have, *ipso facto*, a saltatory mode of head and eye movement. Some birds with short necks and large beaks have smooth head movement, but saccadic exploratory eye movements. That birds of diverse shape and mode of locomotion show similar patterns of head and eye movement should be important to theories of the evolution of visual perception. That the shapes of bones and the points of insertion of muscles and tendons could change during avian evolution while the basic pattern of eye movement remained constant suggests that the neural substrate of this mode of visual exploration must be fundamental to the process of avian vision.

Acknowledgements

The work reported here was carried out in the Department of Psychology, University of Edinburgh.

I am grateful to the S.R.C. for support, to Professor D.M. Vowles for providing laboratory facilities and to Dr. P. Wright for helpful comments and encouragement.

REFERENCES

Bizzi, E. (1974) The coordination of eye—head movements. *Sci. Amer.*, 231 (4), 100—106.
Catania, A.C. (1964) On the visual acuity of the pigeon. *J. exp. Anal. Behav.*, 7, 361—366.
Cowan, W.M. and Powell, T.P.S. (1963) Centrifugal fibres in the avian visual system. *Proc. roy. Soc. B*, 158, 232—252.
Daanje, A. (1951) On the locomotory movements in birds and the intention movements derived from them. *Behaviour*, 3, 48—96.
Duijm, M. (1958) On the position of a ribbon-like central area in the eyes of some birds. *Arch. néerl. Zool.*, 13 (Suppl.), 128—145.
Dunlap, K. and Mowrer, O.H. (1930) Head movements and eye functions of birds. *J. comp. Psychol.*, 11, 99—113.
Gibson, J.J. (1966) *The Senses Considered as Perceptual Systems*. Houghton Mifflin, Boston, Mass.
Graham, C.H. (Ed.) (1965) *Vision and Visual Perception*. Wiley, New York.
Gundlach, R.H. (1933) The visual acuity of homing pigeons. *J. comp. Psychol.*, 16, 327—342.
Hochberg, J. (1972) Perception: space and movement. In J.W. Kling and L.A. Riggs (Eds.), *Experimental Psychology*, Vol. 1, Holt, Rinehart and Winston, New York, pp. 475—550.
Hubel, D.H. and Wiesel, T.N. (1970) Cells sensitive to binocular depth in area 18 of the macaque monkey cortex. *Nature (Lond.)*, 225, 41—42.
Huizinga, E. and van der Meulen, P. (1951) Vestibular rotatory and opkinetic reactions in the pigeon. *Ann. Otol. (St. Louis)*, 60, 927—947.
Kare, M.R. (1965) The special senses. In P.D. Sturkie (Ed.), *Avian Physiology*. Cornell University Press, Ithaca, N.Y., pp. 406—418.
Latour, P.L. (1962) Visual threshold during eye movements. *Vision Res.*, 2, 261—262.
Lettvin, J.Y., Maturana, H.R., McCulloch, W.S. and Pitts, W.H. (1959) What the frog's eye tells the frog's brain. *Proc. Inst. Radio Engr*, 47, 1940—1951.

Lettvin, J.Y., Maturana, H.R., Pitts, W.H. and McCulloch, W.S. (1961) Two remarks on the visual system of the frog. In W.A. Rosenblith (Ed.), *Sensory Communication.* M.I.T. Press, Cambridge, Mass., pp. 757—776.

McGill, J.I. (1964) Organisation within the central and centrifugal fibre pathways in the avian visual system. *Nature (Lond.),* 204, 395—396.

Maturana, H.R. (1964) Functional organization of the pigeon retina. In R.W. Gerard and J.W. Duyff (Eds.), *Information Processing in the Nervous System.* pp. 170—178.

Maturana, H.R. and Frenk, S. (1963) Directional movement and horizontal edge detectors in the pigeon retina. *Science,* 142, 977—979.

Maturana, H.R. and Frenk, S. (1965) Synaptic connections of the centrifugal fibres in the pigeon retina. *Science,* 150, 359—361.

Maturana, H.R., Lettvin, J.Y., Pitts, W.H. and McCulloch, W.S. (1960) Anatomy and physiology of vision in the frog. *J. gen. Physiol.,* 43 (Suppl.), 129—175.

Nye, P.W. (1973) On the functional differences between frontal and lateral visual fields of the pigeon. *Vision Res.,* 13, 559—574.

Pearson, R. (1972) *The Avian Brain.* Academic Press, London.

Pumphrey, R.J. (1961) The sensory organs: hearing in birds. In A.J. Marshall (Ed.), *Biology and Comparative Physiology of Birds, Vol. 2.* Academic Press, New York, pp. 55—68.

Revzin, A.M. (1967) Unit responses to visual stimuli in the nucleus rotundus of the pigeon. *Fed. Proc.,* 26, 656 (abstract 2238).

Revzin, A.M. (1969) A specific visual projection area in the hyperstriatum of the pigeon (*Columba livia*). *Brain Res.,* 15, 246—249.

Reynolds, G.S. (1961) Attention in the pigeon. *J. exp. Anal. Behav.,* 4, 203—208.

Rogers, L.J. and Miles, F.A. (1972) Centrifugal control of the avian retina: V. Effects of lesions of the isthmo-optic nucleus on visual behavior. *Brain Res.,* 48, 147—156.

Sillman, A.J. (1973) Avian vision. In D.S. Farner and J.R. King (Eds.), *Avian Biology, Vol. 3,* Academic Press, New York, pp. 349—387.

Sperling, G. and Speelman, R. (1965) Visual spatial localisation during object motion and image motion. *J. opt. Soc. Amer.,* 55, 1576.

Tansley, K. (1964) Vision. In A.L. Thomson (Ed.), *A New Dictionary of Birds.* Nelson and Sons, London, pp. 859—865.

Trevarthen, C.B. (1968) Two mechanisms of vision in primates. *Psychol. Forsch.,* 31, 299—337.

Trevarthen, C.B. and Tursky, B. (1969) Recording horizontal rotations of head and eyes in spontaneous shifts of gaze. *Behav. Res. Meth. Instruc.,* 1, 291—293.

Walls, G. (1942) *The Vertebrate Eye and its Adaptive Radiation.* Cranbrooke, Bloomfield Hills, Mich.

Whiteside, T.C.D. (1967) The head movements of walking birds. *J. Physiol. (Lond.),* 188, 31P—32P.

Whitteridge, D. (1960) Central control of eye movements. In J. Field, H.W. Magoun and V.E. Hall (Eds.), *Handbook of Physiology, Sect. 1, Vol. 2.* Amer. Physiol. Soc., Washington, D.C., pp. 1089—1109.

Yarbus, A.L. (1967) *Eye Movements and Vision.* Plenum Press, New York.

11

AROUSAL AND ORIENTATION FUNCTIONS OF THE AVIAN TELENCEPHALON

ERIC A. SALZEN and DENIS M. PARKER

Department of Psychology, The University of Aberdeen (Great Britain)

INTRODUCTION — BEHAVIOUR OF DECEREBRATE BIRDS

It has become customary to consider telencephalic functions in terms of learning, and most modern avian forebrain studies follow this approach. We feel it might be more profitable to consider forebrain functions in terms of orientation and particularly in terms of levels of complexity and precision of orientation. The levels of orientation we tentatively recognise are — orientation of the whole body to areas of the environment, orientation of body parts to specific foci of stimulation *i.e.* to objects, and orientation of specific body structures to aspects or features contained within an object or focus of stimulation. We propose to analyse the functions of the avian telencephalon in terms of these orientation processes. We shall give a brief account of a lesion study of chicks which we have analysed in terms of arousal and orientation deficits. In the analysis we shall include a review of lesion, biochemical and stimulation studies of the striatal regions of the avian brain. Finally we shall outline an hypothesis of the functions of the avian striatal areas in terms of an hierarchy of mechanisms for orientation behaviour. In giving this outline we shall consider relevant anatomical and physiological knowledge of the avian telencephalon.

In 1962 Åkerman *et al.* used a high energy proton beam to decerebrate pigeons. The results obtained with this twentieth-century technique were essentially the same as those obtained with less sophisticated methods by many investigators during the previous one hundred years. This earlier work has been reviewed by ten Cate (1936) and has established that decerebrate birds can balance, walk and avoid obstacles, and, if thrown into the air, can fly and land. They may also show typical body movement patterns that include shaking of the head and body, stretching of limbs, wing-flapping, head-scratching, preening, and, in response to holding, struggling and escape behaviour. These somatomotor responses of decerebrate birds are described in the more recent study of Phillips (1964). Most reports emphasise that decerebrate birds are unresponsive to all but strong and crude forms of extero- and proprioceptive stimulation. Strong auditory stimulation may produce arousal and there are a few reports (*cf.* ten Cate, 1936; Pearson, 1972) of visually elicited behaviour including approach to light of moderate intensity and pecking at highly contrasted spots. All reports agree that decerebrate birds show no spontaneous behaviour. It would be reasonable to conclude from this limited responsiveness and lack of spontaneity of behaviour that the avian telencephalon has at least an arousal or facilitatory action on those response mechanisms that can be shown to be still present in the decerebrate preparation. It seems that the latter consist of locomotory patterns and the crudest of approach and avoidance responses. From their study of decerebrate pigeons, Åkerman *et al.* (1962) concluded that "The subtelencephalic structures seem to be able to control not only all important reflexes, including the basic motor patterns of walking and flying, but also the basic visual control of the behaviour associated with locomotion.". In fact Åkerman *et al.* were able to induce a variety of movements in their decerebrate pigeons by electrically stimulating the anterior thalamus, but the behaviour obtained consisted of simple turning movements of the body, general movements of the head, and locomotory actions of the limbs. A more integrated action, escape in response to the experimenter's approaching hand, has been reported by Arduini *et al.* (1948) following the application of strychnine to the tectum of the decerebrate pigeon. It is likely, therefore, that the telencephalon of the bird facilitates generalised sets for adient (approach) and abient (withdrawal) patterns in visceral and somatic response mechanisms in the brain stem. A purely non-specific arousal function has been proposed by Aronson (1967) for the forebrain of fish and Kaplan and Aronson (1969) have emphasised that this control of arousal involves both excitation and inhibition in brain stem response mechanisms. In the case of the avian telencephalon we use the term arousal to mean this facilitation of both excitatory and inhibitory elements of relatively non-specific response sets.

If the arousal action of the telencephalon is not response-specific, it must, nevertheless, be stimulus-specific if arousal is not to occur to any and every stimulus. Thus the telencephalon must also be involved in the selection and discrimination of stimuli, *i.e.* in the processes of attention and recognition. The behaviour of decerebrate birds is consistent with this, and Åkerman *et al.* (1962), after considering their own findings and those of the earlier workers concluded that, ". . . the telencephalon of the bird is concerned with the recognition of stimuli related to some categories of instinctive behaviour, such as feeding, drinking, escape behaviour and social behaviour and with the orientation of the bird and its movements in relation to these stimuli.". In the intact bird orientation in relation to specific stimuli involves more complex response patterns than can possibly be elicited in the decerebrate by non-specific arousal. The most complex response claimed for the latter is the escape response obtained by Arduini *et al.* and this is probably no more than a looming response involving relatively simple stimulus—response mechanisms. Hence the telencephalon must facilitate brain stem response mechanisms in a specific and directive manner so as to produce patterned movements in relation to specific selected stimuli, *i.e.* for orientation of the body and body parts to specific foci of stimulation or objects. This use of the term 'orientation' coincides precisely with its usage in the quotation from Åkerman *et al.* and refers to the behaviour described therein. By orientation we do not mean the 'investigatory reflex' of Pavlov (1927) or the 'orienting reflex' of Sokolov (1963), although this reflex behaviour is clearly an essential part of the more general behaviour which we designate as orientation. Rather we use orientation to mean the performance of specific turning and approach—withdrawal movements akin to the kineses and taxes of Fraenkel and Gunn (1961). Although decerebrate birds may show some crude bodily orientations to the environment these are not discriminating orientations to potential consummatory stimuli for particular motivational or drive states, *i.e.* they do not involve recognition of stimuli related to the categories of instinctive behaviour listed in the quotation from Åkerman *et al.* We suggest that the telencephalic orientation mechanisms are particularly concerned with limb, neck, and mandible actions, since these operate in the most specific manner on the most specific aspects of the environment. Indeed the evolution of the telencephalon to some extent parallels the evolution of the use of limbs, jaws and mobile neck in exploring and manipulating objects in the environment, *i.e.* in the evolution of increasingly complex and specific orientation behaviour.

The establishment and maintenance of a specific orientation to a specific stimulus is the essence of the concept of attention. It is interesting, therefore, that Savage (1968, 1969a and b and 1971) has evidence of both arousal and attention deficits in forebrainless fish and, in addition, has suggested that the forebrain, by maintaining attention to a stimulus, makes possible the association of the stimulus with a subsequent response pattern. In terms of orientation behaviour we would say that the establishment of specific orientations to specific stimuli is the process of discrimination learning, and a maintained orientation enables association of a distant stimulus with a subsequent proximal stimulus and its proximal response, *i.e.* associative learning.

TELENCEPHALIC FUNCTIONS IN NEWLY HATCHED CHICKS

A study of the effects of forebrain lesions

The following study, which is being published in full elsewhere, looked at the effects of aspiration lesions of regions of the telencephalon on the following four behaviour patterns in domestic chicks.

(1) The visuomotor orientation of social approach shown in imprinting.

208

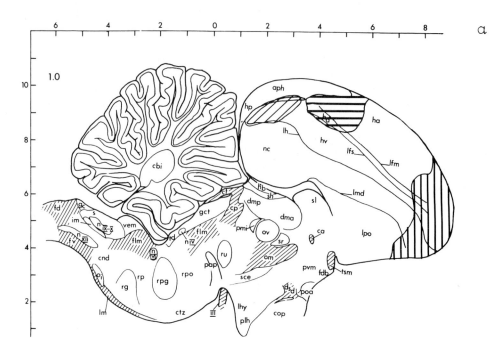

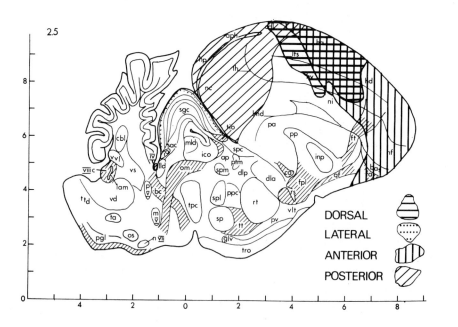

DORSAL

LATERAL

ANTERIOR

POSTERIOR

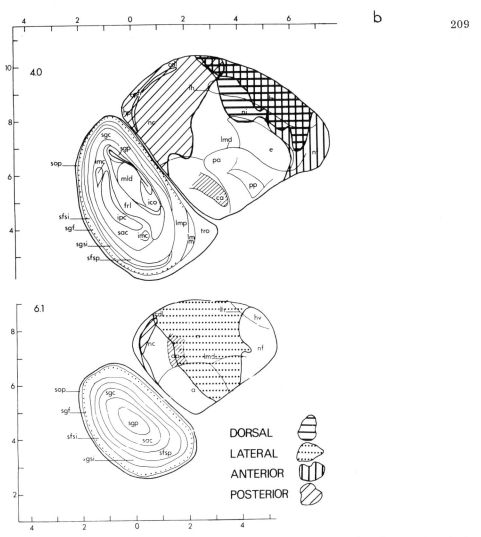

Fig. 1a and b. The diagrams represent a series of sagittal sections taken from a stereotaxic brain atlas for domestic chicks (Salzen and Williamson, unpublished). Stereotaxic coordinates are in millimetres. The shaded areas indicate the tissue removed bilaterally in each chick and combined for each experimental group. The groups, each of 8 chicks, involved anterior, dorsal, posterior, and lateral area lesions made by aspiration two days after hatching. Brains were taken for histology three weeks after hatching. Key to abbreviations of forebrain structures: a, archistriatum; aa, nucleus archistrialis anterior; aph, area parahippocampalis; bas, nucleus basalis; ca, commissura anterior; cdl, area corticoidea dorsolateralis; e, ectostriatum; fa, tractus frontoarchistrialis; fpl, fasciculus prosenphali lateralis; ft, tractus frontothalamicus and tractus thalamofrontalis; ha, hyperstriatum accessorium; hd, hyperstriatum dorsale; hp, hippocampus; hv, hyperstriatum ventrale; inp, nucleus interpeduncularis; lfm, lamina frontalis suprema; lfs, lamina frontalis superior; lh, lamina hyperstriatica; lmd, lamina medullaris dorsalis; lpo, lobus parolfactorius; nc, neostriatum caudale; nf, neostriatum frontale; ni, neostriatum intermedium; pa, palaeostriatum augmentatum; pp, palaeostraiatum primitivum; qf, tractus quintofrontalis; sl, nucleus septalis lateralis and tsm, tractus septomesencephalicus.

(2) The visuomotor orientation and coordination of food-pecking.

(3) The arousal and responses of withdrawal to repeated visual looming and contraction.

(4) The arousal and responses of startle and freezing to a repeated sudden noise.

Chicks were lesioned on the second day after hatching and then reared singly in visual isolation cages. There were five experimental groups consisting of four classes of bilateral extirpation: (1) anterior, which involved the whole of the anterior poles of the telencephalon; (2) dorsal, which involved the middle superficial regions of the cerebral hemispheres; (3) posterior, which involved the whole posterior poles of the hemispheres and (4) lateral, which involved the extreme lateral surfaces of the hemispheres — and a single control group with a craniotomy but no brain damage. Testing began three days and ended ten days after surgery, when the chicks were five and 12 days old respectively. The order of testing was as follows — first imprinting test at five days, first pecking accuracy test at six days, second imprinting test at eight days, second pecking test at nine days, looming and startle tests at 12 days. The animals were sacrificed at 13 days of age. The lesions are shown in Fig. 1a and b in summary form on standard stereotaxic atlas sections from an unpublished atlas prepared by Salzen and Williamson. The lesion areas plotted are the total areas damaged bilaterally in each chick summed for each experimental group.

Imprinting was obtained by hanging either a small green ball or a small yellow rectangular slab in each chick-rearing cage. Under these conditions chicks soon come to treat this object as a social partner. The chicks were tested for imprinting in two tests, each of ten 2-min trials, in a two choice situation requiring approach and social responses to the familiar of the two test objects which were, of course, the two objects used for imprinting. Both the imprinting and the test procedures have been well described by Salzen and Meyer (1968). The results of the two tests were similar and the combined scores are displayed in Fig. 2a as the numbers of correct choices, incorrect choices and failures to make a choice. All the lesioned groups showed more no-choices than the controls but only the lateral and dorsal groups showed significantly fewer correct choices than the control group, and only the lateral group results were not significantly different from a random choice distribution. We conclude that the lateral group showed no evidence of imprinting and that the dorsal group was less strongly imprinted than the control group. Pecking accuracy was tested using round millet seed in two tests. In each test 25 of the chick's attempts to feed were classified as either inaccurate pecking (miss, failure to strike seed; or hit, failure to lift seed) or as accurate pecking (seize, failure to swallow seed; or swallow, successful ingestion). The results of the two tests are combined and shown in Fig. 2b. All groups were affected by comparison with controls, but the most severely affected group was the lateral one which was significantly worse than the

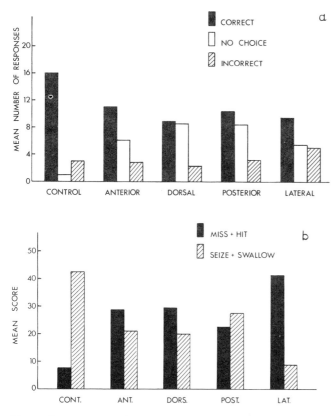

Fig. 2. Imprinting and pecking responses of chicks with lesions of the indicated forebrain areas. Group n = 8. a: approach and social acceptance responses made by imprinted chicks to the imprinting object and a strange one in a simple two choice situation. The data are combined from two separate tests, each of ten 2-min trials, made three and ten days after surgery. The lateral and dorsal groups show fewer correct choices than controls ($P < 0.025$ and 0.041, Mann—Whitney U-test). Laterals respond randomly (χ^2-test). b: accurate and inaccurate pecking attempts to feed on round millet seeds. Each chick was scored for 25 pecks in each of two tests made four and seven days after surgery, and the data have been combined. Inaccurate pecks include both misses and hits without lifting the seed, while accurate pecks include both lifting the seed but dropping it and complete lifting with swallowing. Lateral group is significantly worse than anterior ($P = 0.032$, Mann—Whitney U-test). All groups are worse than controls.

anterior group despite the expectation that the latter would be most affected because of damage to the nucleus basalis which Zeigler and Karten (1973a and b) have implicated in feeding and mandibulation response control mechanisms. The lateral group was the only lesion group that showed no improvement in accuracy on the second test. The posterior group appeared the least impaired.

Looming responses were tested using an alternately expanding and con-

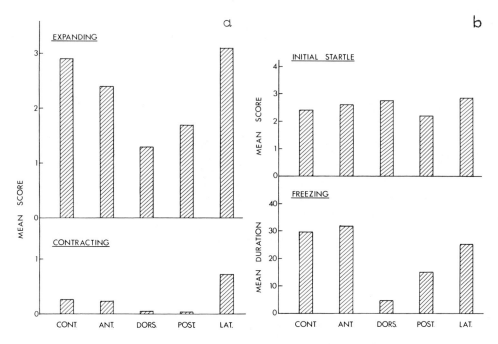

Fig. 3. Looming and startle responses of chicks with lesions of the indicated forebrain areas. Group n = 8. a: mean withdrawal scores in response to an expanding (looming) and contracting (receding) shadow stimulus. Each chick was exposed to an alternating sequence of ten of each type of stimulus ten days after surgery. The responses were rated on a five point scale (from no response score of zero to full flight score of five). Posterior and dorsal groups are less responsive than controls ($P < 0.041$ and 0.001, Mann—Whitney U-test) to the expanding shadow. b: mean startle and freezing scores in response to a loud noise repeated ten times at 1-min intervals ten days after surgery. Initial startle responses were scored on a five point rating scale (from no response score of zero to a short run score of five). Poststartle response freezing behaviour is shown as the mean times in seconds. Dorsal group shows less freezing than controls ($P < 0.001$, Mann—Whitney U-test).

tracting shadow on a screen in front of the chick. The test consisted of ten sequences of expansion followed by contraction, with pauses for recovery between each stimulation and between sequences. The responses of the chicks were scored on a simple rating scale — 0 = no visible response, 1 = head flinch, 2 = body flinch, 3 = jump backwards, 4 = flight. The results of the looming test are shown in Fig. 3a. The posterior and dorsal groups were significantly less responsive to the expanding shadow than the control group. The contracting shadow gave much weaker responses but the same group differences were apparent. There were no apparent differences in the tendency for the looming response to habituate.

Startle responses were tested by dropping a standard metal weight at 1-min intervals for ten trials. Responses were scored on a 4-point scale similar to that for the looming response although it should be noted that bodily movement was forwards or sideways and not backwards as in the looming response. The duration of post-startle freezing behaviour was recorded as the time taken to resume any movement of body or body parts after the initial startle response. The results of the tests are given in Fig. 3b. There were no differences in startle response magnitude, but the duration of post-startle freezing was significantly shorter in the dorsal group than in the control group, and nearly so in the posterior group. No differences in tendency for habituation of either response were apparent.

Imprinting and discrimination learning

The foregoing data implicate the extreme lateral edge of the telencephalic hemispheres in the visual discrimination learning of imprinting with perhaps a contributing influence of a dorsal central region which included the posterior portion of the hyperstriatum. The only other example of a lesion affecting imprinting is that of Hess (1959). He claimed that decerebrate chicks followed an imprinting object but subsequently gave no clear indication of imprinting. Small dorsolateral lesions had no effect, but lesions of the posterior poles of the cerebral hemispheres prevented imprinting. Hess, both then and since, has given no details of his study. It is possible that his posterior lesions were large enough to invade the posterolateral region damaged in the present study. More recently Bateson *et al.* (1973) have found specific biochemical changes in the anterior part of the forebrain roof in chicks exposed to a flashing light as an imprinting stimulus. The rate of incorporation of tritiated uracil into macromolecules during a fixed exposure period on the second day was inversely related to the amount of exposure to the same stimulus on the previous day. This relationship was specific to the anterior forebrain roof area. The series of studies leading to this finding has been reviewed by Horn *et al.* (1973). Gervai and Csányi (1973) have implicated the Wulst in imprinting by their finding that puromycin injected into this region prevented the development of approach responses to an imprinting stimulus and impaired subsequent discrimination but not subsequent approach behaviour. Codish (1971) injected actinomycin-D into the hippocampus of chicks 30 min after imprinting training and found it blocked imprinting responses tested three days later. Injections one day after training did not have this effect. More recently Mayor (1973) has shown that puromycin injected into the dorsal region of the forebrain of quail interferes with acquisition and retention in successive discrimination learning of colour and pattern problems. It is difficult to estimate the significance of the localisation of functions in these injection studies because, at least in the case of puromycin, abnormal electrical activity is produced and may affect widespread areas; *cf.* Cohen *et al.* (1966). Furthermore, Flexner and Flexner (1968) have found that signs of learning may reappear some time after having been apparently 'abolished' by puromycin. Nonetheless the biochemical studies of imprinting seem to implicate the roof of the forebrain in imprinting and discrimination learning whereas in the present study the anterior forebrain lesions had least effect on imprinting, and the dorsal region lesions, while producing some impairment, by no means prevented successful discrimination of the imprinting object. The lateral lesions did appear to prevent discriminative imprinting and perhaps biochemical studies should be directed to this hitherto untested region.

It is of considerable interest that the same lateral region of the chick's forebrain has been shown by Benowitz (1972) to be crucial for the learning and retention of a taste avoidance response. Benowitz used a simple pecking suppression response to a lure dipped in methyl anthranilate *cf.* Lee-Teng (1969) and Lee-Teng and Sherman (1966). Earlier Lee-Teng and Sherman (1969) reported that the sides of the chick forebrain may be implicated in one-trial taste avoidance learning. It is interesting too that Mark and Watts (1971) and Watts and Mark (1971) were able to interfere with short-term and long-term memory of single trial taste avoidance in chicks, using methyl anthranilate, by injecting inhibitors of the sodium pump and of protein synthesis into the centres of the telencephalic hemispheres via the lateral surface. We have some preliminary data for the same taste avoidance behaviour but using drinking rather than pecking, and with a colour discrimination rather than simple response suppression. Preliminary data for six chicks; two control, two with lateral and two with dorsal lesions of the kind already described seem to support the finding of Benowitz *op cit.* that the lateral region of the telencephalon, including the lateral portion of the archistriatum, is crucial for the acquisition and retention of taste avoidance learning. It is well established by Shaklee (1921) that the drinking pattern remains intact in decerebrate birds so this response is a useful one to employ in forebrain lesion studies. Damage to the forebrain feeding circuit, described by Zeigler *et al.* (1969) as including the nucleus basalis and its afferent quintofrontal tract and efferent frontoarchistriatal tract, should not affect the drinking pattern. In our chick study the lateral lesions damaged the frontoarchistriatal tract. In Benowitz' study an anterolateral lesion group received damage to all three forebrain structures involved in the feeding circuit and was not affected in its avoidance learning. It is unlikely, therefore, that the effects of posterolateral forebrain lesions on taste avoidance learning are due to the interruption of the known pathways for taste and beak sensation. It is possible that lateral lesions of the forebrain involve, at least marginally, the periectostriatal belt. This region probably has a visual function, for Parker and Delius (1972) have recorded visual evoked potentials from it in the pigeon. Furthermore, Hodos and Karten (1970) have reported that lesions of this area in the pigeon impaired retention of visual intensity and pattern discriminations in a simultaneous learning situation. The degree of impairment was proportional to the amount of damage and slow relearning occurred. In his taste avoidance study Benowitz had a mediolateral forebrain lesion group which also showed an impairment in learning that was related to the degree of ectostriatal damage. However, he interpreted this deficit as being due to the interruption of afferents from the hyperstriatum because another experimental group with extensive hyperstriatal damage showed a similar impairment. This interpretation is supported by the fact that the evoked potentials recorded in the periectostriatal belt by Parker and Delius were of too-short latency to belong to the rotundoectostriatal visual projection pathway (the tectofugal pathway of Karten, 1969) and might, therefore, represent a projection from the hyperstriatal visual area. Benowitz and Lee-Teng (1973) have reported that damage to the lateral forebrain region also affects conventional visual discrimination learning. They used a simultaneous discrimination of a horizontal *versus* vertical bar with a warm—cold airflow as reinforcement for young chicks. Posterolateral lesions, which included the lateral archistriatum and may have affected the region lateral to the ectostriatum, interfered with the early stage of learning so that there was a slower start to learning but no difference in number of trials to criterion. No other areas of the neo- or archistriatum have been reported as affecting the three kinds of visual discrimination learning — imprinting, taste avoidance, and conventional reinforced simultaneous choice — that have been considered here.

The dorsal region of the telencephalon and particularly the hyperstriatum has been extensively studied with regard to discrimination learning. The structures lesioned, usually in uncertain combinations, include the hyperstriatum accessorium, hyperstriatum dorsale and their intercalated nucleus and the hippocampal and parahippocampal areas. In his study of taste avoidance learning, Benowitz also included a group with damage restricted to the dorsomedial hyperstriatal region and this group was impaired on acquisition only. Our own results for the anterior lesion group of chicks, with destruction of the anterior hyperstriatum, except its most medial parts, indicated no impairment of imprinting. The dorsal group, in which again the most medial parts of the hyperstriatum were spared, showed some effect in terms of smaller number of choices but when a choice was made it was usually correct. In our taste avoidance study, damage to the anteromedial and mediolateral regions of the hyperstriatum gave no impairment of discrimination learning. In their study of simultaneous pattern discrimination learning in chicks Benowitz and Lee-Teng (1973) found no evidence of impairment when the dorsomedial hyperstriatum and the hippocampus were damaged. Hodos *et al.* (1973) have also reported no effect on simultaneous brightness and pattern discrimination in pigeons with accessory hyperstriatal damage. Earlier, Stettner and Schultz (1967), using bob-white quail, had reported that lesions of the Wulst with adjacent hippocampal and parahippocampal areas had no effect on acquisition of a striped pattern simultaneous discrimination. In two papers Macphail (1969 and 1971) has claimed that small lesions within the anterior hyperstriatum did not affect simultaneous brightness discrimination, and lesions of the Wulst which spared the accessory hyperstriatum did not affect simultaneous brightness discrimination in pigeons. Stettner (1974) has recently reviewed the neural basis of discrimination and reversal learning in birds and has included some results from unpublished reports and theses. It is probably fair to conclude from the studies reviewed that there is no good evidence of any serious effects of hyperstriate lesions on simultaneous discrimination learning of colour, brightness and pattern.

There are two studies which seem inconsistent with the preceding conclusion in that they report deficits in visual discriminations following hyperstriatal damage. Zeigler (1963a) found that hyperstriate lesions in pigeons gave deficits in both acquisition and relearning of brightness and pattern problems. Pritz *et al.* (1970) found that Wulst lesions in pigeons produced deficits in retention of brightness and pattern discriminations but had no effect on colour discriminations. As in the case of Zeigler, the control birds showed savings in postoperative relearning that were not shown by the lesioned birds, and the pattern problem was relearned more slowly than the brightness problem. It is significant that both these studies used a successive discrimination procedure in which the animal must learn to withhold responses to the negative stimulus. It is important, therefore, to recall that Macphail (1971), using a successive presentation procedure for an operant response study of pigeons, found that birds with hyperstriate lesions gave more pecks to the negative stimulus in a VI schedule than did control birds. Macphail noted that his lesioned birds showed positive contrast so that their tendency to continue responding to the unrewarded negative stimulus was not simply due to an impairment of emotional reactions associated with non-reward and non-rewarded responses. He concluded that these birds were specifically impaired in their ability to withhold non-rewarded responses. Our own limited study of taste avoidance discrimination is an example of a successive discrimination in which hyperstriate lesions did not seem to interfere with learning or the withholding of responses while also leaving the 'emotional' or 'disgust' reactions unaffected. We used a simple colour discrimination and Pritz *et al.* also found no impairment for a colour problem, so that it is likely that there is an interaction with the inherent distinctiveness of cues. Certainly in the case of the normal pigeon, Jones (1954) has pointed out that pigeons have increasing difficulty with position, colour, and pattern problems. Task variables cannot be neglected in the study of the effects of brain lesions in discrimination learning.

The suggestion that impairment of successive discrimination learning is due to difficulties in withholding responses provides at least a conceptual link with a number of avian forebrain lesion studies of learning that we have not yet considered. These are studies of reversal learning and they have all implicated the dorsal region of the forebrain. Stettner and Schultz (1967), using bob-white quail and a pattern discrimination, established that lesions of the Wulst interfered with reversal learning. Stettner (1974) has also cited unpublished studies with similar findings. Extensive damage to the Wulst does not seem to be necessary for reversal learning to be affected since Macphail (1971) obtained deficits in spatial reversal in pigeons with discrete lesions within the Wulst. In a previous study (1969) he found that rather small lesions of the anterior hyperstriatum accessorium did not affect brightness reversal learning. Earlier Stettner (1966) had noted that corticoid damage alone was not responsible for reversal learning defects. Recently Benowitz and Lee-Teng (1973) found that lesions of the dorsomedial hyperstriatum and hippocampus of chicks impaired their learning of a pattern discrimination reversal. So far we have not tested reversal learning in our chicks, either of the imprinting stimulus or of the taste avoidance learning, but there is no doubt that the experiments would be of interest.

It would appear that reversal learning is undoubtedly affected by lesions of the hyperstriatum and that no other telencephalic region has been shown specifically to have this property. Furthermore, as Stettner (1974) has pointed out, there is a direct relationship between the ease of learning both a multiple spatial reversal problem and a successive brightness discrimination and the relative size of the hyperstriatum in a number of species of birds. Gossette (1967 and Gosette et al., 1966) has shown that the ease of learning these two problems decreases with the species tested according to their taxonomic order — Passeriformes, Psittaciformes, and Galliformes respectively. According to Portmann and Stingelin (1961) and the recent survey by Pearson (1972) this is also the order of decreasing relative size of the Wulst. In assessing the significance of these findings for telencephalic functions one should ask what functional requirement successive discrimination and reversal learning have in common, since it is likely that this will be the function performed by the hyperstriatum rather than any particular form of learning at present fashionable in learning theory. We may, for the moment however, conclude that no telencephalic area other than the posterolateral region (neo- and archistriatum) has been shown to be crucial for simultaneous visual discrimination learning and no area other than the anterodorsal region (hyperstriatum) seems necessary for successive discrimination learning and for reversal learning.

Pecking accuracy and visuomotor learning

The present results for pecking accuracy also implicate the lateral region of the telencephalic hemisphere. These lateral lesions damaged the tractus frontoarchistriatalis so that it might be thought that the impairment was a feeding impairment resulting from interference with the feeding circuit described by Zeigler and Karten (1973a and b). They believe that this system deals with touch information relayed from the bill via the sensory trigeminal nucleus to the nucleus basalis which in turn projects through the tractus frontoarchistriatalis and controls mandibulation, *i.e.* transfer of food from the bill tip to the pharynx for swallowing. This process of mandibulation, which we consider to be a tongue action, was not impaired in any of our lesion groups since instances of seeds being picked up and then dropped rather than swallowed were not common and were indifferent among the groups. The impairment in the present study was not, therefore, a mandibulation deficit. Indeed, it is surprising, in view of Zeigler and Karten's notion of the feeding circuit for mandibulation, that the anterior lesion group of the present study, in which the nucleus basalis and its two associated tracts (quintofrontal and frontoarchistriatal tracts) had been included in the lesions, did not show a high incidence of 'pick-up and drop' (seize) errors. In all our experimental groups any seed that was actually lifted off the bench was usually successfully swallowed. All the lesion groups showed some impairment of feeding and this impairment was of the same nature; namely, an error in the ballistic aiming of the pecking action which resulted in a complete miss or in hitting the seed with insufficient accuracy to permit seizing. Such an effect can account for the frequent early reports, evident in the review by ten Cate (1936), of the inability of partially or totally decerebrate birds to feed successfully. Our own chicks regained the ability to feed themselves by pecking into a mass of fine crumb or meal contained in hoppers so that food was seized 'more by luck than by judgement'. One would like to know whether the pigeons in the study by Zeigler and Karten had difficulty in 'mandibulating' because of tongue—bill incoordination or because seed was not grasped in a suitable position for easy transfer to the pharynx.

It may seem strange that such a basic sensorimotor coordination is not handled entirely by brain stem mechanisms. Our suggestion is that it is the fine adjustment of the aiming system that is resident in the forebrain and that this fine adjustment involves learning. Certainly studies of the development of pecking accuracy in chicks have all shown that there is an improvement during the first few days after hatching and that this improvement is accelerated by practice. Perhaps the best of these studies is that of Cruze (1935) who reviewed the earlier studies and showed experimentally that practice affects the efficiency of the whole feeding pattern, primarily by affecting the accuracy of the pecking so that the incidence of hitting without successfully seizing is drastically reduced. Once a seed is successfully seized the act of transferring the seed to the pharynx appears to be a reflex, as is the final act of deglutition. By using wedge prisms, Hess (1956) claimed to have shown that chicks could not adjust the direction of their pecking in relation to a target, although they did show the normal improvement in consistency of aim by achieving a closer clustering of pecks. However, Rossi (1968) has since shown that a reduction in error does occur while wearing such prisms and that overcompensation occurs when the prisms are removed. We have observed similar adjustments in chicks fitted with monocular occluders. Since the act of pecking involves the integration of vision with proprioception and motor functions of the head, neck, limbs and total body it is not surprising that, as shown in the present study, lesions in any region of the forebrain seem to affect pecking accuracy at least temporarily. All the lesion groups of the present study except the lateral group showed signs of recovery, and presumably recovery occurs because undamaged parts of the system can compensate for the damaged part. A ballistic action like pecking can operate from a large number of possible sets of starting positions of head and body. There is great scope, therefore, for the adaptation and readjustment of these sets when damage is sustained. A somewhat similar kind of readjustment occurred in our chicks with monocular occluders in that they launched their pecks from a position with the head cocked to the seeing side instead of from the normal frontal binocular viewing position. In the case of brain damage a possibly comparable case might be that of a skilled darts player afflicted by a stroke who resumes accurate throwing once his throwing action has adapted to a new body stance enforced by paralysis of the side opposite to his throwing arm. Although proprioceptive systems may adjust in this way, distortions or inaccuracies of reporting by the visual system cannot be corrected or adjusted themselves (*cf.* Hay and Pick, 1966; Walls, 1951). It is possible, therefore, that the lateral forebrain lesions in the present chick study interfered with the visual elements in the pecking accuracy mechanism since the lateral lesion group was the only lesion group that showed no signs of recovery of accuracy. The other forebrain areas may be concerned with proprioceptive elements of the pecking adjustments. We suggest that our posterior lesions may have caused a temporary impairment of pecking accuracy by affecting postural coordinations that form a background to the specific visually guided ballistic head and neck action of pecking and that the dorsal and anterior lesions may have affected the head and neck motor coordinations more specifically.

It is interesting and necessary to compare this analysis of lesion studies with evidence on the localisation of feeding mechanisms in the forebrain obtained from brain stimulation. In one of the earlier studies Boyce and Warrington (1898) stimulated the lateral aspects of the cerebral hemispheres in the pigeon and obtained deglutition, pecking movements, and head rotation. Rogers (1922b) obtained bill and pecking movements from stimulating the same area in the pigeon and Kalischer (1905) found bill and tongue movements after stimulating a similar region in the parrot. The relationship of this lateral region to the adjacent archistriatum is indicated by the finding of Phillips (1964) that electrical stimulation of the lateral archistriatum in ducks gave feeding movements of the head, neck and bill. Putkonen (1967) also described bill movements after stimulating this region of the archistriatum in chickens. However, stimulation studies have produced feeding movements or the movements of feeding organs from a wider area of the forebrain. From the anterior neostriatum Delius (1971) has elicited food-paddling and fish-handling movements in herring gulls, and from the deeper neostriatum and palaeostriatum Harwood and Vowles (1966) have produced a facilitation of feeding in ring doves. In more dorsal regions Kalischer (1905) identified the dorsal cortex of the parrot as a motor cortex with tongue, jaw, and voice represented anteriorly, and Cohen and Pitts (1967) have produced contralateral and vertical head movements by stimulating the hyperstriatum of the pigeon. Even the posterior cortex of the pigeon has been stimulated bilaterally by Bremer *et al.* (1939) to produce pecking movements. Clearly, as Delius and Bennetto (1972) point out, a variety of forebrain structures must contribute to the patterning of feeding behaviour. We have already suggested that in our own lesion study the anterior, dorsal and posterior lesions may have resulted in some impairment of pecking accuracy simply because they are involved with the general sensorimotor coordinations that contribute to the specific visually guided ballistic action of pecking and seizing. The kind of compensatory adjustment that we have postulated must take place, when areas involved in general body coordinations are damaged, seems to have occurred in a study by Cohen (1967). He reported that pigeons pecked accurately after receiving lesions of the accessory hyperstriatum and that reflexes, including visual placing, were apparently normal. But he noticed that there was a suggestion of "...a 'slight awkwardness' of head movements which seemed to result from their frequent accompaniment by movements of the neck and shoulder girdle". The available stimulation as well as lesion data, therefore, are not inconsistent with the conclusions we have drawn from our own lesion study. In particular there seems good evidence of a specific visuomotor defect in pecking accuracy resulting from lesions of the extreme lateral telencephalon. This defect could result from damage to resident data processing systems or to interruption of information flowing through this region.

In view of the taste avoidance studies that we have reviewed it is possible that the lateral region of the telencephalon is involved in both the recognition of the object to be pecked and the aiming of the peck. In both cases this area seems to be concerned primarily with handling visual information. This apparent close relationship between a forebrain feeding mechanism and food recognition recalls the work of Shaklee (1921). He concluded from an examination of decerebrate pigeons that the centres for seizing and transferring food to the pharynx were both in the forebrain and that, in the absence of gustatory or touch information from the bill to these centres, a subcerebral ejection reflex operated. Shaklee speculated that the evolution of forebrain centres for feeding in birds was advantageous in allowing adaptation to changing foods. In contrast, such adaptibility had not been a factor in the evolution of drinking mechanisms which have remained entirely subtelencephalic, as is evident from the normal drinking shown by decerebrates. Of course, this is not necessarily true since avoidance of bad tasting or electrified water with a discriminable associated visual stimulus certainly occurs (*cf.* Shettleworth (1972) and the previously cited taste avoidance studies). Nonetheless, the main point of Shaklee's speculation stands and the present data also suggest a common forebrain region for the integration of visual information concerning food and the adjustment of pecking for feeding.

Arousal and responsiveness

Both the dorsal and posterior lesion groups of the present study gave weaker withdrawal responses to visual looming than did the control group. It is interesting to recall the previously cited work of Arduini *et al.* (1948) in which a type of looming response was evoked by an approaching hand in decerebrate pigeons after application of strychnine to the optic tecta. The present results, therefore, could be explained if the posterior and dorsal regions of the telencephalon of the normal bird facilitate the brain stem mechanisms for the looming response. In view of the fact that the tendency to withdraw from a contracting visual stimulus was also reduced in birds with lesions of these two areas we must conclude that this facilitation is not specific to visual looming but applies to any rapid changes in the area of retinal illumination.

None of the lesions in the present study seemed to affect the startle reflex response to a sudden loud auditory stimulus. It is most probable that the startle reflex in mammals is mediated by brain stem mechanisms and it has been seen in human anencephali, *cf.* Jung and Hassler (1960). In his original study of the startle reaction, Strauss (1929) recognised a primary reflex response involving the brain stem and a secondary more variable emotional or voluntary response that involved cortical mechanisms. The startle response scored in the present study of chicks corresponds with the primary response and it would seem that, in birds also, this response is a brain stem mechanism. It may be recalled from the Introduction, that totally decerebrate birds are aroused by strong auditory stimulation. Whether the telencephalon has any effect on the startle reflex in birds might best be decided by response thresholds rather than by measuring response strength as was done in the present study. What is clear from the present study, however, is that the poststartle freezing response was affected by forebrain lesions. In fact the two lesion groups, dorsal and posterior, that showed reduced looming responses also showed reduced duration of the poststartle freezing response. We suggest that this response corresponds with the secondary acoustic startle responses of mammals and similarly is an 'emotional' and 'variable' response that depends on telencephalic influences on brain stem processes.

Both freezing in response to sudden strange stimulation and withdrawal or flight from a rapidly approaching strange object are clearly elements of what is commonly called fear behaviour in birds. It is relevent, therefore, to consider other studies of the role of the telencephalon in fear responses. Phillips (1964) using mallard ducks and Zeier (1968) using the pigeon have shown that lesions of the archistriatum can drastically reduce fearful behaviour. Electrical stimulation of the archistriatum commonly results in alarm, escape and flight responses (see Phillips (1964) for the mallard; Åkerman (1966b) for the pigeon, and Putkonen (1967) and Phillips and Youngren (1971) for the chicken). The posterior lesions of the present chick lesion study never involved bilateral damage to the archistriatum but usually included some unilateral damage. It is possible that the reduction in the looming and freezing responses of the posterior lesion group was due to some archistriatal damage, but we do not think this probable. Our view is strengthened by the fact that the dorsal lesion group was affected even more than the posterior group and there was no possibility of archistriatal damage in the dorsal lesions. There is no evidence from lesion or stimulation studies that this dorsal region is essential for fear and flight responses as seems to be the case for the archistriatum.

There is, however, some evidence in the literature of a reduction in responsiveness to stimulation in birds with hyperstriatal damage. Cohen (1967), in his study of lesions of the dorsal and accessory hyperstriatum in the pigeon, noted that responsiveness to light flashes and to slowly moving stimuli was reduced and that intense visual or tactile stimulation was required to elicit escape behaviour. In the case of auditory stimulation, unfortunately, both experimental and control animals responded poorly so that no differences were apparent. Earlier, Zeigler (1963b) had reported that pigeons with hyperstriatal lesions showed a reduced and constant level of locomotor activity when kept in an activity cage under constant conditions. Under the same conditions normal pigeons and ones with other forebrain region damage showed a decline in activity over time. In the present chick study it was the posterior rather than the dorsal lesion group which showed least tendency for habituation of both looming and freezing to repeated stimulation, but these effects were not significant and have not been presented here. We do not believe that the reduced responsiveness seen in the present study was due to hyperstriatal damage because the anterior lesion group had more extensive destruction of the hyperstriatum than did the dorsal group yet it was not significantly different from the control group in looming and freezing responses.

In addition, there is evidence implicating the hyperstriatum in orientation of the head rather than in the facilitation of fear and flight responses. Thus Adamo and Bennett (1967) found that chickens with hyperstriatal lesions made errors in orienting the head towards sources of sound and failed to show habituation of this orienting between test sessions. Cohen and Pitts (1967) have shown that electrical stimulation of the hyperstriatum in pigeons produces a variety of head orientation movements. In a subsequent paper, Cohen (1967) suggests that the hyperstriatum mediates somatic orienting to visual stimuli by instigating the head turning that leads into body turning. We shall later present a rather different interpretation of the head orienting functions of the hyperstriatum. For the present, however, it should be apparent that neither archistriatal nor hyperstriatal damage can account for the observed reduction in the two 'fear' responses measured in our chick lesion study.

It is most important to note that the same two lesion groups (posterior and dorsal) were also least responsive in the imprinting tests in that they scored 'no-choice' more frequently than any of the other groups. This suggests that there was a reduction in responsiveness to visual and perhaps auditory stimulation in general rather than any specific effect on the response mechanisms of fear and flight. We conclude this consideration of our chick lesion data in relation to arousal and responsiveness by pointing out that the dorsal and posterior neostriatal areas involved in the lesions of the dorsal and posterior groups probably contribute to the tractus archistriatalis dorsalis and may project to the archistriatum by this pathway. Neostriatal areas immediately above the archistriatum may also project into it by local pathways. It is possible that these neostriatal areas affect approach and withdrawal responsiveness by their action on the archistriatum, which is known to be involved in fear and flight behaviour and which we shall suggest later may also be involved in approach responding.

FUNCTIONS OF THE AVIAN STRIATAL BODIES

In this section we survey the functions of the avian striatal regions in terms of the two functional concepts we have used in interpreting the results of our chick lesion study; namely, the arousal and facilitation of brain stem response mechanisms and the sensorimotor coordination of specific orientation behaviour.

The neostriatum

Stimulation of the neostriatum either produces elements of feeding behaviour or facilitates feeding and preening patterns, and the studies concerned have been cited already in the section on pecking accuracy and visuomotor learning in the chick. We suggest that the neostriatum has the sensorimotor coordinating circuits that guide and adjust the movements of body parts, particularly the action structures of neck, bill and limbs, in relation to appropriate foci of external stimulation or objects. These coordinations include the fine adjustments involved in feeding and in preening, and it is interesting that the anterior neostriatum is most highly developed in birds which have bills adapted for manipulation, such as the Psittaciformes (*cf.* Stingelin and Senn, 1969). Thus there may be a crude arrangement of these coordinating orientation mechanisms with the neck, head and bill systems in the central and frontal neostriatum. In our chick lesion study the anterior and dorsal region lesions impaired pecking accuracy somewhat more seriously than did the posterior lesions. Both the anterior and dorsal lesions involved the hyperstriatum bilaterally as well as the underlying neostriatum.

Nevertheless we suggest that it was the damage to the neostriatum rather than the hyperstriatum that was responsible for the pecking inaccuracy and that the trend to recovery of aim occurred because of the unilateral nature of the neostriatal damage. The extreme anterior neostriatum that was damaged bilaterally we suggest is concerned with bill manipulatory functions rather than with the neck and head aiming system which we have postulated. As we shall indicate in a later section, the overlying hyperstriatum may also be involved in the control of head orientations and body postures, but in a more restricted and specific manner than is the case in the system we are postulating for the neostriatum.

The orientations and adjustments that are involved in the responses to looming and startle were affected by posterior and dorsal lesions in our chicks. This is consistent with the location of systems involving limb and body actions in the posterior and central parts of the neostriatum. Of course, such orientations and adjustments of body and limbs are concerned not only with the defensive responses of withdrawal and freezing, but also with social behaviour involving approach orientations such as occur in filial responses and in courtship. The involvement of the posterior and central neostriatum in approach responses to an imprinting object is indicated by the reduced responsiveness of the posterior and dorsal lesion groups of our chick study. The importance of the lateral neostriatum in the visuomotor orientations of feeding and imprinting has also been shown by the deficits in pecking accuracy, taste learning and imprinting already described in chicks with lateral forebrain lesions. The lesion studies of Rogers (1922a) and Beach (1951) have previously shown that extensive damage to the neostriatum is required for substantial impairment of the reproductive behaviour patterns of courtship, mating and parental care in pigeons.

There is little doubt that the neostriatum receives all the information necessary for the orientation of body parts to specific objects. Auditory projections to the mediocaudal neostriatum have been identified by the neurophysiological recording studies of Biederman-Thorson (1970a and b), Erulkar (1955), Harman and Phillips (1967), and Leppelsack and Schwartz-kopff (1972). Auditory responses were also detected in the frontal neo-striatum by Harman and Phillips, and Bolton (1971b) has cited reports of responses in the ectostriatum and in nucleus basalis. It is clear from the papers of Karten (1968) and Boord (1969) that the anatomically established auditory projection ends in Field L of Rose in the mediocaudal neostriatum. Visual projections to the neostriatum have been indicated by visually evoked potentials recorded by Revzin and Karten (1967) in the ectostriatum, by Kuhlenbeck and Szekely (1963) and by Parker and Delius (1972) in the neo-striatum caudale, and by Parker and Delius (1972) in the periectostriatal belt. Anatomical visual projections to the ectostriatum have been described by Karten and Hodos (1970). Somaesthetic evoked responses to stimulation of the face and beak were reported by Erulkar (1955) to occur in the neostriatum caudale, but more recently Witkovsky *et al.* (1973) have localised beak and buccal cavity sensation in the nucleus basalis.

In the case of the visual projection to the ectostriatum, Revzin (1970) reported that the receptive fields of ectostriatal cells are very extensive and are suited to the detection of movement rather than to the registration of static detail. In the auditory projection Field L, Biedermann-Thorson (1970b) found that units tended to be driven by hisses, clicks and squeals rather than by pure tones. Both these physiological findings are not inconsistent with an object detection and orientation function for the neostriatum. The neostriatum and associated structures of ectostriatum and nucleus basalis have the sensory areas necessary for the functions we have postulated, *i.e.* for the initiation and organisation of the orientating actions of head, bill and limbs in relation to specific objects in space. The processes of initiating and organising specific orientation movements (*e.g.* adduction or abduction) to specific stimuli (*e.g.* stimulus A but not stimulus B) are the processes of discrimination and discrimination learning. It is interesting to recall our review of lesion studies of discrimination learning which showed that simultaneous visual discrimination was seriously affected only by lateral forebrain lesions involving the neostriatum and the ectostriatal region. Simultaneous presentation of positive and negative stimuli in separate spatial locations requires the orientation behaviour systems that we have tentatively assigned to the neostriatum.

The palaeostriatum

We consider the palaeostriatum to be the motor coordinating system which integrates the neostriatally organised 'body-part to object' orientation patterns with the 'whole body posture' control systems so as to allow co-ordinated 'whole animal to object' interactions. The lesion studies of Muskens (1929 and 1931) have indicated palaeostriatal sites where damage results in rearing backwards, falling forwards, circus movements, and rolling actions in pigeons. Putkonen (1967) has produced rearing backwards in chickens by electrical stimulation of palaeostriatal sites and Goodman and Brown (1966) have similarly produced turning movements in pigeons. Electrical stimulation in this area in pigeons has been reported by Åkerman (1966a and b) to produce organised patterns of attack, flight, and courtship behaviour. It is difficult to interpret the significance of these results because the sites stimulated usually border on the archistriatum or the septal area where stimulation commonly results in fear and flight or attack behaviour, *cf.* Putkonen (1967) and Phillips and Youngren (1971). There is also the problem that so much of the palaeostriatum is crossed by ascending and descending fibre tracts that it is difficult to be sure when the responses to electrical stimulation arise in the area.

The recent study of the organisation and projections of the palaeostriatal complex of the pigeon by Karten and Dubbeldam (1973) supports the traditional view that the palaeostriatum augmentatum is homologous with the mammalian putamen—caudate complex and that the palaeostriatum primitivum resembles the globus pallidus. These workers have also proposed that the nucleus intrapeduncularis is equivalent to the internal division of the globus pallidus. Stimulation of the mammalian globus pallidus and putamen has produced flight or attack behaviour, *e.g.* Ursin and Kaada (1960), and turning movements, *e.g.* McLennan *et al.* (1964) and Montanelli and Hassler (1964). Stimulation of the caudate nucleus in mammals commonly gives turning movements, *e.g.* Forman and Ward (1957), Laursen (1962) and McLennan *et al.* (1964). The kind of function we are proposing for the palaeostriatal system is similar to the type of function traditionally (*e.g.* Wilson, 1928) advocated for the striatal parts of the extrapyramidal system in mammals, *i.e.* the integration of specific somatomotor patterns generated in the cortical system with the general body posture patterns operating in the brain stem. In the case of birds we are proposing that the specific orientation patterns are organised in the neostriatum which in this respect is functioning more like the mammalian cortical system. Karten and Dubbeldam (1973) have indicated that both the neo- and hyperstriatum project topographically on to the palaeostriatum augmentatum, which in turn projects to the palaeostriatum primitivum and the nucleus intrapeduncularis. Thus the palaeostriatum can receive organised patterns for specific actions from the neostriatum. It can receive information on the state of the body via thalamofrontal tracts of the lateral forebrain bundle (*cf.* Bolton, 1971a) and so could integrate the specific actions with body posture states and transmit the overall pattern in guidance of brain stem action systems via the tractus cerebellaris et tegmentalis (which Karten and Dubbeldam have called the ansa lenticularis). Obviously there would be feedback between brain stem and palaeostriatum to generate a continuous integration between body postures and the changing specific patterns of orientation generated in the neostriatum in response to changing exteroceptive input. There is much speculation in these suggestions and experimental data are badly needed for this region of the avian striatum.

The archistriatum

In a recent study of the archistriatum of the pigeon, Zeier and Karten (1971) made a distinction between a 'somatic-sensorimotor' portion forming the anterior two-thirds of the structure and a 'limbic' portion forming the medial and posterior part. The latter is probably equivalent to the mammalian amygdala and outflows to the hypothalamus chiefly by the tractus occipitomesencephalicus. The anterior and intermedium archistriatum both outflow through this same tract to the brain stem reticular formation, particularly to its lateral parts, and down to pontine and even spinal levels.

The archistriatum receives its afferents principally from the neostriatum via the tractus frontoarchistriatalis and the anterior commissure to the anterior archistriatum, and by the tractus archistriatalis dorsalis to the archistriatum intermedium. Apart from Zeier (1971), experimental studies of the archistriatum have not distinguished between the 'limbic' and the 'sensorimotor' subdivisions. Some have differentiated lateral and medial areas which are not easily translated into Zeier and Karten's specific divisions. It is probable that these lateral areas are equivalent to the anterior and part of the intermedium divisions while the medial areas also include part of the intermedium as well as the medial archistriatum.

In general, electrical stimulation of the archistriatum seems to produce the behaviour already described for stimulation of the palaeostriatum but with a greater probability. Both attack and fear responses in the form of sneaking, flight, or panic escape, have been reported by Putkonen (1967) and Phillips and Youngren (1971) in chickens, by Åkerman (1966b) and Goodman and Brown (1966) in pigeon, by Maley (1969) and Phillips (1964) in mallard, and by Phillips (1968) in the peach-faced lovebird. Phillips (1964 and 1968) found that lesions of the archistriatum, and especially its more medial parts, resulted in a loss of fearfulness, *i.e.* of orienting away from and fleeing from sudden or strange stimulation. This finding led Phillips (1968) to ascribe an excitatory or amplifying function to the archistriatum. We are sympathetic to this idea and suggest that it applies to orienting behaviour. Simple turning movements are also commonly obtained from stimulation of the archistriatum (*cf.* Putkonen (1967) for the chicken and Goodman and Brown (1966) for the pigeon). The archistriatum, therefore, could serve to amplify specific turning movements in relation to specific stimuli by its action on brain stem reticular excitatory systems and on supporting hypothalamic autonomic systems. Certainly there seems to be evidence for involvement of the more medial part of the archistriatum in turning away in avoidance and flight from specific forms of stimulation.

In a number of papers, Zeier (1968, 1969 and 1971) has reported the effects of lesions of the archistriatum on operant behaviour in pigeons. We think that his results can also be understood in terms of the facilitation by the archistriatum of orientation tendencies. In the most recent study Zeier found that lesions of the posteromedial part of the archistriatum resulted in increased responding in a differential reinforcement of low rate (DRL) schedule which thus led to an impaired performance of the task. In terms of orientation, these animals failed to turn away and leave the key when not rewarded and instead remained pecking and so further delaying reward. This is another case of medial lesions reducing the tendency to turn away from or leave what has become an aversive stimulus or situation. Zeier also found that birds with lesions of the 'somatomotor' part of the archistriatum showed increased responding on a VI schedule and fewer approach responses on a DRL schedule. Both these effects could be due to a decreased tendency to respond by orientation and approach to distance stimuli not in the existing attentional field. Thus while at the key and pecking in the VI schedule these birds were less likely to reorient to any extraneous stimulation, and when away from the key in the DRL condition they were less likely to reorient back to the key. In our view the archistriatum facilitates any bodily reorientation tendency established in the neostriatum in response to novel or significant stimulation. Lack of archistriatal activity should reduce or abolish the likelihood of reorientation to such stimulation and this could account for the 'monotonous pecking' which Zeier saw in the VI schedule. Some support for this explanation is given by the observation of Zeier and Akert (1968) that electrical stimulation of the archistriatum in pigeons engaged in a key-pecking task produced attentive responses which interfered with the ongoing pecking and led to a reduced response rate. The implication for archistriatal function is that approach orientation if facilitated as well as withdrawal orientation. Perhaps there is some degree of spatial arrangement of these opposite orientation or turning tendencies with approach mechanisms operating in the more lateral part and withdrawal in the more medial part, (the medial part in this sense including some archistriatum intermedium or 'somatomotor' area).

The chick lesion data we have presented earlier are also of interest here because they have shown an impairment of approach responses to an imprinting stimulus and impairment of withdrawal or fear responses to visual looming and acoustic startle. This reduced responsiveness resulted from lesions of dorsal and posterior telencephalic regions that probably project to the archistriatum via the tractus archistriatalis dorsalis. If this is the case, then the impairment in responsiveness is consistent with the hypothesis that neostriatal systems for orientation behaviour, in addition to their direct output via the palaeostriatum, also have a parallel output to the archistriatum which has a facilitatory or amplifying function for these specific patterns of orientation to specific stimuli or objects.

Lesions of the lateral and frontal telencephalon did not have such effects on the responsiveness of chicks. We have suggested that the anterior neostriatum is concerned with the organisation of bill and head orientation and it is possible that with gross stimulation this system was not a significant factor in the responses we tested. This anterior region projects to the anterior archistriatum by the tractus frontoarchistriatalis and this may be the parallel facilitatory pathway for the head and bill orientation system. Certainly there is some evidence that electrical stimulation of the archistriatum gives bill movements in the chicken (Putkonen, 1967) and feeding movements of the neck, head and bill in mallards (Phillips, 1964). In the studies by Rogers (1922b) and Boyce and Warrington (1898) electrical stimulation of the lateral surface of the cerebral hemisphere at the level of the anterior commissure gave head and bill feeding movements in pigeon. It is possible that these effects were of archistriatal origin or induced via its afferent tractus frontoarchistriatalis.

We suggest that the archistriatum facilitates somatic and autonomic motor systems of the brain stem according to the specific orientation movement patterns organised in the neostriatum. These specific orientation actions are contemporaneously operating via the palaeostriatum directly on subtelencephalic action systems. The archistriatal actions are thus specific object-oriented facilitatory actions involving patterned excitation and inhibition and operating through brain stem somatic and autonomic facilitatory systems, especially the lateral reticular formation and the medial hypothalamus. There is some evidence for a separation of approach facilitation in the lateral and withdrawal facilitation in the more medial parts of the archistriatum.

The hyperstriatum

The hyperstriatum receives visual sensory projections independently of the neostriatum, according to the anatomical studies of Powell and Cowan (1961), Karten (1969) and Boord (1969). There is evidence of a short-latency visual projection in the pigeon from the evoked potential studies of Rougeul (1957) and Parker and Delius (1972). Auditory responses have been recorded by Adamo and King (1967) in the hyperstriatum of the chicken and cutaneous responses have been reported by Delius and Bennetto (1972) in the pigeon hyperstriatum. We have already cited the work of Cohen (1967), Cohen and Pitts (1967) and Adamo and Bennett (1967) as suggestive evidence for the role of the hyperstriatum in specific orientations to specific stimuli. It is interesting, therefore, that the visual fields of the Wulst studied by Revzin (1969) and by O'Flaherty and Invernizzi (1972) tend to be small and specific in contrast to those reported for the ectostriatum. Revzin found that movement gave the optimal responses but also found two 'horizontal line' detectors. This is the only evidence for specific visual fields within the forebrain that could serve as a basis for detailed pattern vision. The anatomical findings of Hunt and Webster (1972) are most pertinent in this respect for they show that parts of the dorsolateral retinal visual input project bilaterally to the Wulst. This organisation of retinal projection fields could serve the binocular field function which in mammals is served by ipsilateral retinal visual projections and/or callosal projections. Perisic *et al.* (1971) have recorded visually evoked potentials both contra- and ipsilaterally in the pigeon's Wulst.

In behavioural studies Nye (1973) has examined functional differences between the frontal and lateral fields in the pigeon and hypothesised "that pigeons do not possess the neural capability required to learn to use the information contained in laterally located stimuli to directly control pecking behaviour". In fact Nye found that pigeons performed poorly in key-pecking tests for the discrimination of motion, colour and luminance as well as of image structure when the stimuli were located in the lateral visual fields. On the other hand, studies in which the pigeon is required to approach the stimulus from a distance have obtained discrimination performances adequate for the determining of visual acuity, *e.g.* Chard (1939) and Blough (1971). Friedman (Chapter 10 in the present volume) gives observations on the way doves will turn and approach food objects whose image must fall initially on the lateral retina. It is significant that where the lateral visual field is employed whole body orientation is involved, and, according to our hypothesis, this orientation is a function of the neostriatum. We suggest that precise, local and sustained orientation to restricted aspects of visual forms located within the frontal binocular field are mediated by the hyperstriatum. This precise orientation could also apply to auditory functions because the owl, which uses accurate acoustic localisation for prey capture, has a highly developed hyperstriatum (*cf.* Stingelin and Senn, 1969). We propose that the hyperstriatum organises a fine degree of orientation and produces arousal to fine stimulus differences because it is receiving detailed information of more restricted and subtle aspects of the environment, namely, the detailed features within a stimulus complex or object. In an interesting study of cortical function in tree shrews, Jane *et al.* (1972) have used stimulus patterns composed of a small array of lights requiring localised spatial orientation and attention. They interpreted the observed deficits in transfer of learning following cortical damage to an impairment of what they have called 'intra-modality attention'. Tectal lesions in the same species appeared to give deficits in attention to objects or in what Jane *et al.* term 'intermodality attention'. This approach to cortical function is similar to our approach to the avian striatal bodies and to the 'interobject' and 'intraobject' orientation functions for the neostriatum and hyperstriatum.

Orientation to specific details of an object or stimulus complex entails maintenance of attention to the whole complex and gives the opportunity for responding differentially to changes within the stimulus complex over time. This is the kind of function that appears to be displayed in a limited form by the teleost forebrain in the studies of Savage previously cited. We suggest that the avian hyperstriatum is concerned with the holding of orientation to the same locus of stimulation over a period of time during which consummatory responding may not be possible and in which alternative orienting behaviour must be withheld. Minor readjustments or reorientations within the hyperstriate orientation or attention field do not involve changes in general body orientation. In this way specific aspects of a stimulus complex, which may be temporally separate, can be organised in relation to a single response pattern. Inability to sustain orientation or attention could account for the deficits seen in successive discrimination learning performance in birds with lesions of the hyperstriatum. Successful learning in this situation requires maintained orientation to the stimulus source from trial to trial while giving differential orientation responses to the specific stimulus being presented at the same stimulus position. This is the function we are proposing for the hyperstriatum, as a natural consequence of its local sensory and action fields.

Reversal learning deficits, like successive learning inabilities, are character-istic of birds with hyperstriate lesions. At first sight reversal learning difficulties in such birds might seem best explained by postulating a difficulty in withholding a previously rewarded response *i.e.* by a response perseveration hypothesis (*cf.* Stettner, 1974). Macphail (1971) has reported a decrease in withholding non-rewarded responses in pigeons with hyperstriate lesions, but earlier (1969) he had also reported an increase. Stettner (1974) has re-examined some of the evidence from an earlier study of reversal learning in quail with lesions of the Wulst (Stettner and Schultz, 1967) and found that there was a tendency for the lesioned birds to make more errors than control birds in the trials made before they had reached the 50% correct response level following reversal. This would be expected on a response withholding difficulty hypothesis. However, by far the greater number of errors was made after this chance response level had been reached and these errors were due to continued position responding by the lesioned birds. This indicates that the most severe impairment in reversal is not caused by perseveration (*i.e.* failure to withhold the previously rewarded response), but rather it is caused by a failure to maintain a general orientation or attention to the significant aspects of the stimulus display after the 'specific stimulus aspect—specific response' relationship or orientation is disrupted by non-reward. Stettner makes essentially the same point in terms of attention by quoting the suggestion of Cohen (1967) that the reversal learning deficits reported by Stettner and Schultz (1967) in birds with hyperstriate lesions were caused by their inability to shift attention from one aspect of the stimulus complex to another.

A detailed explanation of the reversal learning deficit due to hyperstriate lesions might be given in terms of our orientation hypothesis as follows. The initial simultaneous discrimination learning establishes a particular general orientation response to the significant stimulus aspect and this is a neostriatal function which hyperstriate damage does not impair. Reversal of the significance of this stimulus aspect then destroys the neostriatally organised orientation pattern and the bird reverts to a position response orientation. In doing this it converts the problem into a successive discrimination problem, *i.e.* by tending to respond to one position it can only show discrimination by developing a 'go—no-go' response. This, in our terms, is a differential orientation response to the temporally changing aspect of the stimulus, and this is a hyperstriatal function. When the appropriate orientation to the changing stimulus aspect is established, it can operate in either stimulus position and the reversal is established. Further reversals are accomplished because the hyperstriatal organisation maintains general orientation to the stimulus aspect which is changing and organises the local specific reorientation required by the changing stimulus significance.

Any system for maintained orientation to a particular localised area must involve maintained postures or at least a very restricted range of movement and so must involve massive patterned inhibition. It is important, therefore, that the main direct telencephalofugal projection of the hyperstriatum is the tractus septomesencephalicus, (*cf.* Adamo, 1967) and that this tract appears to reach the medial reticular formation (*cf.* Karten and Dubbeldam, 1973) and so could be responsible for the kind of patterned inhibition we have described. An inhibitory function for the posterior hyperstriatum has already been suggested by Macphail (1969). The tractus septomesencephalicus is probably equivalent to the mammalian fornix system (*cf.* Karten and Dubbeldam, 1973) and is also the efferent tract for the avian hippocampus. The suggestion that it has an inhibitory function recalls hypotheses of inhibitory functions for the mammalian hippocampus, *e.g.* Kimble (1968 and 1969) and Douglas (1967). We are suggesting a double outflow from the telencephalon to the brain stem; one inhibitory to general body orientation from the hyperstriatum (and perhaps involving the hippocampus too) and the other facilitatory to general orientation responses from the archistriatum (probably in part equivalent to the amygdala). This is remarkably like the dual hippocampal—amygdaloid functions in orientation behaviour postulated for the mammal by Douglas and Pribram (1966) and Douglas *et al.* (1969), although arrived at by a different theoretical approach to different evidence. Pribram *et al.* (1969) and Spevack and Pribram (1973) have recently attempted to analyse the nature of reversal learning impairments in monkeys with combined hippocampal and amygdaloid lesions, and have emphasised the reversion to position responding that occurs. Our notion that hyperstriatal inhibition serves to restrict general orientation may also bear comparison with the findings of Vanderwolf (1971) that the mammalian hippocampus seems to be active when the animal is engaged in automatic consummatory behaviour, during alert static behaviour, and in sudden halting (all involving restricted general movements and a maintained posture) and is inactive, as judged by its regular slow rhythmic electrical activity, during voluntary free movements which involve gross bodily movements and orientation behaviour.

Of course, the avian Wulst must be excitatory in so far as it organises and maintains orientations within a stimulus complex. Macphail (1969) has also suggested, from his lesion work, that the anterior hyperstriatum has a facilitatory action. It seems possible that the anterior hyperstriatum, because of its close relationship with the anterior neostriatum and the head orientation system we have postulated for this region, is indeed the hyperstriatal area that serves the active intraobject orientation functions. The patterned outflow from this area probably projects by local paths to the related neostriatal orientation systems to enable interaction with head orientation and bill activities and sensations. We can use this hypothesis of hyperstriatal functions to account for some of the otherwise puzzling effects of hyperstriate lesions in pigeons reported by Reynolds and Limpo (1965) for operant behaviour, just as we have earlier tried to explain the operant behaviour phenomena following archistriatal damage described by Zeier. Reynolds and Limpo found no change in responding during a fixed-ratio schedule but during fixed-interval periods the lesioned birds tended to shorten or miss the usual initial pause following reward and instead pauses occurred several times during high rates of fixed-interval responding. The occurrence of these latter pauses could be due to the lack of hyperstriatal inhibition of general orientation helping to maintain orientation to the key by suppressing any tendency to reorient to extraneous stimulation. The absence of the initial pause in key-pecking following reward in the fixed-interval period could be due to the impairment of a hyperstriatal system for temporal aspects of the stimulus — response organisation and the free uninhibited action of the neo- and archistriatal orientation and facilitating systems producing immediate reorienting to the key.

In summary, then, the avian hyperstriatum may be excitatory in maintaining orientation within a stimulus field or object and inhibitory in suppressing gross orientation tendencies facilitated by the archistriatum in response to extraneous stimulation. We recognise that the system we propose takes no account of the obvious subdivisions of the hyperstriatum and cannot be an adequate account of hyperstriatal function. Nor can it give wholly adequate explanations for the existing studies of the effects of hyperstriatal damage since these too take no account of the various parts of the hyperstriatum. Nonetheless, we feel that our approach to the functions of the hyperstriatal structures may suggest a number of new experimental approaches that could be combined with more specific anatomical and physiological approaches.

CONCLUSION — AROUSAL, ORIENTATION AND THE TELENCEPHALON

The arousal and orientation hypothesis of telencephalic function in birds is summarised in Fig. 4. This is a parasagittal section characteristic of the telencephalon of the pigeon and the domestic chick, cf. Karten and Hodos (1967) and Salzen and Williamson (unpublished atlas). This particular section has been chosen because it shows, at least in part, all the primary striatal bodies and interconnections which we have considered in our experimental study and theoretical discussion. In addition we have indicated the functions we have postulated for the various striatal bodies, the interrelationships of these bodies and their efferent pathways and brain stem effects. The diagram and the hypothesis need additional content and comment if they are to apply to the telencephalon rather than to the striatal bodies only. In this concluding section we attempt to provide this additional content and comment, although there is little reliable experimental information on the functions of non-striatal telencephalic structures in birds.

The palaeostriatal (p.) region shown in Fig. 4 is part of the p. augmen-

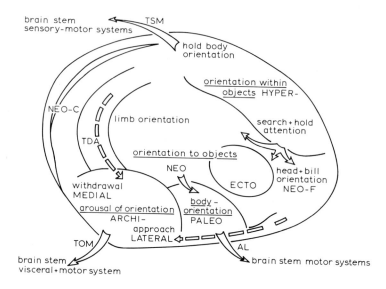

Fig. 4. The diagram is a semi-schematic representation of the structures and functions of the avian telencephalon. It is based on a parasagittal section of the pigeon and domestic chick brain and shows three levels of orientation systems corresponding with palaeo-, neo-, and hyperstriatum, and a parallel facilitatory system operating through the archistriatum. Some degrees of differentiation of function within these systems are indicated. These include a local orientation and inhibition of general reorientation in the hyperstriatum, a head and limb orientation separation in the neostriatum, and a separation of approach and withdrawal facilitation in the archistriatum. The chief interaction pathways and the efferent operations to brain stem systems are also indicated. The orientation hypothesis of telencephalic functions summarised in this diagram can only be fully appreciated in conjunction with the textual description.

tatum. Experimental data are needed before any hypothesis as to the different functions of p. augmentatum, p. primitivum and nucleus entopeduncularis can be formulated. Our discussion and hypothesis treat the palaeostriatum as a functional unit and we have no reason to believe other than that the anatomical subdivisions may correspond with subfunctions of the general function which we have postulated. We have not considered the lobus parolfactorius which Karten and Dubbeldam (1973) included in their study of the palaeostriatum. These authors noted that the efferent flow of this body is via the medial forebrain bundle to the hypothalamus. Electrical stimulation of this area in the chicken has been found by Putkonen (1967) to elicit fear and panic, much as in the case of the medial archistriatum. The lobus parolfactorius may, therefore, be part of the negative orientation facilitation system which we have proposed for the medial archistriatum.

In the case of the archistriatum our discussion and hypothesis have taken into account the anatomical and functional subdivisions for which there is reasonable evidence. This subdivision has been indicated schematically in Fig. 4 because the actual area in this section is part of the lateral archistriatum only.

The neostriatum is not easily subdivided anatomically although frontal and caudal regions are sometimes distinguished. In Fig. 4 we have indicated the frontal and caudal areas in relation with their hypothesised functions. Nucleus basalis is not shown in the diagram but in our hypothesis it is subsumed with the extreme frontal neostriatum in the orientations involving bill manipulations. The ectostriatum is included in Fig. 4 but in our hypothesis it is subsumed under the orientation system of the neostriatum. Its specific function has been suggested implicitly in our discussion as being the visual sensory input of the orientation system probably comparable with the auditory input in Field L of Rose, which lies more medially than the plane of the parasagittal section of Fig. 4.

The several subdivisions of the hyperstriatum (ventral, dorsal, accessory and intercalated) are not shown in Fig. 4 and have not been differentiated in our hypothesis and discussion of hyperstriatal functions. There is no significant and substantial experimental evidence on which any differentiation of function corresponding with these anatomical subdivisions can be based. Experimental data for subdivisions of all the striatal bodies, but particularly for the hyperstriatum, are needed and such studies might profitably seek specific functional aspects of orientation mechanisms in specific subdivisions.

The cortical and olfactory areas are not indicated in Fig. 4 and have not been dealt with in our hypothesis. Anatomical studies of these areas are summarised in the review of the avian brain by Pearson (1972) but apparently critical experimental data for these areas are lacking. Usually lesions of striatal areas also involve damage to the overlying corticoid areas. In our own chick lesion study the dorsal and posterior lesions must also have included dorsal and dorsolateral cortex and it is possible that this contributed to the observed losses in responsiveness. The lateral lesions may also have damaged dorsolateral cortex, and the normal responsiveness of this experimental group makes it unlikely that the reduced responsiveness of the dorsal and posterior groups was due solely to the cortical damage.

The parahippocampal cortex and to some extent the hippocampus are commonly involved in lesions of the hyperstriatum. It is not possible to separate the functions of these areas on the experimental evidence that is available. These areas appear to share the tractus septomesencephalicus for their efferent pathways and could well be involved, perhaps in different ways, in the inhibition of general body reorientation which we have postulated as the function of the posterior hyperstriatum. We have already commented on the parallels between this function and similar ones evident in studies and hypotheses of the mammalian hippocampus. For the present, therefore, we must implicate the avian hippocampus together with the hyperstriatum in the inhibition of reorientation.

We have not considered olfactory structures and associated cortex. Apart from specific olfactory functions these structures might also be involved in arousal and orientation processes, since Wenzel and Salzman (1968) found that damage to the primary olfactory system in pigeon gave deficits in learning to feed from a hopper, in learning to peck a coloured key for food, and in the transfer of this response to a second coloured key.

The afferent tracts of the telencephalon are not apparent in Fig. 4. The primary afferent tracts are in the lateral forebrain bundle and include tractus thalamofrontalis and tractus quintofrontalis (cf. Bolton, 1971a). We have reviewed the telencephalic areas that receive sensory projections from this afferent system and shown that both neostriatum and hyperstriatum receive information independently. We have not considered the extent and distribution of thalamic afferents conveying proprioceptive and interoceptive information to the telencephalon. This important area needs to be investigated and included in any theory of telencephalic functioning as this information is necessary for determining the occurrence, intensity, and sign (i.e. adient or abient) of the orientation that is organised in the telencephalon. In more conventional terms, we have not considered the contributions of internal state, drive, and motivation to telencephalic orientation processes. We do not know if they, like exterosensory information, are represented at all striatal levels, i.e. the palaeo-, neo-, and hyperstriatum, or not at all but rather in the medio-basal region involving lobus parolfactorius, septal and preoptic areas.

The main efferent tracts of the telencephalon are indicated schematically in Fig. 4, although none are in reality present in such a parasagittal section. We have followed the recent summary of telencephalic efferent tracts given by Karten and Dubbeldam (1973) and have used their term ansa lenticularis instead of tractus striocerebellaris et tegmentalis. The medial forebrain bundle has not been included in Fig. 4 but we have already indicated that it carries efferents from lobus parolfactorius to the hypothalamus.

Most of the intrinsic connections of the telencephalon are also shown schematically or pictorially in Fig. 4, so that their origins, destinations and functions should be clear. The chief commissural connection is the anterior commissure which, according to Zeier and Karten (1971) connects the anterior archistriatum of either side and also projects into the neostriatum. Interhemispheric interactions could operate through this commissure and the orientation hypothesis may be applied along the following lines. The anterior archistriatum facilitates turning towards the side in which the neostriatum has organised a positive orientation and at the same time through its commissural projections the archistriatum inhibits any similar ipsilateral positive orientation by the opposite archistriatum and neostriatum. The reader may like to apply this deduction from the orientation hypothesis to the experimental findings presented by Zeier in Chapter 9 of this volume.

We also believe that the system we have proposed is consistent with other studies presented in this book, including the study of avian visual orientation behaviour by Friedman (Chapter 10), the study of amygdaloid and hypothalamic interactions in feeding behaviour by Wright (Chapter 15) and the study of brain stem mechanisms by Andrew (Chapter 13). In the case of the study by Andrew we would point out that the brain stem pathway which he has correlated with the facilitation of attention in chicks is also part of the tractus occipitomesencephalicus projection system which we have postulated as being involved in facilitating patterned orientation responses to stimulation. We recognise that our statements are frequently gross oversimplifications but ask the reader to bear with this and to follow the general direction of our thoughts and to attempt to apply our approach to the specific details with which he is himself familiar. If the hypothesis has any value it will be in facilitating the detailed analysis of the neurological systems involved in specific behaviour patterns in birds. We hope this facilitation will occur through studies designed to test predictions derived from the hypothesis as well as by the reinterpretation of existing studies. It will also serve a purpose if it only leads to alternative essays with fewer theoretical inadequacies and less special pleading to handle the wide ranging and varied data on the avian telencephalon.

Acknowledgements

We wish to acknowledge the work of Alison Williamson, Departmental Research Officer, who carried out the behavioural tests and the histological analysis of the chick lesion study. She also prepared the drawings and graphs of the present paper and assisted with the library and photographic services involved. As a joint author of the experimental paper that will report this work she agreed to our use of the data for the presentation made at this conference. We also wish to record the excellent facilities in this Department that have made this work possible, including the technical assistance provided by Linda MacRae and the help of the Department's technical and secretarial staff.

REFERENCES

Adamo, N.J. (1967) Connections of efferent fibers from hyperstriatal areas in chicken, raven, and African lovebird. *J. comp. Neurol.*, 131, 337—356.
Adamo, N.J. and Bennett, T.L. (1967) The effect of hyperstriatal lesions on head orientation to a sound stimulus in chickens. *Exp. Neurol.*, 19, 166—175.
Adamo, N.J. and King, R.L. (1967) Evoked responses in the chicken telencephalon to auditory, visual, and tactile stimulation. *Exp. Neurol.*, 17, 498—504.
Åkerman, B. (1966a) Behavioural effects of electrical stimulation on the forebrain of the pigeon. I. Reproductive behavior. *Behaviour*, 26, 323—338.
Åkerman, B. (1966b) Behavioural effects of electrical stimulation on the forebrain of the pigeon. 2. Protective behaviour. *Behaviour*, 26, 339—350.

Åkerman, B., Fabricius, E., Larsson, B. and Steen, L. (1962) Observations on pigeons with prethalamic radiolesions in the nervous pathways from the telencephalon. *Acta physiol. scand.*, 56, 286—298.

Arduini, A., Moruzzi, G. e Zanchetti, A. (1948) Ricomparsa delle reazioni visive di disesa nel Piccione talamico acuto in seguito a stimolazione chimica del tetto ottico contro-laterale. *Boll. Soc. ital. Biol. sper.*, 24, 584.

Aronson, L.R. (1967) Forebrain function in teleost fishes. *Trans. N.Y. Acad. Sci.*, 29, 390—396.

Bateson, P.P.G., Rose, S.P.R. and Horn, G. (1973) Imprinting: lasting effects on uracil in-corporation into chick brain. *Science*, 181, 576—578.

Beach, F.A. (1951) Effects of forebrain injury upon mating behavior in male pigeons. *Behaviour*, 4, 36—59.

Benowitz, L. (1972) Effects of forebrain ablations on avoidance learning in chicks. *Physiol. Behav.*, 9, 601—608.

Benowitz, L. and Lee-Teng, E. (1973) Contrasting effects of three forebrain ablations on discrimination learning and reversal in chicks. *J. comp. physiol. Psychol.*, 84, 391—397.

Biederman-Thorson, M. (1970a) Auditory evoked responses in the cerebrum (field L) and ovoid nucleus of the ring dove. *Brain Res.*, 24, 235—245.

Biederman-Thorson, M. (1970b) Auditory responses of units in the ovoid nucleus and cerebrum (field L) of the ring dove. *Brain Res.*, 24, 247—256.

Blough, P.M. (1971) The visual acuity of the pigeon for distant targets. *J. exp. Anal. Behav.*, 15, 57—67.

Bolton, T.B. (1971a) The structure of the nervous system. In D.J. Bell and B.M. Freeman (Eds.), *Physiology and Biochemistry of the Domestic Fowl*. Academic Press, New York, pp. 641—673.

Bolton, T.B. (1971b) The physiology of the nervous system. In D.J. Bell and B.M. Free-man (Eds.), *Physiology and Biochemistry of the Domestic Fowl*. Academic Press, New York, pp. 675—705.

Boord, R.L. (1969) The anatomy of the avian auditory system. *Ann. N.Y. Acad. Sci.*, 167, 186—198.

Boyce, R. and Warrington, W.R. (1898) Observations on the anatomy, physiology and degenerations of the nervous system of the bird. *Proc. roy. Soc. B*, 64, 176—179.

Bremer, F., Dow, R.S. and Moruzzi, G. (1939) Physiological analysis of the general cortex in reptiles and birds. *J. Neurophysiol.*, 2, 473—487.

Chard, R.D. (1939) Visual acuity in the pigeon. *J. exp. Psychol.*, 24, 588—608.

Codish, S.D. (1971) Actinomycin D injected into the hippocampus of chicks: effects upon imprinting. *Physiol. Behav.*, 6, 95—96.

Cohen, D.H. (1967) The hyperstriatal region of the avian forebrain: a lesion study of possible functions, including its role in cardiac and respiratory conditioning. *J. comp. Neurol.*, 131, 559—570.

Cohen, D.H. and Pitts, L.H. (1967) The hyperstriatal region of the avian forebrain: somatic and autonomic responses to electrical stimulation. *J. comp. Neurol.*, 131, 323—336.

Cohen, H.D., Ervin, F. and Barondes, S.H. (1966) Puromycin and cycloheximide: dif-ferent effects on hippocampal electrical activity. *Science*, 154, 1557—1558.

Cruze, W.W. (1935) Maturation and learning in chicks. *J. comp. Psychol.*, 19, 371—409.

Delius, J.D. (1971) Foraging behaviour patterns of herring gulls elicited by electrical forebrain stimulation. *Experientia (Basel)*, 27, 1287—1289.

Delius, J.D. and Bennetto, K. (1972) Cutaneous sensory projections to the avian fore-brain. *Brain Res.*, 37, 205—221.

Douglas, R.J. (1967) The hippocampus and behavior. *Psychol. Bull.*, 6, 416—442.

Douglas, R.J. and Pribram, K.H. (1966) Learning and limbic lesions. *Neuropsychologia*, 4, 197—220.

Douglas, R.J., Barrett, T.W., Pribram, K.H. and Cerny, M.C. (1969) Limbic lesions and error reduction. *J. comp. physiol. Psychol.*, 68, 437—441.

Erulkar, S.D. (1955) Tactile and auditory areas in the brain of the pigeon. *J. comp. Neurol.*, 103, 421—458.

Flexner, L.B. and Flexner, J.B. (1968) Intracerebral saline: effect on memory of trained mice treated with puromycin. *Science*, 159, 330—331.

Forman, D. and Ward, J.W. (1957) Responses to electrical stimulation of caudate nucleus in cats in chronic experiments. *J. Neurophysiol.*, 20, 230—244.

Fraenkel, G.S. and Gunn, D.L. (1961) *The Orientation of Animals*. Dover, New York.

Gervai, J. and Csányi, V. (1973) The effects of protein synthesis inhibitors on imprinting. *Brain Res.*, 53, 151—160.

Goodman, I.J. and Brown, J.L. (1966) Stimulation of positively and negatively reinforcing sites in the avian brain. *Life Sci.*, 5, 693—704.

Gossette, R.L. (1967) Successive discrimination reversal (SDR) performance of four avian species on a brightness discrimination task. *Psychon. Sci.*, 8, 17—18.

Gossette, R.L., Gossette, M.F. and Riddell, W. (1966) Comparisons of successive discrimination reversal performances among closely and remotely related avian species. *Anim. Behav.*, 14, 560—564.

Harman, A.L. and Phillips, R.E. (1967) Responses in the avian midbrain, thalamus and forebrain evoked by click stimuli. *Exp. Neurol.*, 18, 276—286.

Harwood, D. and Vowles, D.M. (1966) Forebrain stimulation and feeding behavior in the ring dove (*Streptopelia risoria*). *J. comp. physiol. Psychol.*, 62, 388—396.

Hay, J.C. and Pick, H.L. (1966) Visual and proprioceptive adaptation to optical displacement of the visual stimulus. *J. exp. Psychol.*, 71, 150—158.

Hess, E.H. (1956) Space perception in the chick. *Sci. Amer.*, 195, 71—80.

Hess, E.H. (1959) Imprinting. *Science*, 130, 133—141.

Hodos, W. and Karten, H.J. (1970) Visual intensity and pattern discrimination deficits after lesions of ectostriatum in pigeons. *J. comp. Neurol.*, 140, 53—68.

Hodos, W., Karten, H.J. and Bonbright, J.C. (1973) Visual intensity and pattern discrimination after lesions of the thalamofugal visual pathway in pigeons. *J. comp. Neurol.*, 148, 447—468.

Horn, G., Rose, S.P.R. and Bateson, P.P.G. (1973) Experience and plasticity in the central nervous system. *Science*, 181, 506—514.

Hunt, S.P. and Webster, K.E. (1972) Thalamo-hyperstriate interrelations in the pigeon. *Brain Res.*, 44, 647—651.

Jane, J.A., Levey, N. and Carlson, N.J. (1972) Tectal and cortical function in vision. *Exp. Neurol.*, 35, 61—77.

Jones, L.V. (1954) Distinctiveness of colour, form and position cues for pigeons. *J. comp. physiol. Psychol.*, 47, 253—257.

Jung, R. and Hassler, R. (1960) The extrapyramidal motor system. In J. Field, H.W. Magoun and V.E. Hall (Eds.), *Handbook of Physiology, Section 1, Neurophysiology*, Vol. 2. American Physiological Society. Washington D.C., pp. 863—927.

Kalischer, O. (1905) Das Grosshirn der Papageien in anatomischer und physiologischer Beziehung. *Abhandl. preuss. Akad. Wiss. Physic-Math.*, 4, 1—105.

Kaplan, H. and Aronson, L.R. (1969) Function of forebrain and cerebellum in learning in the teleost *Tilapia heudelotii macrocephala*. *Bull. Amer. Mus. nat. Hist.*, 142, 141—208.

Karten, H.J. (1968) The ascending auditory pathway in the pigeon (*Columba livia*). II. Telencephalic projections of the nucleus ovoidalis thalami. *Brain Res.*, 11, 134—153.

Karten, H.J. (1969) The organization of the avian telencephalon and some speculations on the phylogeny of the amniote telencephalon. *Ann. N.Y. Acad. Sci.*, 167, 164—179.

Karten, H.J. and Dubbeldam, J.L. (1973) The organization and projections of the palaeostriatal complex in the pigeon (*Columba livia*). *J. comp. Neurol.*, 148, 61—90.

Karten, H.J. and Hodos, W. (1967) *A Stereotaxic Atlas of the Brain of the Pigeon (Columba livia)*. Johns Hopkins University Press, Baltimore, Md.

Karten, H.J. and Hodos, W. (1970) Telencephalic projections of the nucleus rotundus in the pigeon (*Columba livia*). *J. comp. Neurol.*, 140, 35—52.

Kimble, D.P. (1968) Hippocampus and internal inhibition. *Psychol Bull.*, 70, 285—295.

Kimble, D.P. (1969) Possible inhibitory functions of the hippocampus. *Neuropsychologia*, 7, 235—244.

Kuhlenbeck, H. and Szekely, E.G. (1963) Evoked potentials from tectum mesencephali and telencephalon of the chicken after unilateral optic stimulation. *Anat. Rec.*, 145, 332.

Laursen, A.M. (1962) Movements evoked from the region of the caudate nucleus in cats. *Acta physiol. scand.*, 54, 175—184.

Lee-Teng, E. (1969) Retrograde amnesia in relation to subconvulsive and convulsive currents in chicks. *J. comp. physiol. Psychol.*, 67, 135—139.

Lee-Teng, E. and Sherman, S.M. (1966) Memory consolidation of one-trial learning in chicks. *Proc. nat. Acad. Sci. (Wash.)*, 56, 926—931.

Lee-Teng, E. and Sherman, S.M. (1969) Effect of forebrain lesions on acquisition and retention of one-trial learning in chicks. *Proc. 77th Ann. Convention, APA*. pp. 203—204.

Leppelsack, H.-J. und Schwartzkopff, J. (1972) Eigenschaften von akustischen im kaudalen Neostriatum von Vögeln. *J. comp. Physiol.*, 80, 137—140.

Macphail, E.M. (1969) Avian hyperstriatal complex and response facilitation. *Commun. Behav. Biol.*, 4, 129—137.

Macphail, E.M. (1971) Hyperstriatal lesions in pigeons: effects on response inhibition, behavioral contrast, and reversal learning. *J. comp. physiol. Psychol.*, 75, 500—507.

Maley, M.J. (1969) Electrical stimulation of agonistic behavior in the mallard. *Behaviour*, 34, 138—160.

Mark, R.F. and Watts, M.E. (1971) Drug inhibition of memory formation in chickens. 1. Long-term memory. *Proc. roy. Soc. B*, 178, 439—454.

Mayor, S.J. (1973) Puromycin's effects on long-term memory and the acquisition of two successive visual discrimination tasks in Japanese quail. *Physiol. Psychol.*, 1, 33—36.

McLennan, H., Emmons, P.R. and Plummer, P.M. (1964) Some behavioral effects of stimulation of the caudate nucleus in unrestrained cats. *Canad. J. Physiol. Pharmacol.*, 42, 329—339.

Montanelli, R.P. and Hassler, R. (1964) Motor effects elicited by stimulation of the pallido-thalamic system in the cat. In W. Bargmann and J.P. Schadé (Eds.), *Lectures on the Diencephalon, Progress in Brain Research*, Vol. 5. Elsevier, Amsterdam, pp. 56—66.

Muskens, L.J.J. (1929) The tracts and centers in the pigeon dominating the associated movements of the eyes (and other movable parts) in the sense of lateral deviation in the horizontal and of rotation in the frontal plane. *J. comp. Neurol.*, 48, 267—292.

Muskens, L.J.J. (1931) On tracts and centres involved in the upward and downward associated movements of the eyes after experiments in birds. *J. comp. Neurol.*, 50, 289—331.

Nye, P.W. (1973) On the functional differences between frontal and lateral visual fields of the pigeon. *Vision Res.*, 13, 559—574.

O'Flaherty, J.J. e Invernizzi, G. (1972) Organizzazione funzionale dei campi recettivi di singole unita' visive del "Wulst" nel piccione. *Boll. Soc. ital. Biol. sper.*, 48, 137—138.

Parker, D.M. and Delius, J.D. (1972) Visual evoked potentials in the forebrain of the pigeon. *Exp. Brain Res.*, 14, 198—209.

Pavlov, I.P. (1927) *Conditioned Reflexes*. Oxford University Press, London.

Pearson, R. (1972) *The Avian Brain*. Academic Press, New York.

Perisic, M., Mihailovic, J. and Cuénod, M. (1971) Electrophysiology of contralateral and ipsilateral visual projections to the Wulst in pigeon. *Int. J. Neurosci.*, 2, 1—8.

Phillips, R.E. (1964) 'Wildness' in the mallard duck: effects of brain lesions and stimulation on 'escape behavior' and reproduction. *J. comp. Neurol.*, 122, 139—155.

Phillips, R.E. (1968) Approach—withdrawal behavior of peach-faced lovebirds, *Agapornis roseicolis*, and its modification by brain lesions. *Behaviour*, 31, 163—184.

Phillips, R.E. and Youngren, O.M. (1971) Brain stimulation and species-typical behaviour: activities evoked by electrical stimulation of the brains of chickens (*Gallus gallus*). *Anim. Behav.*, 19, 757—779.

Portmann, A. and Stingelin, W. (1961) The central nervous system. In A.J. Marshall (Ed.), *The Biology and Comparative Physiology of Birds*. Academic Press, New York, pp. 1—36.

Powell, T.P.S. and Cowan, W.M. (1961) The thalamic projection upon the telencephalon in the pigeon (*Columba livia*). *J. Anat. (Lond.)*, 95, 78—109.

Pribram, K.H., Douglas, R.J. and Pribram, B.J. (1969) The nature of nonlimbic learning. *J. comp. physiol. Psychol.*, 69, 765—772.

Pritz, M.B., Mead, W.R. and Northcutt, R.G. (1970) The effects of Wulst ablations on color, brightness and pattern discrimination in pigeons (*Columba livia*). *J. comp. Neurol.*, 140, 81—100.

Putkonen, P.T.S. (1967) Electrical stimulation of the avian brain. *Ann. Acad. Sci. fenn.* A5, 130, 1—95.

Revzin, A.M. (1969) A specific visual projection area in the hyperstriatum of the pigeon (*Columba livia*). *Brain Res.*, 15, 246—249.

Revzin, A.M. (1970) Some characteristics of wide-field units in the brain of the pigeon. *Brain Behav. Evol.*, 3, 195—204.

Revzin, A.M. and Karten, H.J. (1967) Rostral projections of the optic tectum and the nucleus rotundus in the pigeon. *Brain Res.*, 3, 264—276.

Reynolds, G.S. and Limpo, A.J. (1965) Selective resistance of performance on a schedule of reinforcement to disruption by forebrain lesions. *Psychon. Sci.*, 3, 35—36.

Rogers, F.T. (1922a) Studies of the brain stem. 6. An experimental study of the corpus striatum of the pigeon as related to various instinctive types of behaviour. *J. comp. Neurol.*, 35, 21—60.

Rogers, F.T. (1922b) A note on the excitable areas of the cerebral hemispheres of the pigeon. *J. comp. Neurol.*, 35, 61—65.

Rossi, P.J. (1968) Adaptation and negative after effect to lateral optical displacement in newly hatched chicks. *Science*, 160, 430—432.

Rougeul, A. (1957) *Exploration Oscillographique de la Voie Viseille du Pigeon*. Foulon, Paris.

Salzen, E.A. and Meyer, C.C. (1968) Reversibility of imprinting. *J. comp. physiol. Psychol.*, 66, 269—275.

Salzen, E.A. and Williamson, A.J. Unpublished stereotaxic atlas of the brain of the domestic chick.

Savage, G.E. (1968) Temporal factors in avoidance learning in normal and forebrainless goldfish (*Carassius auratus*). *Nature (Lond.)*, 218, 1168—1169.

Savage, G.E. (1969a) Telencephalic lesions and avoidance behaviour in the goldfish (*Carassius auratus*). *Anim. Behav.*, 17, 362—373.

Savage, G.E. (1969b) Some preliminary observations on the role of the telencephalon in food-reinforced behaviour in the goldfish, *Carassius auratus. Anim. Behav.*, 17, 760—772.

Savage, G.E. (1971) Behavioural effects of electrical stimulation of the telencephalon of the goldfish, *Carassius auratus. Anim. Behav.*, 19, 661—668.

Shaklee, A.O. (1921) The relative heights of the eating and drinking arcs in the pigeon's brain, and brain evolution. *Amer. J. Physiol.*, 55, 65—83.

Shettleworth, S.J. (1972) Stimulus relevance in the control of drinking and conditioned fear responses in domestic chicks (*Gallus gallus*). *J. comp. physiol. Psychol.*, 80, 175—198.

Sokolov, Y.N. (1963) *Perception and the Conditioned Reflex*. (Translated by S.W. Waydenfeld) Pergamon Press, London.

Spevack, A.A. and Pribram, K.H. (1973) Decisional analysis of the effects of limbic lesions on learning in monkeys. *J. comp. physiol. Psychol.*, 82, 211—226.

Stettner, L.J. (1966) Effect of end brain lesions on discrimination and reversal learning in bob-white quail. Paper presented at American Psychological Association, New York, cited in Stettner (1974).

Stettner, L.J. (1974) The neural basis of avian discrimination and reversal learning. In I.J. Goodman and M.W. Schein (Eds.), *The Bird: its Brain and Behavior*. Academic Press, New York. In press.

Stettner, L.J. and Schultz, W.J. (1967) Brain lesions in birds: effects on discrimination acquisition and reversal. *Science*, 155, 1689—1692.

Stingelin, W. and Senn, D.G. (1969) Morphological studies on the brain of Sauropsida. *Ann. N.Y. Acad. Sci.*, 167, 156—163.

Strauss, H. (1929) Das Zusammenschrecken. *J. Psychol. Neurol.*, 39, 111—231.

ten Cate, J. (1936) Physiology of the central nervous system of birds. *Ergebn. Biol.*, 13, 93—173.

Ursin, H. and Kaada, B.R. (1960) Functional localization within the amygdaloid complex in the cat. *Electroenceph. clin. Neurophysiol.*, 12, 1—20.

Vanderwolf, C.H. (1971) Limbic-diencephalic mechanisms of voluntary movement. *Psychol. Rev.*, 78, 83—113.

Walls, G.L. (1951) The problem of visual direction. *Amer. J. Optom.*, 28, 55—83, 115—146, 173—212.

Watts, M.E. and Mark, R.F. (1971) Drug inhibition of memory formation in chickens. 2. Short-term memory. *Proc. roy. Soc. B*, 178, 455—464.

Wenzel, B.M. and Salzmann, A. (1968) Olfactory bulb ablation or nerve section and behaviour of pigeons in non-olfactory learning. *Exp. Neurol.*, 22, 472—479.

Wilson, S.A.K. (1928) *Modern Problems in Neurology*. Arnold, London.

Witkovsky, P., Zeigler, H.P. and Silver, R. (1973) The nucleus basalis of the pigeon: a single-unit analysis. *J. comp. Neurol.*, 147, 119—128.

Zeier, H. (1968) Changes in operant behavior of pigeons following bilateral forebrain lesions. *J. comp. physiol. Psychol.*, 66, 198—203.

Zeier, H. (1969) DRL-performance and timing behavior of pigeons with archistriatal lesions. *Physiol. Behav.*, 4, 189—193.

Zeier, H. (1971) Archistriatal lesions and response inhibition in the pigeon. *Brain Res.*, 31, 327—339.

Zeier, H. and Akert, K. (1968) Transient changes in operant behavior of pigeons during bilateral electrical forebrain stimulation, *Physiol. Behav.*, 3, 293—296.

Zeier, H. and Karten, H.J. (1971) The archistriatum of the pigeon: organization of afferent and efferent connections. *Brain Res.*, 31, 313—326.

Zeigler, H.P. (1963a) Effects of endbrain lesions upon visual discrimination learning in pigeons. *J. comp. Neurol.*, 120, 161—182.

Zeigler, H.P. (1963b) Effects of forebrain lesions upon activity in pigeons. *J. comp. Neurol.*, 120, 183—194.

Zeigler, H.P. and Karten, H.J. (1973a) Brain mechanisms and feeding behavior in the pigeon (*Columba livia*). 1. Quinto-frontal structures. *J. comp. Neurol.*, 152, 59—82.

Zeigler, H.P. and Karten, H.J. (1973b) Brain mechanisms and feeding behavior in the pigeon (*Columba livia*). 2. Analysis of feeding behavior deficits after lesions of quinto-frontal structures. *J. comp. Neurol.*, 152, 83—102.

Zeigler, H.P., Green, H.C. and Karten, H.J. (1969) Neural control of feeding behavior in the pigeon. *Psychon. Sci.*, 15, 156—157.

12

BRAIN ORGANIZATION AND NEUROMUSCULAR CONTROL OF VOCALIZATION IN BIRDS

RICHARD E. PHILLIPS and FRANK W. PEEK

Department of Animal Science University of Minnesota, St. Paul, Minn. and Department of Biology, Nazareth College, Rochester, N.Y. (U.S.A.)

Behavior is movement and movement is motor activity. Thus, a major brain function is to organize motor outputs and to switch in, or select, the appropriate outputs for the given environmental conditions (inputs). A study of motor output, then, is a relatively direct way of analyzing what the brain is doing and how it does it. This idea is certainly not new. Sherrington (1906) in his studies of spinal reflexes and Baerends and Baerends-van Roon (1950) in their studies of the behavior of cichlid fish were explicitly looking at motor output with the goal of understanding brain organization and function. More recently, Wilson (1967) pointed out, with his analyses of insect locomotor controls, that a study of outputs yields knowledge of what the central machinery must achieve and of likely constraints on proposed mechanisms. The studies of Bentley (1969) of cricket song, and of Willows (1967) of sea slugs are further outstanding recent examples of analyses of motor organization.

An important feature of the invertebrate studies is their concentration on relatively simple behavioral systems involving small numbers of nerves and muscles. The explicit rationale underlying these studies, that understanding of brain—behavioral mechanisms will be achieved most quickly by analysis of simple systems which can then serve as models for the study of more complex ones, has been amply justified in the history of science.

Studies of neuromotor control of vertebrate behavior are usually hampered because the large number of muscles involved in most movements makes measurement and analysis difficult or impossible. What is needed is a system involving a small number of muscles that is also interesting behaviorally. Avian calls satisfy these requirements admirably. Calls are significant behaviors both as indicators of internal states and as signals in communication. The basic variable controlled, air flow, is regulated by two mechanisms: a power source, the respiratory muscles, and a control device, the syrinx and its associated muscles (the pharynx and buccal cavities may be a third set of

244

controls, but no data on their functions in calling are currently available). The two known mechanisms can be studied separately for convenience because of their anatomical independence. Syringeal mechanisms are especially suitable for study because, according to the species, as few as three pairs of muscles are involved, and because these muscles are controlled directly by the brain rather than by way of spinal neurons. These features allow study of the vocal control signals in sufficient detail to ask meaningful questions about how the signal patterns are generated, questions that are potentially answerable with current neurophysiological techniques.

The present paper reviews studies of the control of avian calls. The mechanisms of sound production, from muscle and motoneuron activity through successively higher levels of the central nervous system, are examined in detail, and their significance for brain programming of behavior is discussed. Finally, the results are compared with those from members of the other classes of vertebrates.

PERIPHERAL CONTROL OF SOUND PRODUCTION

Syrinx

The sound producing apparatus for avian vocalizations is the syrinx, a bony and membranous box at the junction of the trachea with the bifurcation of the bronchi (Fig. 1). This contrasts with the location of the sound producing larynx in mammals located at the upper end of the trachea. The syrinx lies deep within the thorax enclosed within the interclavicular air sac. Its structure is basically that of a bifurcating tube, part of the walls of which are composed of very thin membranes supported by bony half rings. In most species the membranes are in the medial walls of the bronchial portions of the syrinx, but, in chickens and their relatives, the membranes are in the lateral walls just at, and rostral to, the junction of the two bronchi. In either case, sounds are produced by outward air flow through the bronchi and trachea causing these membranes to vibrate, and, thus, to set the air column into vibration (Greenewalt, 1968). The tension on the membrane is apparently controlled by extrinsic muscles. In chickens and many non-passerines there are just three: a pair of muscles which run along the sides of the trachea from the larynx to the syrinx (musculi tracheolateralis), a second pair (musculi sternotrachealis) originating on the sternum and running through the space of the interclavicular air sac to insert on the bronchial end of the tracheolateralis muscles, and a pair of ypsilotrachealis muscles that originate on the clavicles and insert, in chickens (*Gallus gallus*), on the skin of the ventral part of the neck about halfway from head to breast and then continue to insert also on the upper end of the trachea and pharynx. These muscles are innervated by the hypoglossal nerve, particularly the ramus cervicalis decendans superior (c.d.s) (Conrad, 1915). The right and left hypoglossal nerves run the length of the trachea, giving off fibers to the tracheolatera-

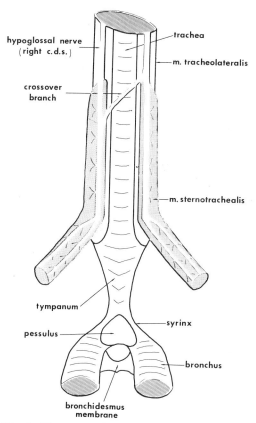

Fig. 1. Diagrammatic representation of the trachea and syrinx and associated muscles and nerves of the chicken.

lis muscles along the way. A branch of the right hypoglossal splits off to the sternotrachealis muscles of that side; the left branch does the same but a twig also crosses over and partially innervates the right sternotrachealis muscle as well as the left (see Fig. 1). In no case does the right nerve innervate the left side. The vagus nerve can be traced down to the crop, but we have not seen a recurrent branch to the syringeal muscles in ducks or chickens, although it has been described by others (Conrad, 1915; Watanabe, 1960). Nottebohm (1971) was also apparently unable to see it in chaffinches, and Fedde *et al.* (1964) seem to have drawn it from the literature rather than from their own observations.

Respiratory pump

The respiratory 'pump' provides the power (the air flow) for vocalization. The avian respiratory system consists of a pair of lungs and several (usually eight or nine) membranous air sacs. The airways in the lungs are all open-ended so that air flows through them rather than in and out as in the dead-end alveoli of mammals. Although the details of air flow are still controversial, (King and Molony, 1971) air probably flows through some parts of the lungs on both inspiration and expiration. The lungs are relatively rigid and are not expanded and contracted by a diaphragm as in mammals. Instead, movements of the abdominal musculature and of the sternum and rib cage compress and expand the air sacs to provide the power for air flow in and out of the bird. The major muscles involved seem to be the external and internal intercostal and the abdominal muscles, mostly the latter (Kadono *et al.*, 1963; Fedde *et al.*, 1964; George and Berger, 1966; for details of names). These are innervated by spinal nerves.

Air pressure in interclavicular air sac

Greenewalt (1968) analyzed the physical aspects of sound production in great detail, especially for passerines. He concluded that the increased complexity of musculature in oscines in comparison to other birds results only in an increased frequency range over which the birds can sing rather than in increased complexity in temporal patterning of the song itself. He also concluded that there must be super-atmospheric pressure in the interclavicular air sac (ICAS) in conjunction with the Bernoulli effect of air flow through the trachea to bow the internal tympaniform membrane into the bronchial lumen in order for sound to be produced. This conclusion is based primarily on *in vitro* experiments of Rüppel (1933) and Gross (1964 and 1968) using fresh syrinxes in a chamber in which external air pressure could be varied independently of the air flow through the trachea.

Greenewalt's conclusions concerning air pressure have been challenged by Gaunt *et al.* (1973) and Gaunt and Wells (1973). These authors contend that the experiments with open air sacs indicate only that air flow has bypassed the syrinx. In their studies of air flow and pressure in juvenile starling distress calling, Gaunt *et al.* (1973) confirmed the observation that large holes in the ICAS eliminate calls, but they found that smaller ones had little effect. In birds with the ICAS opened, air flow through the trachea was negligible. These authors concluded that rerouting of air flow rather than reduction of pressure in the air sacs prevented vocalizing after opening the ICAS.

Youngren *et al.* (1974) have made a direct test of this rerouted flow hypothesis. Working with chickens and using electrical stimulation (ESB) of the midbrain, they showed that chickens were muted by opening the air sac if the hole was large (Fig. 2). Closing the hole with a thin piece of plastic immediately reinstated vocalization. They also confirmed the finding of Gaunt *et al.* (1973) that small holes, even up to several millimeters in diameter, were relatively ineffective. To test the possibility that opening the interclavicular air sac merely shunted air flow away from the trachea and syrinx, Youngren *et al.* plugged the openings (secondary bronchi) from the lungs into the air sac. This restored air flow through the syrinx (Fig. 2), but the

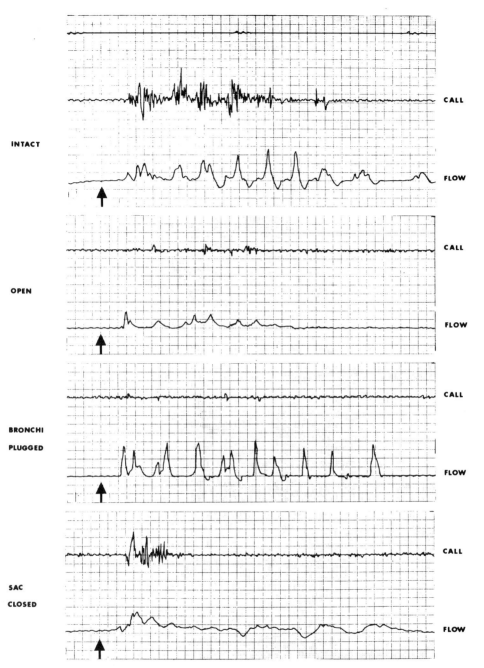

Fig. 2. Relationship between tracheal air flow and sound production. Intact: correspondence between calls and tracheal air flow in an intact bird. Open: loss of both calls and tracheal air flow when the interclavicular air sac is completely opened. Bronchi plugged: restoration of tracheal air flow but not calls by plugging passages from lungs to interclavicular air sac while the latter is open, exposing the tympaniform membranes to atmospheric pressure. Sac closed: restoration of calls even at low tracheal air flow by sealing the interclavicular air sac with thin plastic so that supra-atmospheric pressure can develop within the sac. Arrows mark the onset of the call-evoking stimulus train. The topmost line shows 1-sec time marks.

birds remained mute. Calling returned immediately when the sac was sealed and pressurized by reopening the secondary bronchi. Lockner and Youngren (in preparation) have confirmed these results in adult female and in juvenile mallards (*Anas platyrhynchos*). Adult drakes, whose syringeal anatomy is strikingly more complex than that of females or juveniles (Lockner, personal communication) *could* vocalize with the air sac completely opened. They, however, apparently produce sound by an action similar to that of human vocal cords rather than by means of a vibrating membrane.

Gaunt *et al.* (1973) and Gaunt and Wells (1973) demonstrated that some sounds may be produced passively in anesthetized starlings (*Sturnus vulgaris*). They also found that short sounds in starling and redwinged blackbird (*Agelaius phoeniceus*) alarm calls apparently produced no measurable change in the pressure records. This seems most improbable and needs to be confirmed, since comparable calls in ducklings, chicks and chickens require appreciable internal pressures and flow. Differential measurements of pressure at various places during calling in small birds are less convincing because the very low flow and pressure magnitudes in avian breathing and vocalization may be below instrument sensitivity. Furthermore, the time constants of the pressure system and recording devices may be such as to obscure brief but significant events.

Role of syringeal muscles

Anatomy and action

In chickens the tracheolateralis muscles do not reach the last three or four tracheal rings before the syrinx (Youngren *et al.*, 1974), so when they contract they exert tension on the external tympaniform membranes. Observations of anesthetized birds, both during respiration and during calling, show that the trachea is pulled forward during silent breathing, but that it moves caudally during each call. The sternotrachealis muscles pull the trachea caudally and so relax the external tympaniform membranes and constrict the broncheal passages in the syrinx, when they contract (Youngren *et al.*, 1974; Gross, 1964). The function of the ypsilotrachealis muscles in chickens is obscure, but the muscles are especially well developed in males. In mallards, they are much more highly developed than in chickens and seem to function in calling as do the sternotrachealis muscles in chickens (Lockner and Youngren, in preparation). Our findings do not accord with the statements of Gaunt and Wells (1973) to the effect that in non-oscines membrane tension is increased by relaxing the muscles. This might be true in species where the muscles actually cross the external tympaniform membranes and attach on the tympanum. These authors do not specify which muscles would relax to increase tension, but they apparently were speaking of the tracheolateralis muscles.

Effects of cutting tracheal muscles

Youngren *et al.* (1974) found that cutting the tracheolateralis muscles had little effect on the calls evoked by midbrain stimulation, but that cutting the sternotrachealis muscles markedly altered the sounds produced and sharply limited the range of stimulus intensities that would evoke calls. The range of ESB intensities that evoked the associated pumping body movements remained unaffected. Subsequent section of the tracheolateralis muscles resulted in calls that in sound (and sonogram) appeared more like normal calls. This suggested that much of the sound production in the adults was passive, and that the major role of the tracheolateralis muscles might be to keep enough tension on the tympaniform membranes to prevent vocalizing during normal respiration. Apparently these muscles cannot pull hard enough to prevent calling during intensive vocalizations, even in the absence of the sternotrachealis muscles, for intense midbrain ESB evoked some pulsed calling after bilateral section of the latter muscles. Similar results were obtained with awake birds maintained for several days or weeks after muscle section.

As a partial test of our assumption that our results are pertinent to a wider range of calls than the repetitive, unmodulated bursts that constitute adult chicken alarm calls ('cuks'), we turned to chicks. Collias and Joos (1953), Guyomarc'h (1962), Andrew (1964), and Wood-Gush (1971) indicated that chick calls are largely pure tones whose frequency varies greatly within individual sound pulses. Since frequency modulation is one of the major concerns of students working with bird song (Greenewalt, 1968; Gaunt *et al.*, 1973) we wanted to see if these muscles had a role in pitch modulation. In separate groups of chicks we cut the tracheolateralis muscles bilaterally at the top or the bottom of the trachea or removed the sternotrachealis muscles. Andrew (1973) had reported that such operations eliminated frequency (pitch) modulations in chick calls, but all ten of our chicks with tracheolateralis sections and three of the five with sternotrachealis muscles excised still modulated frequency appreciably although the range was reduced approximately 50% by sternotrachealis section (Fig. 3).

We interpret these data to mean that the sternotrachealis muscles play a role in frequency modulation of chick calls, but that tracheal air flow and interclavicular air sac pressures also contribute importantly. The fact that the tracheal muscles are involved in frequency control increases the likelihood that our findings are pertinent to a broad range of avian vocalizations. The demonstration of air flow control of pitch in chicks suggests some role for flow effects in passerine song, even though they cannot vocalize in the complete absence of muscular control (Peek, 1972; Nottebohm, 1971 and 1972). This is further supported by the large frequency shifts illustrated in Nottebohm's (1971) sonograms of 'wheezing sounds' made by chaffinches during inspiration after bilateral section of the hypoglossus.

Electromyograms of tracheal muscles during calling and breathing

In order to study timing of the action of the tracheosyringeal muscles in

250

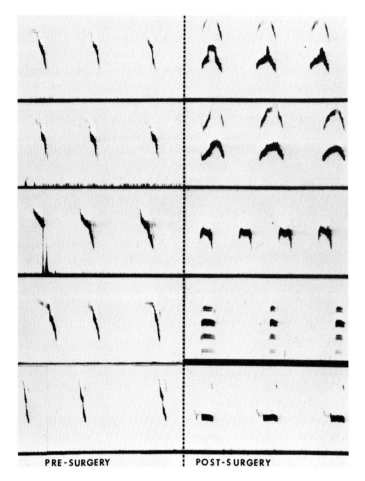

PRE-SURGERY | POST-SURGERY

Fig. 3. Sonograms illustrating the effects of cutting the sternotrachealis muscles bilaterally in chicks. Each row samples one chick; before surgery on the left, after surgery on the right. Control chicks with opened air sacs only or with section of N. XII (nervus hypoglossus) were much less affected.

relation to vocalization we turned to electromyograms (EMGs) of their activity during calling evoked by midbrain ESB and during silent respiration (see Youngren *et al.*, 1974, for details). EMG activity during spontaneous respiration under pentobarbital anesthesia indicated that, at the pharyngeal end of the trachea, musculus tracheolateralis fires during the end of an expiration and on through the subsequent inspiration. A few centimeters caudally, it fires mostly in synchrony with inspiration, but at progressively lower levels it becomes more and more synchronized with expiration. When respiratory activity is driven more strongly by adding CO_2 to the inspired air, tracheolateralis activity at the pharynx remains unchanged, but at more caudal positions more and more EMG activity appears during both inspiratory and expiratory phases of the cycle. The sternotrachealis and ypsilotrachealis muscles are silent during both normal and forced respiration.

During calling evoked by ESB in the midbrain, the tracheolateralis EMG shows activity synchronous with that of the abdominal muscles, and thus with calling (Fig. 4), as well as a longer burst of poststimulus inspiratory (presumed) activity. This synchrony is more marked at caudal than at rostral positions. During calling sternotrachealis and ypsilotrachealis muscles fire in phase with, but slightly after, the start of activity in the abdominal muscles. Sound production itself lags the activity in the abdominal muscles by an amount even greater than that seen in the sterno- and ypsilotrachealis muscles. In summary, the EMG recordings confirm speculations, based on the origins and insertions of the tracheal muscles, that the tracheolateralis mus-

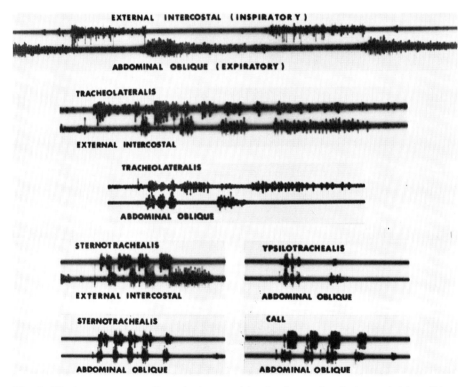

Fig. 4. Electromyograms of respiratory and tracheal muscles during quiet breathing and during calling showing the temporal relationships of the activity of the various muscles. The top line illustrates reciprocal firing of external intercostal and abdominal muscles. The remaining records relate activity in tracheal muscles and sound (call) production to respiratory phase as indicated by the records of respiratory muscle in the lower trace of each record.

cles function with every breath to keep the tympaniform membranes out of the air stream to permit silent breathing, whereas the sternotrachealis muscles pull the trachea caudally, relaxing the tympaniform membranes and permitting them to vibrate.

Neural control of syringeal muscles

Electrical stimulation of the hypoglossal nerves at points along the sides of the trachea resulted in contraction of the tracheolateralis and sternotrachealis muscles. Severing the nerves and stimulating the distal stump was equally effective. Neither contraction nor EMG activity in these muscles resulted from stimulation of either the right or the left vagus, although stimulation of either caused marked deceleration of the heart. These effects were not changed by severing the nerve and stimulating the distal stump. Nottebohm (1971) mentioned, in passing, that he had produced contraction of the 'syringeal muscles' by stimulation of the hypoglossal, but not by stimulation of the vagus in chickens. He was apparently unable to see a recurrent vagal branch to the syrinx in chaffinches (*Fringilla coelebs*). Therefore, any innervation by the vagus (Fedde, 1964; Nottebohm, 1971; Watanabe, 1960; Conrad, 1915), if it exists, must be sensory in chickens, mallards and chaffinches.

Section or cold-block of the hypoglossals eliminated EMG activity both in the sternotrachealis and tracheolateralis muscles during midbrain evoked calls and in the tracheolateralis muscle distal to the section or block during silent respiration. Section or cold-block of the right hypoglossal had no effect on EMGs in the left muscles, but blocking or cutting the left hypoglossal frequently sharply reduced the EMG in the right sternotrachealis as well. This confirms the anatomical finding of the crossover of the left nerve. Innervation of respiratory pump muscles has not been studied in detail, but bursts of activity synchronous with calling and persisting under tubocurarine block of neuromuscular transmission can be recorded from spinal nerves as far caudally as thoracic four (Murphy and Phillips, 1967). The timing of the activity in these muscles relative to syringeal muscle activity and to sound production has been shown in an earlier section. Peek (in preparation) found that like the muscle, the nerve of the external abdominal oblique shows sharp bursts of activity correlated with calls and silence in the intervals between calls. The nerve to the external intercostal (inspiratory) muscle is silent during calling (expiratory) bursts. It may or may not fire between call bursts (it usually does), but during successive intervals between calls of a series, activity tends to increase in this nerve and muscle as the series progresses. Frequently the nerve and/or muscle shows a strong burst of activity following a series of calls. The nerve records were obtained from hyperventilated curarized preparations, so O_2–CO_2 shifts and proprioceptive feedback could not contribute to the patterns that were seen. During silent breathing the inspiratory and expiratory activities overlap to some extent, but during calling they seem to be mutually exclusive. This will be discussed further in the next section.

CENTRAL MECHANISMS OF CALLING

In our studies of central mechanisms we are concentrating on the control of syringeal muscles for several reasons. Firstly, the information to the respiratory pump and to the syrinx seems to be very similar, at least for the repetitive 'cuks' that we have been studying. Secondly, the control information for the syrinx flows over the hypoglossal (see preceding section) rather than passing down the spinal cord, and so the hypoglossal nucleus provides both the motor output and a starting point into central processing. Thirdly, control of syringeal action by only three pairs of muscles, two of which (sternotrachealis and ypsilotrachealis) receive very similar patterns of excitation, offers hope that their relationship might be fully understood. Obviously, the possibilities for interactions and interrelationships increase much more rapidly than the number of muscles, so that it is to our advantage to deal with the simplest possible system for the first attack.

Pons and medulla

ESB evoked calls

To the best of our knowledge, only two studies of electrically evoked calling in birds have dealt with points in the pons and medulla. Peek and Phillips (1971) mapped the distribution of calls from the region using roving electrodes in anesthetized chickens, testing only with currents of 100 μA (peak) or less. The distribution of these points is illustrated in Fig. 5. The points lie ventrolaterally back past the lateral pontine nucleus, then swing up through the area of the motor nucleus of the trigeminal and course caudally in a ventrolateral position back to the level of the vagal and hypoglossal nuclei. There they appear to swing medially to the regions of these two nuclei. This is a tentative tracing at best, for the points do not correspond closely with any well known tracts. Brown (1971) sampled a few points in the medulla and pons of restrained but awake redwinged blackbirds and reported some vocalizations. His points were too few for mapping, and the thresholds for the points he did plot were over 300 μA, so that detailed comparison with our results is not possible. The points that he plotted just lateral to the abducens nucleus seem to be slightly dorsal and medial to effective points in comparable planes to the chicken medulla, but the significance of the difference is not clear.

We attempted to test whether or not ESB could elicit pulsed calling from the medulla after transection of the brain rostral to it, but our experiments were not conclusive. In one bird we were able to evoke a few hoarse 'cuks' after transection. Data from frogs (Schmidt, 1973), fish (Demski and Gerald,

254

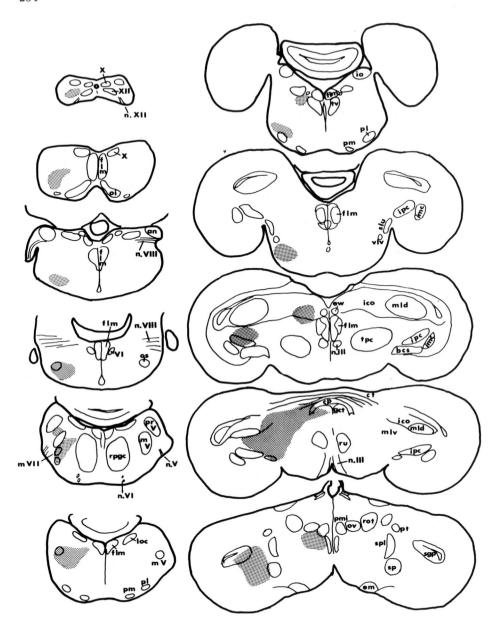

Fig. 5. Brain areas from which calls are most readily and reliably produced (stippled) in chickens. Abbreviations: an, nucleus angularis; bcs, brachium colliculi superioris; cp, commissura posterior; ct, commissura tectalis; em, nucleus ectomamillaris; ew, nucleus of Edinger-Westphal; flm, fasciculus longitudinalis medialis; gct, substantia grisea centralis; ico, nucleus intercollicularis; io, nucleus isthmoopticus; imc, nucleus isthmi, pars magnocellularis; ipc, nucleus isthmi, pars parvocellularis; loc, locus ceruleus; n. III, nervus oculomotorius; n. V, nervus trigeminus; n. VI, nervus abducens; n. VIII, nervus octavius; n. X,

1972 and 1974) and mammals (Kanai and Wang, 1962) indicate that calling can be evoked after such transections, so it would be surprising if they could not be evoked from the area in birds. Two types of vocal responses were evoked from medullary regions: repetitive 'cuks' and single, prolonged sounds that lasted the duration of the stimulus train. These latter were found only in the medulla. A reasonable interpretation of their significance is that the electrode is in the actual control mechanism for pulse formation, and therefore jams it unless current and position are optimal.

Unit recording

Because all excitatory impulses to the syringeal muscles pass over the hypoglossal nerve, we recorded unit activity from the hypoglossal nucleus during calling evoked by midbrain ESB. Electrodes made of 25 μm stainless steel wire insulated except for the cross section of the tip and stiffened with a no. 000 insect pin, isolate units quite well and also allow marking electrode locations by electroplating ferric ions for Prussian blue reaction.

Despite considerable effort, we found no way to stabilize this caudal portion of the brain during calling, so we have taken advantage of the fact that neural correlates of calling persist under curare (Murphy and Phillips, 1967). We can relate activity temporally to calling by simultaneously recording from a nerve to one of the respiratory muscles. We can also study potential respiratory function of units by adding CO_2 to the otherwise hyperventilating gas stream. This effectively drives respiratory rhythms in medullary units.

Responses to electrical stimulation of the hypoglossal nerve are concentrated within the nucleus intermedius of Karten and Hodos (1967). This seems to be the same as the nucleus intercalatus of Ariëns Kappers et al. (1960), and it is the nucleus that these authors describe as the source of the ramus cervicalis decendans superior (c.d.s.) of the hypoglossal nerve. We have not analyzed the response in detail, but they are complex potentials rather than spikes. The complexity, though not the amplitude, of these waves is highly dependent on stimulus frequency and, to a lesser degree, intensity. The frequency dependence strongly suggests interaction between hypoglossal

nervus vagus; n. XII, nervus hypoglossus; mld, nucleus mesencephalicus lateralis, pars dorsalis; mlv, nucleus mesencephalicus lateralis, pars ventralis; oi, nucleus olivaris inferior; os, nucleus olivaris superior; ov, nucleus ovoidalis; pl, nucleus pontis lateralis; pm, nucleus pontis medialis; pmi, nucleus paramedianus internus medialis thalami; prv, nucleus sensorius principalis nervi trigemini; pt, nucleus pretectalis; rpgc, nucleus reticularis pontis caudalis pars gigantocellularis; ru, nucleus ruber; sgp, substantia grisea et fibrosa periventricularis; slu, nucleus semilunaris; sp, nucleus subpretectalis; spl, nucleus spiriformis lateralis; tpc, nucleus tegmenti pedunculopontinus, pars compacta; tv, nucleus tegminti ventralis (Gudden); vlv, nucleus ventralis lemnisci lateralis; X, nucleus nervus vagus; XII, nucleus nervi hypoglossi (all names from Karten and Hodos, 1967).

neurons, presumably in the motor nucleus. Although this nerve is usually thought of as a purely motor path, it remains possible that it contains some sensory fibers. Two other possible explanations for the complex evoked potentials would be collaterals from hypoglossal motor axons activating nearby cells, or electrotonic interaction between the hypoglossal neurons themselves as has been reported for toadfish (Demski et al., 1973). We have not attempted to determine experimentally which of these hypotheses is most likely, but the persistence of the response at stimulus rates of 300/sec suggests that any effect is oligosynaptic. Latencies for the earliest responses to hypoglossal nerve stimulation range from 6 to 8 msec, and were typically 7 msec. This gives conduction velocities of at least 20 m/sec, well within the range of vertebrate myelinated nerve. Part of the variability can be accounted for by different placements of the stimulating electrodes. The difference between the sum of conduction velocities between midbrain and medulla (± 3 msec) and between medulla and muscle (± 7 msec) and the latency to the earliest call bursts in sternotracheal or abdominal oblique muscles (± 40 msec) is evidence that much processing of the stimulus goes on before the bursting activity reaches the medulla.

Our preliminary look at medullary units whose activity was altered by midbrain call site ESB have disclosed several types of responses. One group of cells bursts in phase with call rhythm just like peripheral expiratory cells (Fig. 6). A second group shows a similar rhythm plus the burst of poststimulus firing that characterizes peripheral inspiratory units. In some cases these identifications have been confirmed by recording normal respiratory activity from the same cells. In addition to these two types we find cells that are turned on for the duration of an ESB train but are silent otherwise, spontaneously firing cells that are turned off for the train duration, and cells whose activities either speed up or slow down but are neither turned on nor turned off per se. All of these classes of cells are altered only when the stimulating current in the midbrain equals or exceeds threshold for evoking calling from that site. As stimulus intensities increase over call threshold, additional units are brought in at some sites, as one might expect from the increasing intensity of calling that occurs with increased stimulus intensity. In addition, the firing patterns of other cells are altered by midbrain ESB currents below call threshold. We are temporarily treating these as not related to calling because of the many fiber tracts that pass through the areas of the stimulating and recording electrodes that are surely not directly related to vocalizations.

In summary, a first look at medullary unit activity discloses units whose firing patterns match peripheral activities and a few additional types whose relationship to calling is uncertain.

The firing patterns of some of the units that are influenced by ESB at midbrain sites where calls can be evoked at low thresholds are highly resistant to the effects of CO_2, suggesting that these units probably play no role in normal ventilation.

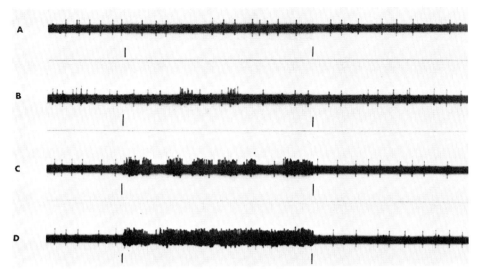

Fig. 6. Bursting activity in a unit located near the obex. A—D is a series showing effects of increasing ESB current in the midbrain call region. Vertical bars below the traces indicate beginning and end of the 2-sec stimulus train (50 pps).

Some of the units in the region of the caudal medulla (obex) are fired in bursts by midbrain ESB and are also activated to fire single spikes by hypoglossal nerve stimulation. More often groups of neurons fire in bursts that are repeated at the call rhythm in response to midbrain stimulation, and evoked potentials to hypoglossal stimulation can be picked up from the same site. These hypoglossal responses, however, seem not to be unit responses. More surprisingly, many of the places where hypoglossal stimulation evokes activity seem to contain no cells that burst at the call rhythm.

Relationships of vocal and respiratory systems in caudal brain stem

The vocal machinery almost certainly evolved from respiratory mechanisms both peripheral and central, so our preliminary hypothesis has been that both systems make use of the same motor neurons and probably also the same neurons in the pons and medulla. The existence of medullary units that fire with both respiratory and calling rhythms supports this hypothesis. Our observations that several medullary units did not show respiratory rhythms even with very high CO_2 levels in the ventilating air stream, and that recordings from the peripheral nerves to expiratory muscles show much larger spikes during call bursts than during expiratory bursts suggest that the vocal and respiratory systems are at least partly independent. These effects can be seen under anesthesia and sufficient tubocurarine paralysis so that no EMG can be detected, thus it is unlikely that these bigger spikes result from proprioceptive neurons in the mixed nerve from which we are recording. The observation that both sterno- and ypsilotrachealis muscles are active only during calling is even stronger evidence for some separation, or in ethological terms 'emancipation', of call from respiratory mechanisms.

Midbrain

The greatest impetus for the study of brain control of avian vocalizations came from Brown's papers in which he showed that electrical stimulation of an anatomically restricted portion of the midbrain evoked more or less natural sounding calls in redwinged blackbirds (Brown, 1965a), and that relatively small lesions in this same area would mute them (Brown, 1965b). Holst and Schleidt (1964) and Saint Paul (1965) made use of electrically evoked vocalizations and even mentioned that they could be produced under anesthesia, but the lack of anatomical information kept these papers from having maximum impact. The constancy of the anatomical representation across species is one of the most striking features that has emerged from the subsequent studies.

Extensive studies in chickens (Murphy and Phillips, 1967; Phillips and Youngren, 1971; Peek and Phillips, 1971), and in Coturnix quail (Potash, 1970a and b), Stellers jays (Brown, 1973), and mallard ducks (Maley, 1969; Phillips and Youngren, 1973) all show very similar anatomical distributions of effective points for the evocation of calls. Vowles and Beazley (1974), however, evoked very few vocalizations even from the MCR (see below) in ring doves (*Streptopelia risoria*). This seems to be correlated with a lack of the repetitive alarm calls which are so common in the other species studied to date. The matches between call site locations in different species are especially good in the midbrain which has been studied most thoroughly. This distribution is illustrated in Fig. 5. The vocal area includes the torus externus of Brown (1971) or nucleus intercollicularis of Karten and Hodos (1967), but not the distinctive central core labeled nucleus mesencephalicus dorsalis (MLD) by the latter authors. Newman (1970) in redwinged blackbirds and Potash (1970b) in Coturnix quail (*Coturnix coturnix japonica*), have demonstrated that the MLD is auditory, but not vocal, in function. The lowest thresholds for calling have been obtained from the anteromedial portions of the nucleus intercollicularis and the parts of the nucleus mesencephalicus lateralis pars ventralis (MLv) (Karten and Hodos, 1967) immediately ventral to it. Electrodes in this region (hereafter referred to as midbrain call area or MCR) frequently evoke calls and little or no other behavior even with intense stimulation (Phillips et al., 1972). Lesions in the MCR also muted both redwinged blackbirds (Brown, 1965a) and chickens (Phillips et al., 1972). The call regions of the left and right sides of the brain lie close to the ends of the tectal commissure, and stimulation in this fiber tract is effective in evoking calls in both awake and anesthetized chickens.

Ventral to this primary area, a broad band of call sites runs laterally then swings medially and caudally through the lateral reticular formation dorsal to the isthmic nuclei, exiting from the midbrain in a ventrolateral position and running to the region of the pontine nuclei described above. Tectal sites are largely ineffective. So are all sites in known sensory projection nuclei such as MLD, rotundus, ovoidalis, spiriformis lateralis, and ectomamillaris. The caudal projections have already been described. These areas are not all equipotent for call production. ESB outside the prime area (MCR) generally requires higher stimulus intensities to produce calls, and the calls are usually accompanied by various signs of behavioral arousal and relatively unpatterned activation of non-vocal motor systems. Calls can also be evoked from the periaqueductal gray near the posterior commissure.

Small electrolytic lesions in the anteromedial nucleus intercollicularis and MLv result in dense degeneration (as revealed by the Nauta technique) bilaterally in the intercollicular nuclei and in parts of the MLv medial and ventral to the lesion. Heavy degeneration is found in the tectal commissure. A dense band of degeneration passes ventrolaterally to the region of the pontine nuclei, and another band crosses the midline ventral to the medial longitudinal fasciculus and turns caudally along the midline. This latter bundle, the crossed tectospinal tract, would not appear to be involved in calling, judging from the plots of points that have evoked vocalization. The ipsilateral path matches the active areas diagrammed in anesthetized chickens (Peek and Phillips, 1971).

Murphy and Phillips (1967) showed that neural bursts at the call rhythm could be recorded from spinal nerves in anesthetized, curarized chickens during ESB of the midbrain call region. Thus the pattern can be generated without timing cues from either distance receptors or proprioception, since the curare prevented muscle activity. Call rhythm thus joins a growing list of behaviors whose basic programming is centrally organized, even in vertebrates. These include swallowing in dogs (Doty and Bosema, 1956), use of deafferented limbs in both skilled and expressive actions by monkeys (Taub and Berman, 1968) and chewing in rabbits (Dellow and Lund, 1970).

The whole of the responsive area in the midbrain receives extensive projections from the archistriatum by way of the occipitomesencephalic tract (Zeier and Karten, 1971), which will be considered in the discussion of more rostral regions.

Zigmond et al. (1973) have added an interesting facet to the study of midbrain controls of vocalizations in their demonstration that the cells of the nucleus intercollicularis selectively take up tritium-labeled testosterone. This suggests that the area might be involved in sensitizing call mechanisms to circulating gonadal hormones.

Diencephalon

Tracing vocal mechanisms rostrally from the midbrain into the diencephalon leads to the second major call area described from the brain stem in the region just ventral to the nucleus *ovoidalis* (Brown, 1971; Peek and Phillips, 1971; Phillips *et al.*, 1972). The nucleus itself does not yield vocalizations, but thresholds at points just below it are almost as low as those in the lateral midbrain, and calling can be almost as free from other behaviors here as from the midbrain (Phillips *et al.*, 1972). The effective area coincides with the tractus occipitomesencephalicus (OM) which distributes extensively to the midbrain calling regions described above (Zeier and Karten, 1971).

Simultaneous threshold stimulation of OM and MCR resulted in more calls than the sum of those produced by stimulating each site alone (Phillips *et al.*, 1972). With slightly more intense stimulation the interaction equalled the sum of the two alone, and with maximal stimulation at both sites, fewer calls were evoked than the sum from each of the sites stimulated alone. This appears to be classical Sherringtonian facilitation and occlusion. A second experimental test involved implanting an electrode in each of the sites, testing for evoked calls, then lesioning the midbrain call area. Such lesions greatly reduced or abolished calling evoked from the subovoidalis site as one would expect if the electrode was stimulating OM and that OM was exerting its effect on neurons of the lateral area (Phillips and Youngren, 1974). To be effective these midbrain lesions had to be large, apparently much larger than those that are effective in eliminating spontaneous vocalizations. This suggests that the muting lesions reported by Brown do not destroy the call coordinating mechanisms but more probably eliminate some sort of input. This hypothesis is supported by observations that, after lateral lesions which abolish OM evoked calls, stimulation 1—2 mm ventral to OM in the hypothalamus will still yield poststimulus calls. These observations do not rule out a possible crossed pathway for the hypothalamic effect, but the effects were not by way of the contralateral MCR or OM, because those areas had also been lesioned as controls prior to the ipsilateral MCR lesions (Phillips and Youngren, 1974). Calls can be evoked from the ventral and caudal portions of the reticular formation lateral and caudal to the lesions after the lesions.

Except for the area ventral to the nucleus ovoidalis, ESB in the avian thalamus has rarely produced vocalizations. Putkonen (1967) plotted a few 'excitement' points in the dorsal thalamus, and Phillips *et al.* (1972) showed a few points in the caudal ventral part of the dorsal thalamus of chickens. Phillips and Youngren (1973) also elicited vocalizations from a few points in the dorsolateral thalamus of mallard ducks, but Maley (1969) found only negative points there. Likewise Brown (1971 and 1973), and Peek and Phillips (1971) evoked no calls from extensive surveys of the thalamus outside the ovoidalis area. Vocalizations have been produced from many points in the hypothalamus and preoptic areas in all species studied (redwinged blackbirds, Brown, 1971; pigeons, Åkerman, 1966a and b; mallard ducks, Maley, 1969, Phillips and Youngren, 1973; and chickens, Peek and Phillips, 1971, Phillips *et al.*, 1972). Although Peek and Phillips (1971) found no calling from the caudal parts of the hypothalamus in anesthetized chickens, both

Maley (1969) and Phillips and Youngren (1973) plotted numerous points in the posterior hypothalamus of awake mallard ducks. Poststimulus calling is readily produced by ESB at points in the posterior hypothalamus ventral to the nucleus ovoidalis and OM in adult chickens, and these points match Andrew's (1973) points that interrupt peeps during stimulation. Andrew (1973) has analyzed in detail the vocalizations that can be elicited from various parts of the hypothalamus in awake domestic chicks. Brown's (1971) conclusion that septal and diencephalic sites yielded more complex and variable calls in redwinged blackbirds than did mesencephalic placements seems to hold in all the studies that have been reported to date. Similar contrasts in variability of behavior evoked from forebrain and hypothalamic sites and between diencephalon and midbrain sites have been reported by Harwood and Vowles (1966 and 1967). The ventral septum and the region surrounding the anterior commissure have also yielded calls in almost all studies.

The existence of call sites in the anterior hypothalamus and preoptic area is of special interest because of the role of these areas as hormone receptors and regulators of pituitary gonadatrophic function. Meyer (1973) demonstrated that the uptake of tritium-labeled testosterone in male chick brains was concentrated in the medial preoptic and periventricular areas. Barfield showed that testosterone implants in preoptic areas restored mating behavior in capons (Barfield, 1969) and in ring doves (*Streptopelia risoria*) (Barfield, 1971). Hutchison (1969 and 1970) also facilitated reproductive behavior in castrated male ring doves with testosterone implants in the anterior hypothalamus and preoptic area, and Komisaruk (1967) found that courtship was suppressed by progesterone implants in this same area. The coincidence of hormone receptors and sites where stimulation evokes calling suggests a possible integrating role for this area in the production of song and other vocalizations that are conditioned by gonadal hormones. Andrew (1973) has discussed the connections between the preoptic and hypothalamic areas and the MCR. These are not well known in birds, but comparative studies strongly suggest that substantial numbers of fine fibers run in periventricular regions caudally to the midbrain periaqueductal gray (the dorsal longitudinal bundle). How this interrelates with the more lateral call area remains to be determined.

Telencephalon

Telencephalic structures have been much less extensively explored for vocalizations than have those in the brain stem. Phillips (1964) obtained fear and escape behavior accompanied by a wide range of vocalizations from medial archistriate sites in mallard ducks. Åkerman (1966a and b) evoked bowing and nest calling from the paleostriatum and parolfactory area and also from a few sites in the caudal neostriatum of pigeons; these vocalizations were components of what he called bowing and nest-demonstration displays. He also reported escape responses and accompanying growling alarm calls from the archistriatum and nearby points in the neostriatum and hyperstriatum. Putkonen (1967) found many 'attention or excitement' reactions, which included vocalizations, from archistriatum and adjacent points in paleostriatum and caudal neostriatum. Other points in the anterior basal forebrain were probably in the preoptic and parolfactory areas and in the anterior borders of the paleostriatum. Maley (1969) had few forebrain points but did evoke aggressive calling from a few archistriate points in mallard ducks. Most of the extensive samples of forebrain points in awake redwinged blackbirds (Brown, 1971) or awake (Phillips *et al.*, 1972) or anesthetized (Peek and Phillips, 1971) chickens yielded no vocalizations. The few exceptions in the redwings were mostly in the archistriatum or occipitomesencephalic tract and a few totally unexpected points near the brain surface in the hyperstriatum. None were evoked from the forebrain in the studies of anesthetized chickens, and only a few in awake chickens. These latter were mostly in the archistriatum, or close to it, and mostly had thresholds of over 100 μA.

The scarcity of active points for vocalization in archistriatum is puzzling in light of the ease with which vocalizations can be evoked from its chief efferent pathway, the tractus occipitomesencephalicus, in the diencephalon. We have tried repeatedly to evoke calls from the archistriatum using large currents, large electrodes and wide interelectrode spacing, but none of these consistently evoked vocalizations. Simultaneous stimulation of points in the archistriatum and in the MCR indicates a role of the archistriatum in vocalizations. Results are very similar to those with combined stimulation of the OM and MCR: facilitation at low levels and occlusion at high intensities. This interaction is not unexpected in light of the OM studies, but it makes the inability to evoke calls from most archistriate sites more puzzling.

Recording calling responses to stimulating electrodes in both the archistriatum and the MCR of the same side, first to each separately, then to the two stimulated simultaneously, demonstrated facilitation. These tests were then repeated successively after lesions in the contralateral MCR, contralateral OM and finally in the ipsilateral OM. Stimulation of the archistriatum increased the number, amplitude, and duration of the calls. Lesions in the ipsilateral OM were most effective in eliminating the interaction between the archistriatum and midbrain. This was expected since OM is on the path from the archistriatum to the MCR. Our interpretation of these data is that the archistriatum contains local inhibitory circuits as well as excitatory output neurons so that the net effect of ESB is only weak activation of the call area. Some tentative support for this comes from a series of birds in which we compared the effects of stimulating various subdivisions of the archistriatum on midbrain evoked calling in an attempt to take advantage of Zeier and Karten's (1971) anatomical findings. The only clear effect that came of this was that medial archistriate sites facilitated calling better than did lateral placements. A cursory look at the anatomy indicates that the output fibers are scattered in the lateral portions of the archistriatum and that they converge in the medial parts so that medial stimulation would hit a higher percentage of output fibers.

Interactions of other forebrain areas with midbrain in evoking calling were tested with similar techniques. Hippocampal, septal and pyriform cortex sites had no effect, but placements in the neostriatum (11 electrodes) and paleostriatum augmentatum (4 electrodes) near the archistriatum significantly increased calling. These latter structures do not send projections out of the forebrain (Zeier and Karten, 1971) but they project to the archistriatum and paleostriatum primitivum, respectively, both of which send axons to the midbrain.

Location and nature of call generator

Fig. 7 summarizes our knowledge of central mechanisms for calling in birds.

The figure follows, from top to bottom, the course of this discussion from the periphery centrally. Inspiratory and expiratory mechanisms represent the controls for normal breathing. They send outputs to both spinal motor neurons and to the hypoglossal nucleus which control the intercostal and abdominal muscles and the tracheal muscles, respectively. Two pathways from nucleus nervus hypoglossus (n. XII) to musculus tracheolateralis are shown because that muscle is active both during silent breathing and during calling. The musculi sternotrachealis and ypsilotrachealis, in contrast, contract only during voiced expirations (calling). The pathway from the 'call generator' (presumably located near the obex) to the inspiratory mechanism is included because inspiratory activity augments during ESB trains even in curarized, anesthetized, hyperventilated preparations where no possible proprioceptive feedbacks or changes in blood or lung gas composition could be involved. 'Affect' next to archistriatum suggests that the latter plays a major role in integrating calls with other aspects of affective behavior. We strongly suspect that both midbrain and hypothalamic inputs are also importantly involved, but data are scarce. The single pathway from archistriatum via the OM to the MCR and thence to the 'call generator' reflects the ability of MCR lesions to block OM evoked calling. The figure depicts one side of the brain only, for clarity; addition of known crossed effects would not change its basic form. The 'call generator' is shown caudal to the MCR because (1) no call-like bursting of neurons has yet been detected in midbrain, (2) calls can be evoked from sites caudal to the MCR, and (3) transection experiments provide some evidence that calls can be evoked after transection of the brain stem at the rostral end of the medulla.

We think that the mechanisms shown in Fig. 7 are pertinent to more complex vocalizations as well as to alarm calls because (1) it is presumably

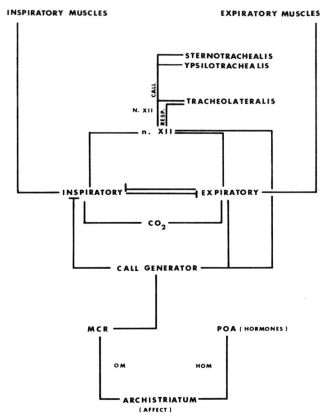

Fig. 7. Diagrammatic summary of control of avian vocalization from periphery (top) to best documented forebrain areas (bottom). See text for discussion. N. XII, nervus hypoglossus; n. XII, nucleus nervus hypoglossus; OM, tractus occipitomesencephalicus; HOM, tractus hypothalamooccipitomesencephalicus; MCR, the midbrain call region described in the text; POA, preoptic and anterior hypothalamic area.

easier to modify existing mechanisms during evolution than to 'invent' totally new ones, thus vocal mechanisms are intimately related with respiratory circuits and probably are derived from them, and (2) brain stimulation and lesions studies have not revealed separate machinery for various calls. The same brain areas are active in all birds studied, and all reports of ESB evoked vocalizations mention the ability to change call parameters considerably by varying stimulus parameters. Potash (1970a) interpreted his results to indicate that different calls were produced by different parts of the midbrain, but he got different calls from single sites and similar calls from different sites (and tested too few points to investigate repeatability of call type—brain structure relationships). Andrew (1973) reported correlations between call types and hypothalamic structure in baby chicks, but as he points out, calls evoked there are parts of more complex responses and may not be closely related to motor coordination mechanisms. Our extensive studies of evoked vocalizations in chickens and in mallard ducks have not yielded clear correlations between anatomy and call type, other than the fact that, as one stimulates more and more rostral structures, the stimulus—response relationships become more variable.

We have reported (Phillips *et al.*, 1972) that a wide range of call parameters can be produced by altering the stimulus parameters at a given midbrain site in chickens. These range from very brief cuks to prolonged squawks with varying intensities and intervals between sound pulses. Furthermore, Peek (in preparation) has shown that respiratory drive, produced by adding CO_2 to the ventilating air stream, strongly interacts with midbrain evoked neural and muscular correlates of calling. In particular, strong inspiratory drive converts a brief call pulse into a series of two or three closely spaced bursts followed by a very prolonged one. This pattern resembled crowing activity so markedly that Peek recorded EMG's from inspiratory and expiratory muscles of a spontaneously crowing rooster. He found that the pattern was very similar. Furthermore, increasing bursts of inspiratory activity began to appear as much as a minute or two before the rooster crowed, and the rooster took an abnormally deep breath in his final preparation for each crow. A similar deep inspiration precedes song in canaries (Calder, 1970).

It is no great surprise that a bird takes a deep breath before it sings, but the similarities between the interaction of inspiratory drive and calling activity produced by midbrain ESB and spontaneous crowing suggest a specific input that appears capable of altering the output of the hindbrain call generator from 'cuks' toward 'crows'. This hypothesis is supported by the results of neuromime and computer simulation of neural oscillators by Harmon (1964) and Wilson (1967), which indicate that even very simple neural nets can be driven to oscillate by unpatterned input, and that the outputs of these nets can vary widely and abruptly with continuous variation of input drive. This means that a single call oscillator has the potential to produce a wide range of vocalizations such as those that are actually observed in birds. This becomes even more likely when the complex, multisensory processing that has been demonstrated for receptors and early stages of sensory pathways in a variety of animals is taken into account. For instance, to produce song, hormone sensitive cells in the preoptic area may discharge in either a patterned or a tonic way to the midbrain (or more caudal sites) where these impulses would interact with other inputs to alter the call oscillator output in the direction of crowing or song. Perhaps this could be tested by simultaneous stimulation in preoptic and midbrain areas to see if crows can be produced.

The demonstrated interactions between respiration and calling raise the possibility that a single oscillator is responsible for both calling and for respiration, but the fact that the sternotrachealis (and ypsilotrachealis) muscles fire only during calling and never during even very intense respiration makes it necessary to postulate at least a semi-independent call system.

CENTRAL CONTROL OF VOCALIZATION IN OTHER VERTEBRATES

A wide enough variety of species have now been studied in detail for meaningful comparisons of the mechanisms in the different groups to be made. In the toadfish (*Opsanus beta*), the sounds are made by contraction of muscles attached to the air bladder (Demski *et al.*, 1973). Apparently these contractions produce vibrations in the air of the bladder which are transmitted through the body wall into the water. In birds, the mechanism is quite different: the expiratory air flow sets the syringeal membranes into vibration, and these vibrating membranes produce alternating compression and rarefaction of the air stream. Apparently the rest of the respiratory system does not appreciably modulate frequency (Greenewalt, 1968). In frogs and mammals the sounds are produced by the expiratory air stream setting vocal cords into vibration, so that little puffs of air are passed periodically to give a basic frequency which is then modified by resonant chambers of the mouth and pharynx (mammals), and probably by the throat sacs (frogs). Despite these wide differences in effector mechanism, the hypoglossal nerve plays a major role in all four classes, controlling the air-bladder muscles in toadfish, the syringeal muscles in birds, and the tongue in frogs and mammals.

Detailed comparisons can best be made in the midbrain where vocal mechanisms have been studied most thoroughly. Magoun *et al.* (1937) explored the midbrain of anesthetized rhesus monkeys and cats under light anesthesia and/or decerebration. Their plots show vocalizations from an extensive area running from the periaqueductal gray at the pretectal level rostrally, laterally and caudally in the gray under the superior colliculus, back ventromedially to the inferior colliculus, and on down ventrolaterally near the facial nucleus. Many of these points produced vocalizations without autonomic effects, much as we have reported from lateral midbrain sites in birds. Kanai and Wang (1962) replicated these studies and produced very similar plots for anesthetized cats. In addition, they described an area just ventral to the medial lemniscus from which vocalizations were evoked without signs of rage and where lesions muted the cats, blocking responses to more rostral stimulation. The calling from the ventrocaudal area was described as 'rhythmic crying' with each expiratory phase; the same type of vocalization could be produced even after transection of the brain stem between the pons and midbrain.

In frogs (Schmidt, 1965) and toadfish (Demski *et al.*, 1973) vocalizations can also be obtained after transections at pontine or even anterior medullary levels, and our results with birds suggest that vocalizations can be evoked from medullary levels after brain stem transection. Furthermore, the lateral pathways plotted by Peek and Phillips (1971) in the ventrolateral medulla are in a very similar position to those Kanai and Wang found just below the medial lemniscus. According to Nauta and Karten (1970), non-mammalian forms have no homolog of the medial lemniscus, so that exact homologies are difficult to determine without more knowledge of the connections of 'call-sites' in all classes. The distribution of points from which vocalizations could be evoked from the midbrain of awake rats (Waldbillig, 1975) is almost identical to those for cats and rhesus monkeys.

Recently, Jürgens *et al.* (1967) and Jürgens and Ploog (1970) have mapped the entire brain of the squirrel monkey for points that yield vocalizations. The midbrain portion of the system that they mapped closely resembles that reported for other mammals. These authors report that, like the call areas in birds (Phillips *et al.*, 1972) and frogs (Schmidt, 1965), mechanical stimulation in midbrain areas of squirrel monkeys produces calls. The plots of 'vocalization' sites from the midbrains of fish (Demski and Gerald, 1972 and 1974) closely resemble those from birds in that the points lie in the medial parts of the torus semicircularis (the intercollicular nucleus of birds and the probable homolog of the inferior colliculus of mammals) and the adjacent reticular formation. In more caudal regions, the effective areas are more ventral and lateral, just as they are in birds and mammals.

Comparisons with frogs are slightly more difficult because points have been presented in horizontal rather than transverse sections and because, for excellent reasons, Schmidt has focused attention more on the minimum tissue that would suffice for sound production rather than on a distribution of active points. Nevertheless, the structures involved appear to be very similar to those involved in fish, birds, and mammals. Schmidt (1965) refers to calling produced by mechanical stimulation of the posterior parts of the optic lobes, and he reports temporary muting resulting from removal of the inferior colliculus (muting lasted as long as six weeks in some animals). On anatomical grounds the inferior colliculus of frogs can be expected to be closely related to tissues homologous to the MCR in birds. The most effective area for electrical stimulation in frogs lies below the inferior colliculus, in, or very closely adjacent to, the secondary visceral nucleus at the level of the main sensory nucleus of the trigeminal nerve. This area lies in the presumed output path from the avian MCR. Neural correlates of vocal activity recorded in this area of frogs persist after total denervation of the isolated brain stem, showing that pattern generation is not dependent on peripheral feedback (Schmidt, 1974a). Demski and Gerald (1972) described their midbrain points as being in the medial part of the torus semicircularis, the ganglion isthmi and the area surrounding the acousticolateral lemniscus, whereas Schmidt speaks of the primary area in frogs as isthmotegmental but also mentions calls from just below the inferior colliculus. The call areas are very likely homologous structures in all four classes of animals. Stimulation of the isthmotegmental area in the totally deafferented (isolated) brain stem of frogs still produces the neural correlates of calling (Schmidt, 1974b), demonstrating the central nature of the pattern generating mechanism even more strikingly than does such persistence in curarized birds.

Demski and Gerald (1972) plotted only one point in the hypothalamus and that was in the midline. In frogs, Schmidt repeatedly evoked mating calling from the preoptic area (Schmidt, 1968). In cocks, anterior hypothalamic and preoptic points produced calls that were indistinguishable from food calls or tidbitting calls to the ear, but we did not analyze them sonographically (Tweeton et al., 1973).

The work in fishes was done in both anesthetized (Demski and Gerald, 1972) and awake (Demski and Gerald, 1974) toadfish, so that in all four classes of vertebrates, midbrain ESB has been found to elicit vocalizations in both anesthetized and awake animals. Demski and Gerald (1974) have proposed that the two kinds of calls they have evoked: grunts and boat whistles, are produced by a single mechanism, much as we have proposed for bird vocalizations. Schmidt (1971) also suggested a single mechanism when he said that: "The afferent vocal center may be basically organized to cause release calling . . . and that this effect may then be modified by preoptic or intense pain input to lead, instead, to mating calling or warning crying". In the same paper Schmidt discussed the close relationship between medullary respiratory and call mechanisms, much as did Peek and Phillips (1971) for chickens.

Jürgens and Ploog (1970) reached totally different conclusions from their work with squirrel monkeys. They interpret their results to indicate that midbrain sites are afferent rather than efferent paths, in contrast to the interpretations of all the other studies. Their arguments include: (1) species-specific behaviors are highly complex and require coordination of several cranial nerves and autonomic activity; (2) the shortest latencies for vocalizations are found in the periaqueductal gray but these are 0.2 sec, far too long to be efferent conduction times; (3) all the vocalizations elicited persist until the end of stimulation and (4) only by lumping different call types is a continuous plot obtained.

Certainly species-specific calls are complex and require coordination of several sets of muscles in monkeys, but this does not mean that, in principle, several pattern generators are needed. Latencies of 0.2 sec are too long to represent efferent conduction times, but not necessarily too long to activate a pattern generating network whose output activates (or is) the actual efferent path. Minimum latencies reported for calls in birds after midbrain ESB range from about 1 sec (estimated from graphs of Brown, 1973) to 0.12 sec (mean latency to first call in 100 trials in five anesthetized chickens). This includes the time to move air through the trachea to produce the call. Latencies to EMG bursts in abdominal muscles were as brief as 40 msec, a very short time if one assumes that the stimulating electrode is in afferent paths.

The strongest reason for considering midbrain areas as efferents rather than afferents is that calls can still be elicited after decerebration and even after transection of the brain between midbrain and medulla in fish, frogs, cats, rhesus monkeys, and, probably, chickens, indicating that the medulla is capable of producing call patterning. Furthermore bursts of nerve spikes patterned just like calling can be recorded from more caudal regions in both frogs (Schmidt, 1974b) and chickens, but have not been found rostral to the caudal medulla in chickens or to the isthmotegmental region in frogs (Schmidt, 1974b). In toadfish (Demski and Gerald, 1972) hindbrain ESB interfered with call patterning. The single, prolonged calls reported by Peek and Phillips (1971) after hindbrain ESB also seem to indicate jamming of a pattern generator, and the placements are similarly in the medulla oblongata.

Finally, lumping various calls may give a more accurate picture than attempts to classify the sounds and to plot their anatomical substrates separately. A look at the sonograms of cackling calls in Fig. 1 of Jürgens and Ploog (1970) shows that all the modulations that are present in the other categories of calls are also found within the cackling category. Since their eight categories are condensed from some 30 categories that Winter *et al.* (1966) distinguished for squirrel monkeys, a number of sounds intermediate to the eight types must exist. This is just what one would expect if a single sound generating mechanism was modulated in various ways by inputs from other brain structures to produce the observed range of sounds. Possibly vocal mechanisms have differentiated much more in squirrel monkeys than in other mammals or in birds, fish or amphibia, but, in view of the striking similarities in the anatomical localization of points within the brain where ESB will reliably yield vocalizations in four of the five vertebrate classes, we think that it is unlikely at brain stem levels.

We think that treating all vocalizations as a single class has already proven itself as a powerful tool in the analysis of their neural mechanisms. Although a distant goal of such analyses is to explain the way in which *all* vocalizations are produced, we firmly believe that progress will be more rapid if we concentrate on the simplest kinds of controls rather than attempting to analyze more complex ones. While we agree with Brown's statement in his 1969 review that "...the central problem in the study of neural vocalization control mechanisms is not to locate brain regions which 'control' vocalizations; it is to determine how the brain as a whole functions during the various vocalizations...", and that "...this is not an isolated problem but must be viewed within the larger context of brain—behavior relations in general.." (Brown, 1969), we disagree with his subsequent statement that "The controls of vocalizations are the controls of sexual, agonistic and other types of behavior in which vocalization play a role...". We think this statement is premature. We do not know in detail how any behavior in any vertebrate is controlled, not even the Mauthner cell-mediated tail-flip of fish and tailed amphibia. Even if it is not true in principle, we submit that it is good tactics to investigate controls as if they were isolated bits of neural functioning for our first analyses, even while trying to integrate the results into a broader context and meaning. Our ultimate goal is to understand brain—behavior relationships, but at present we are a long way from understanding such 'simple' problems as how alarm calls are turned on and off, much less how a thrush derives his complex musical patterns (Nelson, 1973) or a human being speaks.

IN CONCLUSION

This paper has attempted to review studies on the brain mechanisms that control avian vocalizations with the hope that they will shed light on brain—behavior relationships generally, and, more specifically, on brain production of affective states and associated species-typical behavior. It has concentrated on a study of motor output as a direct way to move from observed behavior, which is movement, into the central neuronal control of that behavior. Avian vocalizations were chosen because (1) they have relatively simple effector mechanisms that promise easier analysis than most behaviors, (2) the repeated and rhythmical nature of many vocalizations makes analysis easier, and (3) of the availability of an excellent neurophysiological preparation for this study. The alarm 'cuk' call of chickens was traced back from the respiratory pump and syringeal control muscles into the hindbrain, and, with ever decreasing detail and understanding, into the midbrain, diencephalon, and forebrain.

Probably the most exciting product of this analysis was the realization of the amazingly close similarities between both the peripheral and central control mechanisms for vocalizations in toadfish (if drumming on an air bladder can be called a vocalization), several species of frogs, several species and orders of birds, and three orders of mammals. In frogs and birds, at least, this similarity extends to the central nature of the pattern generating mechanisms. Sufficient information has now accumulated for us to ask increasingly sharp questions, and both the physiological preparations and the neurophysiological tools are available for their answering.

Acknowledgements

Paper #8838, Scientific Journal Series, Minnesota Agricultural Experiment Station. The work on vocalizations of adult chickens and mallards was supported by the Minnesota AES, by research grants GB 2838 and G 9583 from the National Science Foundation, and by U.S. Public Health Service research grants NB 06981 and NS 10341. We gratefully acknowledge the indispensable assistance of Mr. Orlan Youngren over the years. We thank Dr. F. Russel Lockner for his critical reading of and comments on the entire manuscript.

REFERENCES

Åkerman, B. (1966a) Behavioural effects of electrical stimulation in the forebrain of the pigeon. I. Reproductive behavior. *Behaviour*, 26, 323—338.

Åkerman, B. (1966b) Behavioural effects of electrical stimulation in the forebrain of the pigeon. II. Protective behaviour. *Behaviour*, 26, 339—350.

Andrew, R.J. (1964) Vocalization in chicks and the concept of "stimulus contrast". *Anim. Behav.*, 12, 64—76.

Andrew, R.J. (1973) The evocation of calls by diencephalic stimulation in the conscious chick. *Brain Behav. Evol.*, 7, 424—446.

Ariëns Kappers, C.V., Huber, C.G. and Crosby, E.C. (1960) *The Comparative Anatomy of the Nervous System of Vertebrates, Including Man.* Hafner, New York.

Baerends, G.P. and Baerends-van Roon, J.M. (1950) An introduction to the study of the ethology of cichlid fishes. *Behaviour*, Suppl. 1, 1—242.

Barfield, R.J. (1969) Activation of copulatory behavior by androgen implanted into the preoptic area of the male fowl. *Horm. Behav.*, 1, 37—52.

Barfield, R.J. (1971) Activation of sexual and aggressive behavior by androgen implanted into the male ring dove brain. *Endocrinology*, 89, 1470—1476.

Bentley, D.R. (1969) Intracellular activity in cricket neurons during generation of song patterns. *Z. vergl. Physiol.*, 62, 267—283.

Brown, J.L. (1965a) Vocalization evoked from the optic lobe of a songbird. *Science*, 149, 1002—1003.

Brown, J.L. (1965b) Loss of vocalization caused by lesions in the nucleus mesencephalicus lateralis of the redwinged blackbird. *Amer. Zoologist*, 5, 693.

Brown, J.L. (1969) The control of avian vocalization by the central nervous system. In R.A. Hinde (Ed.), *Bird Vocalizations*. Cambridge University Press, London, pp. 79—96.

Brown, J.L. (1971) An exploratory study of vocalization areas in the brain of the redwinged blackbird (*Agelaius phoeniceus*). *Behaviour*, 39, 91—127.

Brown, J.L. (1973) Behavior elicited by electrical stimulation of the brain of the Stellar's jay. *Condor*, 75, 1—16.

Calder, W.A. (1970) Respiration during song in the canary (*Serinus canaria*). *Comp. Biochem. Physiol.*, 32, 251—258.

Collias, N.E. and Joos, M. (1953) The spectrographic analysis of sound signals of the domestic fowl. *Behaviour*, 5, 175—187.

Conrad, R. (1915) Untersuchungen über den unteren Kehlkopf der Vögel. I. Zur Kenntnis der Innervierung. *Z. wiss. Zool.*, Bd. CXIV, 532—578.

Dellow, P.G. and Lund, J.P. (1970) Evidence for central timing of rhythmical mastication. *J. Physiol. (Lond.)*, 215, 1—13.

Demski, L.S. and Gerald, J.W. (1972) Sound production evoked by electrical stimulation of the brain in toadfish (*Opsanus beta*). *Anim. Behav.*, 20, 507—513.

Demski, L.S. and Gerald, J.W. (1974) Sound production and other behavioral effects of midbrain stimulation in free-swimming toadfish (*Opsanus beta*). *Brain Behav. Evol.*, 9, 41—59.

Demski, L.S., Gerald, J.W. and Popper, A.N. (1973) Central and peripheral mechanisms of teleost sound production. *Amer. Zoologist*, 13, 1141—1167.

Doty, R.W. and Bosema, J.F. (1956) An electromyographic analysis of reflex deglutition, *J. Neurophysiol.*, 19, 44—60.

Fedde, M.R., Burger, R.E. and Kitchell, R.L. (1964) Anatomic and electromyographic studies of the costo-pulmonary muscles in the cock. *Poultry Sci.*, 43, 1177—1184.

Gaunt, A.S. and Wells, M.K. (1973) Models of syringeal mechanisms. *Amer. Zoologist*, 13, 1227—1247.

Gaunt, A.S., Stein, R.C. and Gaunt, S.L.L. (1973) Pressure and air flow during distress calls of the starling, *Sturnus vulgaris* (Aves; passeriformes). *J. exp. Zool.*, 183, 241–262.

George, J.C. and Berger, A.J. (1966) *Avian Myology*. Academic Press, New York, pp. 250–270.

Greenewalt, C.H. (1968) *Bird Song: Acoustics and Physiology*. Smithsonian Institution Press, Washington, D.C.

Gross, W.B. (1964) Voice production by the chicken. *Poultry Sci.*, 43, 1005–1008.

Gross, W.B. (1968) Voice production by the turkey. *Poultry Sci.*, 47, 1101–1105.

Guyomarc'h, J.C. (1962) Contribution à l'étude du comportement vocal du poussin de "*Gallus domesticus*". *J. Psychol.*, 3, 283–305.

Harmon, L.D. (1964) Neuromimes: action of a reciprocally inhibitory pair. *Science*, 146, 1323–1325.

Harwood, D. and Vowles, D.M. (1966) Forebrain stimulation and feeding behavior in the ring dove (*Streptopelia risoria*). *J. comp. physiol. Psychol.*, 62, 388–396.

Harwood, D. and Vowles, D.M. (1967) Defensive behaviour and the after effects of brain stimulation in the ring dove (*Streptopelia risoria*). *Neuropsychologia*, 5, 345–366.

Holst, E. von and Schleidt, W.M. (1964) Wirkungen von psychopharmaka auf instinktives verhalten. In P.B. Bradley, F. Flugel and P. Hoch (Eds.), *Neuropsychopharmacology*. Elsevier, Amsterdam, pp. 22–29.

Hutchison, J.B. (1969) Changes in hypothalamic responsiveness to testosterone in male Barbary doves (*Streptopelia risoria*). *Nature (Lond.)*, 222, 176–177.

Hutchison, J.B. (1970) Influence of gonadal hormones on the hypothalamic integration of courtship behaviour in the Barbary dove. *J. Reprod. Fert.*, Suppl. 11, 15–41.

Jürgens, U. and Ploog, D. (1970) Cerebral representation of vocalization in the squirrel monkey. *Exp. Brain Res.*, 10, 532–554.

Jürgens, U., Maurus, M., Ploog, D. and Winter, P. (1967) Vocalization in the squirrel monkey (*Saimiri sciureus*) elicited by brain stimulation. *Exp. Brain Res.*, 4, 114–117.

Kadono, H., Okada, T. and Ono, K. (1963) Electromyographic studies on the respiratory muscles of the chicken. *Poultry Sci.*, 42, 121–128.

Kanai, T. and Wang, S.C. (1962) Localization of the central vocalization mechanism in the brain stem of the cat. *Exp. Neurol.*, 6, 425–434.

Karten, H.J. and Hodos, W. (1967) *A Stereotaxic Atlas of the Brain of the Pigeon (Columba livia)*. Johns Hopkins Press, Baltimore, Md.

King, A.S. and Molony, V. (1971) The anatomy of respiration. In D.J. Bell and B.M. Freeman (Eds.), *Physiology and Biochemistry of the Domestic Fowl*. Academic Press, New York, pp. 93–169.

Komisaruk, B.R. (1967) Effects of local brain implants of progesterone on reproductive behavior in ring doves. *J. comp. physiol. Psychol.*, 64, 219–224.

Lockner, F.R. and Youngren, O.M. (1975) Functional syringeal anatomy of the mallard. I. *In situ* electromyograms during evoked calling. In preparation.

Magoun, H.W., Atlas, D., Ingersoll, E.H. and Ranson, S.W. (1937) Associated facial, vocal, and respiratory components of emotional expression: an experimental study. *J. Neurol. Psychopath.*, 17, 241–255.

Maley, M.J. (1969) Electrical stimulation of agonistic behavior in the mallard. *Behaviour*, 34, 138–160.

Meyer, C.C. (1973) Testosterone concentration in the male chick brain: an autoradiographic survey. *Science*, 180, 1381–1382.

Murphy, R.K. and Phillips, R.E. (1967) Central patterning of a vocalization in fowl. *Nature (Lond.)*, 216, 1125–1126.

Nauta, W.J.H. and Karten, H.J. (1970) A general profile of the vertebrate brain, with sidelights on the ancestry of the cerebral cortex. In F.O. Schmitt (Ed.), *The Neurosciences: Second Study Program*. Rockefeller Univ. Press, New York, pp. 7–26.

Nelson, K. (1973) Does the holistic study of behavior have a future? In P.P.G. Bateson and P.H. Klopfer (Eds.), *Perspectives in Ethology.* Plenum, New York, pp. 281—322.

Newman, J.D. (1970) Midbrain regions relevant to auditory communication in songbirds. *Brain Res.,* 22, 259—261.

Nottebohm, F. (1971) Neural lateralization of vocal control in a passerine bird. I. Song. *J. exp. Zool.,* 177, 229—262.

Nottebohm, F. (1972) Neural laterization of vocal control in a passerine bird. II. Subsong, calls, and a theory of vocal learning. *J. exp. Zool.,* 179, 35—50.

Peek, F.W. (1972) An experimental study of the territorial function of vocal and visual display in the male redwinged blackbird (*Agelaius phoeniceus).* *Anim. Behav.,* 20, 112—118.

Peek, F.W. and Phillips, R.E. (1971) Repetitive vocalizations evoked by local electrical stimulation of avian brains. II. Anesthetized chickens (*Gallus gallus*). *Brain Behav. Evol.,* 4, 417—438.

Phillips, R.E. (1964) 'Wildness' in the mallard duck: effects of brain lesions and stimulation on 'escape behavior' and reproduction. *J. comp. Neurol.,* 122, 139—156.

Phillips, R.E. and Youngren, O.M. (1971) Brain stimulation and species-typical behaviour: activities evoked by electrical stimulation of the brains of chickens (*Gallus gallus*). *Anim. Behav.,* 19, 757—779.

Phillips, R.E. and Youngren, O.M. (1973) Electrical stimulation of the brain as a tool for study of animal communication. Behavior evoked in mallard ducks (*Anas platyrhynchos*). *Brain Behav. Evol.,* 8, 253—283.

Phillips, R.E. and Youngren, O.M. (1974) A brain pathway for thalamically evoked calls in birds. *Brain Behav. Evol.,* 9, 1—6.

Phillips, R.E., Youngren, O.M. and Peek, F.W. (1972) Repetitive vocalizations evoked by local electrical stimulation of avian brains. I. Awake chickens (*Gallus gallus*). *Anim. Behav.,* 20, 689—705.

Potash, L.M. (1970a) Vocalizations elicited by electrical brain stimulation in *Coturnix coturnix japonica. Behaviour,* 36, 149—167.

Potash, L.M. (1970b) Neuroanatomical regions relevant to production and analysis of vocalization within the avian *torus semicircularis. Experientia (Basel),* 26, 1104—1105.

Putkonen, P.T.S. (1967) Electrical stimulation of the avian brain. Behavioral and autonomic reactions from the archistiatum, ventromedial forebrain and the diencephalon in the chicken. *Ann. Acad. Sci. fenn. A5,* 130, 1—95.

Rüppel, W., (1933) Physiology and acoustic of bird voices. *J. Orn.,* 3, 441—453.

Saint Paul, U. von (1965) Einfluss von pharmaka auf die auslosbarkeit von verhaltensweisen durch elecktrische reizung. *Z. vergl. Physiol.,* 50, 415—446.

Schmidt, R.S. (1965) Central mechanisms of frog calling. *Behaviour,* 26, 251—285.

Schmidt, R.S. (1968) Preoptic activation of frog mating behavior. *Behaviour,* 30, 239—257.

Schmidt, R.S. (1971) A model of the central mechanisms of male anuran acoustic behavior. *Behaviour,* 39, 288—317.

Schmidt, R.S. (1973) Vocal mechanisms of release calling in northern leopard frog. *Copeia,* 624—627.

Schmidt, R.S. (1974a) Neural correlates of frog calling. Independence from peripheral feedback. *J. comp. Physiol.,* 88, 321—333.

Schmidt, R.S. (1974b) Neural correlates of frog calling: trigeminal tegmentum. *J. comp. Physiol.,* 92, 229—254.

Sherrington, C. (1906) *The Integrative Action of the Nervous System.* Scribner, London.

274

Taub, E. and Berman, A.J. (1968) Movement and learning in the absence of sensory feedback. In S.J. Freedman (Ed.), *The Neuropsychology of Spatially Oriented Behavior.* Dorsey Press, Homewood, Ill., pp. 173—192.

Tweeton, J.R., Phillips, R.E. and Peek, R.W. (1973) Feeding behaviour elicited by electrical stimulation of the brain in chickens, *Gallus gallus. Poultry Sci.*, 52, 165—172.

Vowles, D.M. and Beazley, L. (1974) The neural substrate of emotional behaviour in birds. In M.W. Schein and I.J. Goodman (Eds.), *Birds: Brain and Behavior.* Academic Press, New York, pp. 221—258.

Waldbillig, R.J. (1975) Attack, eating, drinking, and gnawing elicited by electrical stimulation of rat mesencephalon and pons. *J. comp. physiol. Psychol.*, in press.

Watanabe, T. (1960) Comparative and topographical anatomy of the fowl. VII. On the peripheral course of the vagus nerve in the fowl. *Jap. J. vet. Sci.*, 22, 152—158.

Willows, A.O.D. (1967) Behavioral acts elicited by stimulation of single, identifiable brain cells. *Science*, 157, 570—574.

Wilson, D.M. (1967) Central nervous mechanisms for the generation of rhythmic behaviour in arthropods. *Soc. exp. Biol. Symp.* XV. Academic Press, New York, 199—228.

Winter, P., Ploog, D. and Latta, J. (1966) Vocal repertoire of the squirrel monkey (*Saimiri sciureus*), its analysis and significance. *Exp. Brain Res.*, 1, 359—384.

Wood-Gush, D.G.M. (1971) *The Behavior of the Domestic Fowl.* Heinemann, London.

Youngren, O.M., Peek, F.W. and Phillips, R.E. (1974) Repetitive vocalizations evoked by local electrical stimulation of avian brains. III. Evoked activity in the tracheal muscles of the chicken (*Gallus gallus*), *Brain Behav. Evol.*, 9, 393—421.

Zeier, H. and Karten, H.J. (1971) The archistriatum of the pigeon: organization of afferent and efferent connections. *Brain Res.*, 31, 313—326.

Zigmond, R.E., Nottebohm, F. and Pfaff, D.W. (1973) Androgen: concentrating cells in the midbrain of a songbird. *Science*, 179, 1005—1007.

13

MIDBRAIN MECHANISMS OF CALLING AND THEIR RELATION
TO EMOTIONAL STATES

R.J. ANDREW

Ethology and Neurophysiology Group, School of Biological Sciences, University of
Sussex (Great Britain)

INTRODUCTION

Everyday language is very rich in words which draw a variety of subtle
distinctions between a great many different human emotional states. External signs of such states can also be described in considerable detail by means
of specialised terms (*e.g.* chuckle, grin, frown). There is little doubt that it is
usually possible to identify particular emotional states from direct observation of real sequences of behaviour, and that, once identified, emotional
states can be described to other observers who will understand and agree.

Despite long sustained interest on the part of psychologists the scientific
investigation of emotional states has hardly advanced with any security
beyond the detailed and extensive knowledge which we all share by right of
being human. Indeed, it could be argued that most attempts at systematisa-
tion have been misleading rather than helpful, in that they have disregarded
important aspects of emotional behaviour in their attempts at a complete
description and explanation. To take as an example the area with which I am
most familiar, ethological discussions of displays have emphasised the importance as display components of intention movements, such as incomplete
movements of copulation, attack and locomotion. In this, and the demonstration that the external signs of human emotional states can be regarded as
displays, evolved like the displays of animals for communication, the
ethological approach has made a considerable advance. However, equally
obvious states such as those of general excitement or depression have been
largely ignored by ethological theory, if not by all ethologists.

Experimental work on emotional behaviour has been handicapped by the
fact that those species which have traditionally been used in the laboratory,
such as rats, mice and pigeons do not readily perform displays which are
conspicuous to human observers. In the case of rats and mice, many displays
certainly pass unnoticed, since most of their calls are supersonic, and the

276

specific scent signals which they emit are also not obvious to man. The domestic chick is far more suitable for such work. Its calls are given very readily in every type of situation, and can be divided into a number of categories which offer parallels to various human emotional states.

Before proceeding to discuss the evidence which can be obtained from a study of the relationship between calls and other behaviour, and the disturbances produced in both by the destruction of various midbrain structures, it is necessary to give a few definitions. A group of display components, which include in mammals calls, expressive facial movements, piloerection of back and tail (and probably pupil dilatation and secretion of some cutaneous glands), and in the chick calls (and probably erection of neck feathers and pupil dilatation) are here termed emotional display (ED) components, whenever a general term is needed. The facilitation of urination and defaecation in species such as rats and mice should probably be included; indeed, definitions of 'emotionality' have sometimes been so restricted as to consider only increases of defaecation and decreases in locomotion (*e.g.* Candland and Nagy, 1969). However, a consideration of the large number of studies of emotional defaecation and urination (for review see Archer, 1973) shows these two indices to be ambiguous and variable in significance. This is no doubt partly because they do not involve behaviour which is specific to the state which they are used to define, but merely a change in frequency of performance of behaviour which occurs at a variety of other times.

As a great deal of the evidence cited comes from the calls of the chick, a special term 'calling phase' will be applied to periods of behaviour in which calling occurs: such periods have a number of other special properties, as will be shown.

Finally, a brief description of chick calls is necessary. I shall be concerned here only with the main sequence of chick calls, and not with crows and a variety of patterns of calling which become more common after testosterone (Andrew, 1969). The main sequence was divided by Collias and Joos (1953) into ascending 'pleasure calls' and descending 'distress calls'. Subsequently, the terms 'twitter' and 'peep' were suggested (Andrew, 1963a). Recent studies (Andrew, 1967 and 1973) of the relation between intracranial self-stimulation and calling have confirmed that it is best to avoid terms which presuppose a relatively direct relationship between a hypothetical central state such as pleasure and a particular call. It is now known that both ascending twitters and short descending calls of similar length occur together in the same situations (*e.g.* during feeding), although the latter tend to occur at points of maximum frequency of calling, and so may be regarded as slightly higher intensity responses. In view of this, and the convenience for computer recognition of a classification based solely on length of call and of the gaps preceding and following it (Andrew, 1973), both twitter and short descending call are now classed together as 'short calls' (calls <100 msec in duration, and separated by gaps >90 msec.).

These are sharp calls about as long as the sound produced by a light tap, and with the same quality of discreteness. 'Peeps' are defined as calls longer than 135 msec (and separated by gaps >90 msec). Such calls are heard as slightly elongated and descending calls, with a clearly musical quality, like that of a short whistle. 'Middle calls' are of intermediate length (100—135 msec); approximately the same calls have also been termed 'short peeps' in the past.

Short calls, middle calls and peeps form one continuous series in form as well as duration. A somewhat separate type of call will also be considered: the 'short trill', which is made of components no longer than 100 msec and separated by gaps so short (<50 msec) that the whole call sounds like a trill. Finally, peeps may intergrade with 'warbles', which occur in bouts separated by short gaps. They are variable in length but nearly all bouts contain a call longer than 100 msec, and this has been made part of their definition; both ascending and descending limbs are commonly well developed.

It may help to give a general impression of the kind of situations in which chicks are likely to call if a number of examples are briefly given at this point. The opportunity will also be taken to compare chick calling phases with what, it will be argued, are corresponding human emotional states.

The main points in this argument have already been set out in part (Andrew, 1962) for a number of species of primate (*e.g. Lemur, Galago*). These are, firstly, that at least some of the calls of each species form a series of responses which can all be evoked by the same stimulus or situation; calls of lowest intensity are evoked by brief exposure or a relatively ineffective stimulus, and calls of highest intensity by prolonged exposure to a very effective stimulus or situation. Secondly, the range of situations effective in producing any particular call is wide. Three important examples are perception of a strange object (which evokes mobbing in arboreal species such as *Lemur* and *Galago*), recognition of a sought-after stimulus such as food with or without an obstacle to its requisition, and being placed in a strange environment (which involves separation from social fellows or their equivalent). The same calls (in *Lemur* these are, in order of intensity, grunts, wails and shrieks) can be evoked in all three situations (as well as in a range of purely social encounters: *e.g.* greeting at a distance, in close bodily contact, persecution by a superior). The second situation requires some elaboration. Low intensity calls may occur both on seeing food and on beginning to eat. If an obstacle is placed in the way such as a closed door which makes it impossible to reach the food (or water or a social companion or to emerge for evening exercise), then calls of progressively higher intensity are performed, sometimes passing up to those of greatest intensity which otherwise may be heard commonly only during attack by a superior. Specific stimuli (*e.g.* from social fellows) are thus not essential for the performance of these calls, and it is difficult to explain them in all their occurrences as caused by drives such as fear or aggression (Andrew, 1962 and 1972b). This is not to say that specific stimuli are not involved in the causation of some calls (including ones which fall into the main series just discussed) and associated ED; the complications introduced by these and other causal factors are briefly considered at the end of this chapter.

Very much the same holds for chick calls in the short call—peep series and for some human sounds and associated facial expressions. The situations which will be specially considered here include the following.

(1) The sudden recognition of a significant stimulus (*e.g.* food or a conditioned stimulus (CS) for food) may evoke EDs, which are also evoked by a startling stimulus, even if the significant stimulus is not in itself conspicuous. Thus the chick may give short trills both when suddenly touched or on suddenly seeing a strange object, and on catching sight of food (or CS for food presentation). A comparable response in man is the sharp indrawing of breath, which is given both in startle response, and on finding a long sought-after object.

When chicks are running towards a goal or are feeding hastily and excitedly (which, because a chick pecks up its food, involves a series of responses to different stimuli), they give frequent short calls. Human responses are obviously much more complex. In particular, emotional vocalisations are complicated by the superimposition of speech: an excellent example is 'no', which can be heard in man as part of a chuckle, a breathy aggressive pant, a friendly grunt or a wail (Andrew, 1963b). However, young children sometimes smile and even chuckle on catching sight of a favourite food or toy. Smiles and interjections which may include chuckles may continue during hasty and excited play with the toy. Adult equivalents will be obvious to the reader.

(2) Attempts to respond (*e.g.* attempts to mount to copulate, pecks at a key for reinforcement) are accompanied by short calls in the chick, which are replaced by higher intensity calls if reinforcement is withheld. Warbles are more usual than peeps, and may develop directly out of short calls. This is probably because the warbles often occur in place of peeps when the chick becomes unresponsive and immobile in a situation which evokes intense calling, and such unresponsiveness is the commonest final result of withholding reinforcement to key-pecking in the young chick (Andrew, 1973 and unpublished). Duncan and Wood-Gush (1972) have described intense alarm calls in frustration situations in the adult fowl.

Comparable human behaviour is once again much more variable and complex. However, expletives are common in response to obstacles and are usually delivered as sharp grunts. If frustration is sustained and intense, wails and crying may develop, together with a cessation of responding, resembling the condition during chick warbles, or alternatively screams and a full tantrum may appear.

(3) In a strange environment chicks peep continuously, looking around in all directions. Under similar circumstances, children may wail and weep as

they look around; calling out is obviously in part an attempt at verbal communication, but its quality often suggests a component of emotional vocalisation.

If the exposure to the situation is for a long period, unresponsiveness tends to develop with warbles in chicks and crying in children.

(4) Finally, when fleeing from a very frightening object chicks usually peep and children often scream. Both show the vocalisation of the highest intensity of all to painful stimuli (long irregularly trilled peeps or trills in the chick, violent screams in man).

A variety of parallels can thus be drawn between chick and man. Comparable evidence from other primates has already been mentioned. Many other mammals appear to possess vocalisations with similar causation (*e.g.* Artiodactyls and Perissodactyls; Kiley, 1972). Parallels in the cat and dog are important, since much of the anatomical and physiological evidence which will be considered below comes from these species. Miaows or barks to closed doors and other obstacles are a matter of common experience. Dogs also bark at novel objects either singly as a part of a sudden startle, or persistently in a response like mobbing; cats tend to be silent at such times although they show other EDs (such as piloerection of back and tail). Miaows do occur with nervous scanning movements. Dogs howl when lost. Both species give intense vocalisations in fear or pain.

In the sections which follow it will be argued that activation of a single functional system explains the performance of a series of calls, and associated responses of the type just discussed, and some attempt will be made to establish the properties of such a system.

THE EFFECTS ON GENERAL BEHAVIOUR OF MUTING LESIONS IN THE CHICK

Data which bear on this problem were obtained from chicks in which the intercollicular area (ICo) or its medial edge was destroyed bilaterally (Fig. 1). This area was already known to yield calls at low threshold, even under deep anaesthesia, in the chick (De Lanerolle, 1972) as in a number of other birds (Brown, 1965b; Peek and Phillips, 1971). Stimulation in the conscious unrestrained chick yields little but calling and alerting, unlike stimulation in the central mesencephalic grey, for example, from which both low threshold calls and escape can be evoked (De Lanerolle 1972; Allan, 1971).

Lesions of the anterior ICo induce total muteness in the chick (De Lanerolle and Andrew, 1975), just as was earlier described by Brown (1965a) for the redwinged blackbird. However, in addition, it was found that all the phases of behaviour in which the chick would have been calling were also absent.

Full details are given in Andrew and De Lanerolle (1975) and Andrew

280

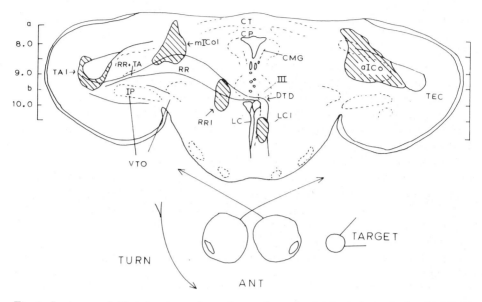

Fig. 1. Lesions and fibre tracts are here shown for a single plane through the chick brain as figured in Andrew (1973), where stereotactic coordinates and procedures are fully described. Depth is shown in millimetres below a skull roof reading of 0 mm in the ear—ear plane. Lateral measurements are to the same scale. The anteroposterior plane is titled in relation to the plane figured: a is 2.00 mm behind the ear—ear line, and b 1.65 nm. The full extent of lesions is shown in De Lanerolle and Andrew (1975) and Andrew (1975). In each case one of the smallest fully effective lesions is shown: in the case of TAl this is a reconstruction since the lesion figured had a dorsal extension into the tectum which was absent in the three other TA birds and so was omitted in the figure. The raphé route (RR), tectal afferents (TA) and main ventral tectal outflow (VTO) are shown diagrammatically, since RR and VTO break up into fascicles, which are widely scattered at some points (*e.g.* middle stretch of RR); in the case of TA, the medial origin is unknown, and no attempt is made to show its connections to deep tectal layers. A typical lesion of the anterior intercollicular area (aICo) is shown on the right hand side of the figure, and one of the medial edges of the intercollicular area (mICo) on the left. Lesions with no extensions above the plane of the tectal ventricle were effective in the mICo series. Amongst lesions shown on the left of the figure, mICol would block turning to the target shown as presented to the contraversive eye, whereas RRl and LCl (lesion of nucleus linearis caudalis) would tend to depress pecks at the target but promote exaggerated turning in the direction shown (*i.e.* ipsiversive to LCl and contraversive to RRl). Abbreviations: aICo, anterior intercollicular area; CT, tectal commissure; CMG, central mesencephalic grey; CP, posterior commissure; DTD, dorsal tegmental decussation; IP, nucleus isthmi pars principalis parvocellularis; LC, nucleus linearis caudalis; mICo, medial edge ICo; RR, raphé route; TA, tectal afferents; TEC, tectum; VTO, ventral tectal outflow; III, oculomotor nerve. The suffix l = lesion.

(1975). It will suffice here to consider a limited number of examples. Firstly, short calls given during feeding are most frequent at the onset of feeding and diminish in frequency during each bout, and between bouts of a feeding session. Calling is at its most frequent when the chick has not seen food for some little time: periods of over ten minutes are sufficient to ensure a good rate of calling. The first bout of feeding after such a period is typically longer than usual and is followed by a resumption of feeding after a very short gap. Lesioned birds give a short first feeding bout even after deprivation of food for three hours or more, and follow it with a long gap. They behave, in this respect, very like completely undisturbed chicks which are continually exposed to food: it is significant that such normal chicks also eat in almost or complete silence.

Short calls are also frequent when a chick is examining a novel object of about chick size, and showing intention movements of pecking or full pecks. In the lesioned chicks, examination of such an object is cursory, and far briefer than that shown by normal chicks; in particular, intention pecks and pecks are absent.

The same deletion of the whole calling phase of behaviour is shown in open field tests. Lesioned birds show neither the persistent scanning which normally accompanies peeps, not the inhibition of locomotion and all other responses which occur during peeping. Instead, locomotion resumes soon after entry to the open field and behaviour typical of an undisturbed chick such as preening or pecking often soon follows.

It is not possible to explain the complex of changes produced by the lesion by a primary effect specific to a single system of responses. The effects in the open field test, for example, could be ascribed to a depression of fear responses (assuming that scanning and the inhibition of locomotion can be so classified), but this would not explain the changes in feeding behaviour, or the unresponsiveness to novel objects (or a number of other effects: Andrew and De Lanerolle, 1975). Indeed, apart from the deletion of calling phases of behaviour, the chicks appear largely unaffected.

They are still capable of a wide repertoire of responses, including vigorous copulation (which is given in silence by normal chicks). They peck accurately, and feed freely if without excitement, so that normal growth is sustained. In addition, primary perceptual deficits do not seem to be involved (Andrew and De Lanerolle, 1975). The most obvious possibility is an effect on hearing due to damage to the adjacent nucleus mesencephalicus lateralis, but totally deaf chicks show none of the changes induced by ICo lesions.

THE PROPERTIES OF CALLING PHASES OF BEHAVIOUR

Calling phases, whether defined from normal behaviour or from the changes induced by anterior ICo lesions, have a number of basic properties. Firstly, during such phases the chick is commonly either treating a visual stimulus as though it were conspicuous or even startling (even if it is intrinsically inconspicuous) or, alternatively, scanning as though greatly affected by the absence of a stimulus (*e.g.* features of the home cage). Conversely, following ICo lesions, conspicuous visual stimuli no longer evoke ready targetting, fixation and pecking. This was systematically tested (Andrew, 1975) by the presentation of a small metal ball in the lateral visual field. In controls this almost always evoked targetting and pecking. Such targetting and pecking were almost invariably absent to stimuli presented to the eye contralateral to an ICo lesion. Pecking was also depressed when targets were presented in the binocular field of birds with bilateral lesions. The same effect is almost certainly involved in the loss of pecking and associated responses of attack in lesioned birds which have received testosterone; controls following such treatment all attacked vigorously when exposed to the visual stimuli presented by a thrusting hand.

A system of fibres running to the deep layers of the tectum from the ICo area or areas medial and rostral to it (as is discussed further below) degenerates following ICo lesions. Extratectal effects on the responsiveness of tectal units are now well established in mammals and Anura. Wickelgren and Sterling (1969) have shown in the cat that reversible inactivation of the visual cortex causes collicular units to become unresponsive; ablation of the visual cortex results in a broadening of the types of stimuli which would drive units. Evidence that behavioural unresponsiveness in the cat following such a lesion is due to the removal of a facilitatory input to the tectum has recently been presented by Sherman (1974). In frogs and toads the effects of input from the pretectal area has been investigated. Ewert (1970) found that removal of this input caused the prey-catching responses (targetting and striking with the tongue), which are initiated from tectal units when an appropriate visual stimulus is present, to be evoked more readily and by a far wider range of stimuli. Ingle (1973) showed that this was accompanied by a parallel change in the responsiveness of the units themselves.

It is known that in most vertebrates the optic tectum mediates visual targetting. Resemblances between the organisation of pecking in the chick and prey-catching in frogs and toads will be considered later; they are close enough, taken with the fact that accurate pecking persists in decerebrate chicks (Andrew, unpublished), to suggest that pecking, as well as targetting, is a tectal reflex in the chick. If this is assumed to be the case, then the following hypothesis may be suggested for further investigation: during calling phases extratectal mechanisms, which involve a path passing through the ICo, facilitate targetting, and the tectal reflexes which follow targetting. If this effect were to be topographically organised, and involve areas concerned with the recognition of visual stimuli, it would allow an inconspicuous stimulus, which would normally be ineffective, to evoke such responses. A general facilitation would explain persistent scanning.

The possible counterparts in human emotional states would be the ability of a particularly highly valent stimulus to attract and sustain gaze, and in the case of scanning, the ready and repeated targetting to any slight movement or other conspicuous stimulus shown in nervousness, frustration and states of excitement.

A second property of calling phases is that they are usually accompanied by motor facilitation. A very close coincidence of calling with the onset of responses like pecking has already been noted. When calling occurs in a period in which locomotion to a goal or (in extinction sessions, for example) restless and apparently undirected locomotion occurs, calling begins as locomotion begins and is often confined to times when locomotion is occurring.

A third property, which may appear more unexpected, is that of the inhibition of response. This is clearest in the chick during peeping, when it is unusual for any reponse to be given to a particular stimulus, despite the continuous scanning of the environment, and the high postural tonus and other suggestions of readiness to respond. The association of inhibition of response with peeping is made particularly clear by the fact that if a peeping chick does fixate a localised stimulus and show signs of response (*e.g.* of pecking), then at the same time peeping almost invariably shows a decrease in intensity, and is often replaced by short calls.

It is likely that inhibition is also important during periods of short calls, when the chick is completely ready to perform a response but must for the moment wait. Two examples may make this clearer. During a hasty excited search (*e.g.* for food), when calling is at its most frequent, repeated decisions are required as to whether a stimulus is a food grain, is near enough to peck or is as accessible as alternative targets; during decisions pecking must be inhibited. Secondly, when attempting to reach a goal (*e.g.* when attempting to mount for copulation, at which times calls are frequent), the animal is clearly on the edge of performing the response. Intention movements are not unusual (*e.g.* crouching and treading without full mounting), which suggests that only an opposing inhibitory effect prevents premature full performance. In periods of emotional behaviour an immediate readiness to perform a particular response may be thought of as balanced by opposing inhibition of response. Human states of hysteria, such as may arise in fear or great excitement, are interesting examples of an extreme condition of this sort in which, just as in periods of peeping, no complete response to any particular stimulus is carried out despite the appearance of great responsiveness.

The final property of calling phases and of emotional behaviour which should be mentioned here are the special responses which characterise them. Calls, piloerection, pupillary dilatation and comparable responses have been used throughout the preceding discussion as indices of emotional states. A second category of response is suggested by the hypothesis that certain tectal reflexes such as visual targetting are facilitated during calling phases. Evidence will be considered shortly which suggests that some components of attack and defence may also fall in this category.

THE MIDBRAIN MECHANISMS INVOLVED IN CALLING PHASES

All four of the properties of calling phases considered above are abolished or depressed by ICo lesions. It seems likely, therefore, that they are mediated by the fibre connections which are interrupted by such lesions. These may either pass through the ICo or begin or end within it. Some preliminary data concerning the midbrain mechanisms which might be involved have been obtained by interrupting these connections separately from each other. However, before discussing these results, it is necessary to suggest some homologies between avian and mammalian midbrain structures so that evidence from mammalian studies can also be introduced.

The subcollicular area lying below the inferior colliculus and lateral part of the superior colliculus resembles the ICo both anatomically and functionally. A full discussion of the evidence has been given elsewhere (De Lanerolle and Andrew, 1975; Andrew, 1975) and so will only be summarised here.

Firstly, exaggeration of a condition already present in some mammals (*e.g.* ungulates, Huber *et al.*, 1943) in which the anterior pole of the inferior colliculus underlies the superior colliculus, coupled with a lateral extension of the aqueduct below the superior colliculus and above the inferior colliculus, would bring the subcollicular area to lie in the avian position. Secondly, in the cat, Kanai and Wang (1962) found sites from which only vocalisation was elicited in the lateral part of the subcollicular zone, and sites yielding both vocalisation and defence more medially, as the central mesencephalic grey (CMG) was approached. A similar distribution of sites occurs in the chick, with the CMG yielding calls and escape. Lesions of the subcollicular area produce muteness in cats (Skultety, 1958; Berntson, 1972). In addition, they may also cause a loss of responsiveness to visual stimuli which are normally very effective in evoking response. The evidence here is complicated by the fact that the superior colliculus itself is either removed during such an operation or its main connections are cut. However, Casagrande *et al.* (1972) showed that removal of the superficial layers of the superior colliculus in the tree shrew had little effect on targetting and prey-catching, whilst the further ablation of deeper layers extending down into the subcollicular area produced severe visual unresponsiveness. This result (although unexpected in that it suggests that visual systems which do not pass through the tectum can control targetting and prey-catching) provides a close parallel with the chick. A variety of other studies confirm that subcollicular damage (with varying involvement of the superior colliculus and tegmentum) produces visual unresponsiveness (in cat: Sprague and Meikle, 1965; and in hamster: Schneider, 1969).

Having established this tentative homology between the subcollicular area and the ICo, we can now turn to the fibre tracts which are cut by ICo lesions in the chick. The first runs medially and somewhat ventrally to decussate below the oculomotor nuclei in the dorsal tegmental decussation and then descends vertically into the area immediately lateral to the midline nuclei of the raphé. At least the majority of fibres terminate close to these nuclei; in the chick it is not yet certain whether terminals enter these nuclei, but this is the case for the corresponding mammalian system (see below).

Exactly the same route is followed by the predorsal component of the efferent outflow from the tectum in the cat (Altmann and Carpenter, 1961). In this case, a small proportion of fibres are known to continue as the tectospinal tract, and the same may be true in the chick since tectospinal fibres have been described in the pigeon (Karten, 1966). The origin of at least some components of this fibre tract in the tectum has been demonstrated for the chick (Andrew, unpublished); however, to avoid possible future confusion it will be termed, for the present, the raphé route.

If the raphé route is cut lateral to the dorsal tegmental decussation and well away from the ICo (Fig. 1) the most obvious change is a loss of inhibition. Of nine birds lesioned unilaterally (Andrew, unpublished), three had faulty placements and showed no effect and four out of the remaining six in which the raphé route was interrupted showed contraversive circling; it was not clear why this effect was absent in the remaining two, but all animals showed progressive recovery so that the two exceptions may represent extreme cases in which the condition developed briefly or not at all. Two of the birds which showed circling were tested with blinkers which occluded vision of each eye separately, without disturbing the chick directly. In these animals circling proved to be greatly depressed, or absent, when the contralateral eye was covered. It could also be shown to be initiated when a target was presented in the lateral field of the contralateral eye. In three out of the four birds which circled, circling was a usual response to target presentation in the affected eye. It looked like an exaggeration of the usual targetting response, in which the body is often not turned at all, although rapid turning on the spot will follow if the target is moved as the chick turns its head. The raphé route may thus mediate an inhibition of locomotion during targetting to a conspicuous stimulus, despite the fact that a command to turn by locomotion is initiated.

The involvement of an inhibitory system is further confirmed by lesions of the major midline nucleus, nucleus linearis caudalis, near or in which much of the raphé route terminates (Andrew, unpublished). Three birds with symmetrical destruction of the nucleus showed persistent forward locomotion in which large obstacles were often ignored. Nine birds with asymmetric damage showed persistent circuits of the home cage with ipsiversive turning. Both findings can be explained if it is supposed that each half of the nucleus (or the zone immediately lateral to it) tends to inhibit turning ipsiversive to the side on which the stimulus is presented. This must be, in part at least, dependent on an input from the contralateral tectum along the raphé route (Fig. 1). The more posterior raphé nucleus appears to be similarly organised (two birds).

It is not possible to explain the results by postulating that the same inhibitory system lying laterally to the raphé nuclei is injured by both raphé route and midline lesions since the two have opposite effects (the former facilitating turning contraversive to the lesioned side, and the latter ipsiversive; Fig. 1).

Lesions of the raphé route also depress targetting and pecking, although to a less extent than do ICo lesions. Instead of targetting and pecking (which still sometimes occurs) a chick with a raphé route lesion will often give some other response: it may either locomote (forward or contraversive circling) or show a response to visual stimuli other than the target (*e.g.* by looking or pecking at the floor).

Voneida (1970) has described a rather similar syndrome in the cat following a midline section of the tegmentum: controls showed that damage to the CMG or eye muscle nuclei were not responsible. As in the chick, a slightly asymmetric placement produced ipsiversive circling. Inability to avoid large obstacles suggests the persistent pushing against obstacles shown by chicks with symmetrical midline lesions. In both cases, this could be another example of loss of inhibitory effects initiated from the tectum following the presentation of appropriate visual stimuli. Depression of striking with the forepaw at a moving target in the cat is rather similar to the depression of targetting and pecking in the chick. The cats also showed a marked and persistent loss of responsiveness to hand thrusts which produced striking and withdrawal in normal animals. Attack to hand thrusts has unfortunately not yet been studied in chicks with raphé or raphé route lesions. In the rat, Lynch and Crain (1972) have demonstrated both greatly increased spontaneous locomotion (such as occurs in chicks with raphé or raphé route lesions), and hyperresponsiveness to handling following lesions of nucleus reticularis gigantocellularis and the associated nucleus reticularis pontis caudalis. Injury to descending fibres of the predorsal route could be involved in this disinhibitory effect.

Voneida (1972) ascribes the changes described above for the cat to section of crossed tectal efferents. This seems even clearer in the case of the chick where the raphé route has been cut laterally, as well as at its central destination; here, at least, the effects are probably specifically due to interruption of the avian equivalent of the predorsal route.

Three possible different types of effect of interruption of this route may be conveniently distinguished. The first is the interruption of a motor command route. Voneida notes that section of the tectospinal tract might directly prevent 'orientation'. In the chick, it may be that there is a specific interruption of tectospinal commands to turn the head to a conspicuous target. Spontaneous head-turning, including turning to peck marks on the cage wall, persists and appears to be normal in ICo chicks, and so a very specific route would have to be postulated. More indirect motor routes are also possible via the medial reticular cells which lie just lateral to the final descending limb of the raphé route (*e.g.* nucleus reticularis gigantocellularis: Rossi and Brodal, 1956).

Following ICo lesions the chick shows an almost complete abolition of pecking in attack directed at a thrusting hand presented binocularly; this clearly resembles the loss of responsiveness to thrusting hands shown by

Voneida's cats. It is possible that this and the loss of defensive attack in the cat both involve interruption of a motor route. However, a second type of effect would also explain these findings. It may be that the raphé route is involved in ascending pathways which initiate organised defence or attack to purely visual stimuli.

The third type of effect that is suggested by the behaviour of chicks with raphé route lesions is the disinhibition of responses such as locomotion (*cf.* also Lynch and Crain, 1972; above) and pecking at the floor, which may compete with targetting and pecking. An effect of this type might explain the fact that Voneida's cats sometimes struck with their paws at a target, but after a long delay.

It has not yet proved possible to cut the tectal afferent route mentioned above, which also degenerates after ICo lesions, independently of the raphé route. However, both can be cut out together well away from the ICo, at a point close to the lateral border of the tectal ventricle (Fig. 1; Andrew, unpublished). Only the four birds with the most complete section of the combined tract will be considered here. In a number of others there was a good correlation between the degree of damage and the magnitude of the effect. All four showed the same changes following contralateral presentation of a target: the chick relaxed, ceased other visual responses such as looking around or at the floor, and became rather still. This condition, which did not involve turning the head towards the target, could sometimes be sustained for a considerable time by keeping the target in continuous slight movement. Ipsilateral presentation to the unaffected eye produced targetting and pecks, just as in normal birds.

The hypothesis already advanced for the function of the tectal afferents provides a possible explanation for the different consequences of cutting the raphé route at the ventricle edge and medially. If a tonic facilitation of the responsiveness of tectal units to a conspicuous stimulus is removed by cutting the tectal afferents (lesion at ventricle edge), then the effectiveness of such a stimulus in producing other responses such as body turns in targetting and locomotion will also be reduced. The reduction or loss of a simultaneous inhibitory command would then have little effect. However, the inhibition of other responses in birds with such a lateral lesion suggests the survival of some inhibitory effects; possibly intratectal effects are important. Further work is needed. It may also be that the raphé route is not fully interrupted by the lateral lesion: some of its fibres may even arise in the ICo. The most important point so far established by this lesion is that interruption of the tectal afferents (admittedly in combination with the raphé route) has an effect consistent with the function suggested here for those afferents.

The depression of attack following ICo lesions, which has already been noted, has special interest here. One of the most obvious features of pecks directed at looming objects, whether with a sudden dart of the head from a defensive crouch or in a sparring attack, is that they are evoked by objects much larger than are usually effective in producing pecks. A change in the responsiveness of tectal units such as has been described for Anura following pretectal lesions (Ewert, 1970; Ingle, 1973), and which might be mediated by tectal afferents, would explain this.

Finally, it must be noted that the depression of targetting and pecking produced by ICo lesions differs yet again from the effects of either of the lesions just discussed. Chicks with such lesions (Andrew, 1975) turn their head away from a target presented in the eye contralateral to the lesion; they may also shake the head or turn bodily away. Since lateral lesions do not produce such a change, it seems probable that it involves other connections of the ICo. In the Anura the pretectum and the tectothalamic border has been shown to mediate avoidance of obstacles quite independently of the targetting and prey-catching initiated from the tectum. A similar independence of the two systems seems likely in the chick (Andrew, 1975). A similar condition may be general in higher vertebrates: in mammals, the rostral superior colliculus appears to initiate avoidance.

The removal of a depressive effect on an anteriorly placed system of this sort by ICo lesions would explain the substitution of avoidance for targetting. Ewert (1970) has shown a comparable shift in dominance, but in the opposite direction, after pretectal lesions in the toad.

Midbrain mechanisms which might mediate enhanced visual responsiveness and inhibition of response have now been considered. The motor readiness also characteristic of calling phases most probably involves motor routes not directly affected by ICo lesions and is not considered here. There therefore remains to discuss the induction of muteness. Fibres descending ventrally from the lesion are probably involved. They coincide with the descending route plotted from sites which yield calls to stimulation (Peek and Phillips, 1971; De Lanerolle, 1972). It may also be significant that large lesions which destroy not only the ICo but also the area ventral to it, no longer give complete muteness: rapid calling may be given during locomotion, although calling is still not evocable by visual stimuli.

ROSTRAL CONTROL OF MIDBRAIN MECHANISMS

In view of the existence of separate mechanisms mediating different aspects of the calling phase of behaviour caudal to the ICo, it might be argued that the various effects of the lesions are associated simply because a number of routes may all, by chance, be interrupted at this point. The best argument against this is the close association of all the properties of behaviour, which are deleted by the lesion, in calling phases of normal behaviour (the appearance of avoidance after ICo lesions is perhaps an exception). This association might depend upon the convergence at the ICo of various inputs which are able to initiate calling phases; alternatively, the ICo might be the terminus of a single system, the activation of which initiates calling phases. The role of possible inputs to the ICo will now be examined.

The medial edge of the ICo is bordered by a lateral prolongation of the CMG. In the rhesus monkey Crosby and Woodburne (1951) have described an extension of the diencephalic periventricular system (as part of the main dorsal longitudinal fasciculus in their terminology), through the CMG to the subcollicular tegmentum. Connections of this sort between the CMG and the ICo or subcollicular area are probably of importance in calling phases and emotional behaviour. CMG lesions abolish calling in the cat, (together with facial expression, Adametz and O'Leary, 1959) and rat (Chaurand et al., 1972). In the chick, lesions confined to the lateral edge of the CMG (so far there is one chick with histology confirmed) appear to abolish calling to startling stimuli (short trills) and to significant stimuli such as food (short calls). Peeps were greatly reduced in number in open field tests, but did develop slowly in this test, and also when the chick was held for a long time or otherwise constrained. Thus, it is probable that in both mammal and chick the CMG provides an input to the ICo or subcollicular area which is important or essential for vocalisation.

There is some evidence that routes of importance in the initiation of vocalisation also converge on the CMG (and perhaps ICo) from lower centres. In the cat, vocalisation and piloerection to tactile stimulation can still be evoked after complete transection of the brain at the midbrain—diencephalic junction (Kelley et al., 1946). In the pigeon, Karten (1967) has noted convergence of cutaneous and auditory information on the ICo and this may explain the ease with which tactile stimuli produce calls. Ascending pain routes are known to reach the CMG caudally, and this route may initiate the trilled calls which follow painful stimuli in the chick (Andrew, 1969).

The descending routes to the CMG which might explain calling following the recognition of complex stimuli by forebrain mechanisms are unfortunately not yet properly established. It is clear from anatomical evidence (Crosby and Woodburne, 1951; Szentágothai et al., 1968) that the main rostral input to the CMG is by the descending part of the diencephalic periventricular system (DPS: anterior limb of Szentágothai et al.). No mammalian studies involving the effects of the interruption of this system on normally evoked vocalisation and associated behaviour are known to me; Fernandez de Molina and Hunsperger (1962) have shown that lesions of the perifornical hypothalamus blocks the evocation of calls and threat by amygdaloid stimulation. In the chick (Andrew and Mitchell, unpublished) lesions of the nucleus paraventricularis magnocellularis have been investigated, which, in addition to cutting the paraventricular tract (Mitchell and Andrew, in preparation) and so specifically depressing copulation, almost certainly also affected the dorsal limb of the DPS. Such lesions significantly depress peeping in open field tests. Other possible effects on calling have not yet been studied.

Special interest attaches to the possibility that the route to the ICo via the DPS and CMG, which is suggested by studies of muting and loss of facial expression, may also be involved in other functions disturbed by ICo lesions.

In the cat, stimulation experiments suggested that vocalisation, piloerection of the back and tail, and pupillary dilatation are closely associated with the DPS—CMG route (Hunsperger, 1963). This association has been described in the past as being due to the mediation of defensive threat (of which the above responses are components) by this route. However, it is worth noting that visual scanning with frequent miaows can be obtained by stimulation in the system, without overt threat or escape. The DPS—CMG route could well be involved in the performance of vocalisation and associated EDs in a wide variety of situations. It has already been noted in primates that situations involving frustration may evoke vocalisations of a wide range of intensity, including high intensity vocalisations such as calls usually given when separated from social fellows. It is probable from the behaviour of pet cats that miawos are produced by obstacles such as closed doors, although experimental data on this point are not known to me. The miaow thus could be a call which is associated with low levels of activation of the DPS—CMG.

In view of this evidence from stimulation it is interesting that the DPS has been identified on the basis of self-stimulation experiments as an important component of the 'punishment system' (Olds and Olds, 1963; Stein, 1964), and is, as a result, supposed to be activated by non-reinforcement, such as occurs in frustration situations. Unfortunately, too little is definitely known of the way in which the hypothetical 'punishment system' functions to allow any conclusion to be drawn beyond the general one that the DPS clearly has important general functions in behaviour over and above the initiation of EDs and defence.

Again there appear to be little data on the general effects of interruption of the DPS. One human clinical study (Cairns et al., 1941), in which a cyst of the third ventricle was repeatedly drained thus reversibly removing pressure on the ventricle walls in which the DPS runs, is of interest. Pressure produced akinetic mutism in which facial expression and loud vocalisation was absent, suggesting, on the present hypothesis, interference with the DPS—CMG route. The patient at the same time was extremely unresponsive, even though ready eye movements to any moving stimulus gave the impression of alertness, and motor tonus remained normal for much of the time. There were signs that words and objects such as food were recognised even though response was rare and feeble. All of the effects diminished or disappeared very rapidly when the cyst was drained.

Routes whose interruption does affect responsiveness certainly traverse the diencephalon. Two types of effect should be distinguished. The first involves unresponsiveness which is accompanied, and perhaps entirely caused, by tonic and intense general inhibition. Fonberg (1972) has described this following lesions of either the medial amygdala or lateral hypothalamus in the dog. Such animals are not only largely unresponsive to all stimuli (although they remain capable of sporadically recognising and responding correctly to a CS), but also are immobile with low postural tonus, giving an appearance of great depression. They resist manipulation. Their characteristic aphagia, which is classic following lesions of the tracts passing through the lateral hypothalamus, may be regarded as only one aspect of their chronic inhibition of all response. Marshall *et al.* (1971) have shown that rats with lateral hypothalamic lesions are unresponsive to stimuli in all modalities (including both touch, with routes ascending to the midbrain, and modalities such as olfaction, with descending routes), and that recovery from the usual aphagia follows the same time course as recovery from the other syndromes.

The conclusion that chronic inhibition is basic to this syndrome is strongly confirmed by the demonstration (Fonberg, 1972) that it is completely reversed in both cases by a further lesion in the lateral amygdala. It would thus seem to depend on a greatly increased dominance of the lateral amygdala and more caudal systems driven from it. A second type of unresponsiveness, in which inhibition seems not to be crucial, appears to be produced by complete removal of the amygdala. Taming has long been known to result from total amygdalectomy in mammals (temporal lobe: Kluver and Bucy, 1938; amygdala alone: cat, Schreiner and Kling, 1953; rat, Blanchard and Blanchard, 1972); Ursin (1965) showed, using feral cats, that the basolateral amygdala is probably the crucial structure in taming. Cutting the outflow from the archistriatum, which includes the avian amygdala, has a taming effect in mallard (Phillips, 1964) and parakeet (Phillips, 1968); in the latter case, lesions in the medial archistriatum were employed, and a great reduction in mobbing calls was noted in addition to reduced withdrawal. The chapter by Wright in this volume (Chapter 15) should also be consulted. In both mammal and bird conspicuous and strange visual stimuli, including other animals, are no longer avoided or threatened. This effect is not accompanied by any sign of general inhibition or depression: the strange stimulus may be investigated, and social partners evoke vigorous responses including frequent vocalisations. Such unresponsiveness is thus somewhat like that shown by chicks with ICo lesions but affects only one class of stimuli.

The lateral amygdala is a complex structure, but it is possible that both chronic inhibition and escape of, or threat at strange stimuli reflect the same aspect of its functioning. It has already been noted that as calling phases become more protracted and intense, inhibition of response becomes more marked. In the chick, a final state may develop out of, or in place of, peeping in which a resting attitude is assumed with reduced tonus and closed eyes. Calling continues, however, as a series of warbles, and if disturbed the chick may open its eyes and return to peeping. This 'warble' state thus appears to represent a variant of the calling phase with very high inhibition.

It could therefore be argued, remembering the effect of unbalanced hyper-activity of the lateral amygdala, that this structure when briefly activated may promote defence or threat directed against whatever visual stimulus is being examined, but that when it is sustainedly activated general inhibition becomes more and more marked.

Exactly this sort of sequence can be seen when an animal is repeatedly attacked or threatened by superiors. Von Holst (1969) has shown in the tree shrew (*Tupaia*) that tail piloerection persists throughout the period of inhibi-tion and immobility which results. Here again, the persistence of an ED component characteristic of DPS—CMG activation suggests that such activa-tion may persist through phases of extreme tonic inhibition.

The arguments advanced here would suggest that such an effect, if originated by the lateral amygdala, would be exerted via the CMG. The route connecting the lateral amygdala and CMG remains to be established. Anatomically, the DPS is the most likely candidate since it is known to reach the CMG; the outflow from the amygdala terminates in both medial and lateral hypothala-mus (Field, 1972).

The exact range of situations which may activate the lateral amygdala also remains to be established. It is possible that visual stimuli recognised as significant but not frightening may also act via this route; if so, it should be possible to abolish short calls to food, for example, by amygdaloid damage in the chick. In the cat, vocalisations are readily evoked by the sight of a social fellow at a distance, following total amygdalectomy (Schreiner and Kling, 1953). The exact behavioural circumstances in which this will occur remain to be defined; the route involved is in any case unknown. The sur-vival of calls to handling is less surprising since tactile routes remain intact.

Other routes from the forebrain to the CMG—ICo almost certainly exist and are probably important. Phillips and Youngren (1974) have argued that the mesencephalic component of the tractus occipitomesencephalicus (OM) may initiate calling via the ICo. It is not known whether, or under what circumstances OM lesions may reduce calling, which is necessary before de-ciding what role the OM may play in normal calling and associated behaviour.

Possible routes which might increase visual responsiveness remain to be investigated. In mammals it is known that the visual cortex (areas 17, 18 and 19) exerts a strong facilitatory effect on the ipsilateral optic tectum (refer-ences in previous section; Sherman, 1974). The route involved remains to be established, although it presumably involves the corticotectal connections described by Huber *et al.* (1943). The equivalent in the chick is quite un-known.

A SUMMARY OF THE MAIN HYPOTHESIS

It may be helpful at this point to summarise the properties which have been postulated for the CMG—ICo and its rostral and caudal connections.

Midbrain mechanisms exist, it is proposed, which can make a visual stimulus more effective in evoking targetting and a range of tectal reflexes such as pecking (visual enhancement), and which can also inhibit response during targetting, or adjustment of position, or more generally ('inhibition'). Such mechanisms are accessible to ascending routes and can mediate a variety of basic responses such as prey-catching to a visual stimulus, or turning to bite or peck at a tactile stimulus. Painful or startling stimuli might subsequently potentiate such responses via visual enhancement.

However, the same mechanisms, if available to forebrain mechanisms, could serve a wider range of functions. Thus the recognition of a significant but inconspicuous stimulus could then be made to initiate the same rapid targetting with inhibition of the other responses that a startling or painful stimulus normally evokes. The initiation of defence may represent only the highest intensity of this process. Checks, obstacles and delays may also activate the same midbrain mechanism (not necessarily via the amygdala, although the DPS seems likely to be involved; see above). Visual enhancement would make persistence in attention to the evoking stimulus more likely, whilst concurrent inhibition would nevertheless prevent response until the check was overcome. In the absence of an appropriate stimulus, the same two effects (again with the participation of rostral mechanisms) would make possible repeated targetting to possible stimuli, coupled with immediate readiness to respond but without premature response.

The vocalisations, facial expression, piloerection and other EDs have evolved, it would be argued, to make conspicuous and explicit the state of these midbrain mechanisms to social fellows. They provide crucial information about readiness to respond, the degree of significance attached to a stimulus, the progressive development of inhibition and a number of other variables which have often been ignored in past discussions of communication. A fuller treatment of this topic has been given previously (Andrew, 1972b; see also below).

Finally, this seems an appropriate time to point out that some purely behavioural aspects of this hypothesis have been set out in earlier papers (Andrew, 1964) in a rather different way. It was there pointed out that if a stimulus is to be capable of evoking short calls in the chick, it must at the same time also evoke signs of marked visual interest (*e.g.* alert responses such as targetting). Stimuli with this latter property might (a) contrast with background stimulation or represent a sudden change in stimulation, or (b) although intrinsically inconspicuous possess significance (*e.g.* as a CS), or (c) constitute a persisting change from a familiar pattern of stimulation.

It was emphasised that no other single property (or complex of properties) could be found which all stimuli that evoked short calls had in common. In particular, explanations in terms of particular drives or combination of drives could be excluded. It was, therefore, proposed that stimuli in categories a—c above, should be said to possess 'stimulus contrast'. This term has proved confusing in that no property of the stimulus itself is involved but rather a characteristic of the reaction of the chick to the stimulus. The term should, therefore, be abandoned, particularly as it can now be replaced by a more concrete set of hypotheses, which relate directly to possible neural mechanisms.

Other phases of behaviour

It is important to emphasise that it is not suggested that the participation of the midbrain mechanisms postulated above is essential for all coordinated response. Indeed, it is postulated that they are inactive in silent chicks (with possible exceptions such as freezing in fear) which nevertheless respond vigorously. The conditions under which silence is most likely are ones which involve a practised response to a stimulus which is recognised unambiguously and immediately. It is important that there be no check or obstacle to response; if the chick is highly motivated, finding itself at a little distance from the goal or even incorrectly oriented to it may be equivalent to a check. Finally, there should be no distracting stimuli which are potentially strange or significant, and may require brief examination before they can be finally recognised.

Under these conditions even responses such as feeding or locomotion, which normally are accompanied by calling, are quite silent (Andrew and De Lanerolle, 1975). They can be vigorously performed and, indeed, copulation is carried out in silence once mounting is over.

No doubt yet other phases of behaviour may be usefully distinguished. Careful investigation of a strange object is often carried out in silence, and at such times it is rare for a response to occur. During such silent periods, the bird may be learning the characteristics of the object rather than preparing to respond. Visual enhancement of the effectiveness of a particular feature might well be a disadvantage at such a time by locking gaze on that feature and interfering with the methodical examination of all features.

The possible differences between this latter hypothetical phase of behaviour and others may be made clearer by an approach set out at greater length by Andrew (1972b). In that paper, it was proposed that recognition of a stimulus inevitably required a formal process of comparison between centrally held specifications of an appropriate type of stimulus and the stimulus itself. A variety of properties of these specifications such as their accuracy, detail and ease of retrieval would therefore affect behaviour. The most important point here is that one can distinguish between a process of comparison in which the animal is responding to match or mismatch between a particular set of specifications and the stimulus, and a process of a quite different sort in which the stimulus is being examined in order to establish new or changed specifications which describe the stimulus accurately. This latter process would dominate the silent investigatory phase whereas calling would occur when match or mismatch between established specifications and the stimulus called for response.

Hippocampal theta rhythm

The arguments advanced above may be more convincing to workers familiar with chick or guinea pig, whose frequent and varied vocalisations divide their behaviour up unambiguously into phases of calling and phases of silence, than to those who use the silent pigeon or (apparently) silent rat. However, there is a large volume of work in mammals on a discrete phase of behaviour which is clearly defined by theta frequency of synchronised EEG in the hippocampus.

Andrew (1969) noted that there was a close resemblance between the situation in which theta rhythm occurs in the cat, rat or guinea pig and those in which the chick calls; a fuller comparison is set out in Table I. There are some uncertainties about the relationship between theta and locomotion, but even these suggest a variable association like that between locomotion and calling. It is probable that theta does not occur during locomotion which is part of a highly practised response and performed by a completely relaxed animal.

Like calling, theta occurs during locomotion to a goal and in attempts to initiate a response (*e.g.* attempts to mount for copulation). It appears when a stimulus is perceived to which the animal is likely to respond, if that response has to be checked while the stimulus is properly recognised, or while the animal decides what to do. If the response is highly practised, the theta phase disappears or is highly abbreviated.

Hippocampal theta rhythm thus provides some evidence that phases of behaviour equivalent to calling phases occur even in animals which do not readily give EDs (or which give ones which we do not readily perceive).

The question of the possible function of the hippocampus in calling phases or emotional states will be examined here only briefly. A number of apparently unrelated functions which are all, in fact, appropriate to a structure which is important in calling phases of behaviour, have been ascribed, by different authors, to the hippocampus. Firstly, and most importantly, Grastyan *et al.* (1959) have shown that the hippocampus is important in controlling the evocation of the 'orienting reflex' (targetting and fixation).

Vanderwolf (1971) argued that the hippocampus is involved in triggering (and presumably sustaining) a cortically controlled motor programme. The evidence for this deserves further examination since it is the largest body of data directly comparable to that available for chick calls. In Vanderwolf's terminology (for references see Table I) theta rhythm occurs during voluntary movements and is absent during 'automatic' responses. Automatic responses appear to include copulation, grooming, eating, drinking and, perhaps, relaxed locomotion. The absence of both calls and theta rhythm during

296

TABLE I

Comparison of situations eliciting calling in chicks with those eliciting hippocampal theta rhythm in mammals.

Eliciting situations			Chick call	Mammal theta rhythm and references
(A)	Startling or conspicuous stimulus	(i) Inducing sudden awakening from dozing	Short trill or short calls	?[1]
		(ii) Moving visual stimulus	Short trill	Theta[9]
		(iii) Inducing freezing	None	None[6]
(B)	Valent stimulus		Short trills or short calls	Theta[2]
(C)	Locomotion	(i) to goal (a) highly practised: animal relaxed	None	?[1,2,9]
		(b) excited; motivated	Short calls	Theta[1,2,5,6,8]
		(ii) sustained (e.g. treadmill)	?	Theta[9]
(D)	Manipulation of object		Short calls	Theta[3,5,6]
(E)	Struggle		Short calls/peeps/shriek	Theta[7]
(F)	Frustration, non-reinforcement	(i) locomotion	Short calls or peeps	Theta[8]
		(ii) stand and scan	Peeps	
		(iii) resting attitude	Warbles	
(G)	Run in escape		Peeps	Theta[7]
(H)	Consummatory acts	(i) feed and drink: hasty and excited;	Short calls	Variable[6,8]
		relaxed and established;	None	
		(ii) copulation attempts;	Short calls	Theta[6]
		copulation	None	None[6]
		(iii) groom	None	None[6]
(I)	Careful investigation without response		None	None ?[6]

Notes.

(A) Theta does not occur to stimuli which waken cats from sleep[1]; however, it is possible that sudden wakening from a light doze would have different effects.

(B) Calls are given to a conditioned stimulus for food or water, when locomotion to copulation, in contrast to their presence during attempts to mount, is particularly striking. Ingestion and locomotion require more careful examination. It has been shown that calls are typical of hasty excited pecking, in which the animals' responses appear to be markedly facilitated and sustained against momentary checks (*e.g.* in order to swallow) and indecisions (*e.g.* when locating the next grain). The same may be true of theta: cats (apparently relaxed and at ease) do not show theta whilst eating or drinking (Whishaw *et al.*, 1972), but theta has been recorded during drinking in deprived rats immediately after a runway approach (Gray and Ball, 1970).

A similar distinction could be drawn between locomotion in an excited rapid approach to a goal, or in frustration (when an animal might be argued to be continuously ready to respond and on the edge of premature performance of some response, even though it is not close enough to an appropriate stimulus to be able to perform the response), and relaxed locomotion in a familiar and highly practised behaviour sequence. Chicks call during the first but are usually silent during the latter (as long as they are fully undisturbed). It is not clear whether theta disappears entirely during highly practised relaxed locomotion. This was reported in cats by Grastyan *et al.* (1959) but Whishaw and Vanderwolf (1973) argue that this observation may have been based on recordings from the dentate gyrus; their own data showed disappearance of theta with practice at such sites (representing an attenua-

---

dispenser follows. Theta is given to conditioned stimuli requiring movement[1−4] but not if immobility is necessary[3,4] (cat, dog, rat).

(C) Calls coincide with locomotion to a goal when behaviour is strongly driven by external stimuli (*e.g.* when goal has just become available); highly practised approach by a completely undisturbed bird is silent. Theta coincides with locomotion under all circumstances tested in rat[9], and in particular during interaction with social partners[5,6,9] (rat, guinea pig). In the cat[1], theta is said to be present during 'non-specific purposeless movement' during training, but to disappear once approach to a goal has become practised[1,2], however, it is present during protracted locomotion on a treadmill[9] (see text).

(D) Calls immediately precede and accompany pecks at a conspicuous object. Light pecks during protracted examination may be given in silence. Theta occurs during lever operation (dog, rat)[3,5,6].

(G) Theta is known to accompany escape induced by posterior hypothalamic stimulation (rat)[7].

(H) (i) Calls occur at the onset of feeding, becoming less frequent as it establishes. Relaxed feeding by a completely undisturbed chick is almost entirely silent. The same distinction may hold for theta. Thus, the rat shows (slow) theta during drinking after approach down a runway[8], but no theta whilst chewing food[6]. There is no theta during drinking or eating in the cat (perhaps whilst relaxed)[9].

(I) There is no theta during alert immobility (rat)[6], but this may be equivalent to freezing rather than investigation.

References 1. Grastyan *et al.*, 1959 (cat). 2. Grastyan *et al.*, 1966 (cat) 3. Black and Young, 1972 (dog). 4. Klemm, 1972 (rat). 5. Vanderwolf, 1971 (rat, guinea pig, gerbil). 6. Whishaw *et al.*, 1972 (rat). 7. Bland and Vanderwolf, 1972 (rat). 8. Gray and Ball, 1970 (rat). 9. Whishaw and Vanderwolf, 1973 (cat, rat).

tion of the state), but its persistence in the layer of Ammon's pyramids, even over very prolonged locomotion on a treadmill. An additional possibility, which is suggested by comparison with chick calls, is that theta might disappear entirely even from the dentate gyrus in completely undisturbed animals, in their home quarters. It should be noted that Whishaw and Vanderwolf also argue that theta occurs in preparation for movement, on the grounds that it is shown by cats during motionless fixation of a moving object: they thus agree that it is not confined to locomotion.

Klemm (1970 and 1972) extended and altered the Vanderwolf hypothesis to cover those instances in which theta occurred despite immobility, and proposed that theta reflects "a general readiness state whereby an animal is prepared to make adaptive responses to biological significant stimuli".

A number of authors have emphasised the inhibitory role of the hippocampus. Thus Douglas (1967) has argued for an 'inhibitory control of attention' by the hippocampus, and Jarrard (1973) has suggested that the hippocampus may inhibit lower level structures. Again, inhibition of response has been shown to be important in calling phases.

Finally, Gray (1970) has claimed that non-reinforcement and frustration situations cause hippocampal theta rhythm, and has presented evidence that, if theta is prevented, some of the normal consequences of frustration do not appear.

None of these theories (at least, as set out briefly here) consider reasons why a forebrain structure of such size and complexity as the hippocampus should be needed for such apparently simple functions. It may, therefore, be worth examining the sort of processes which are likely to be important in calling phases, but which are unlikely to be mediated by the midbrain mechanisms which have been discussed here. In such periods it is necessary to be able to identify stimuli as irrelevant without losing the set to respond to the particular type of stimulus with which the animal is at that time concerned. This must be done during checks and delays which increase the likelihood of distraction and attention shift. A recent review by Bennett (1971) concluded that the hippocampus is important in the mediation of attention shifts, and it seems likely that it is in this role that it participates in calling phases. Midbrain mechanisms exist which would be capable of mediating targetting to a particular stimulus, with associated inhibition and motor readiness, and these might be available to the hippocampus; the hippocampus itself is more likely to be directly involved in processes such as retrieval or temporary storage of central representations of stimuli, so as to allow the recognition of one stimulus without the loss of the representation of the one to which the animal was ready to respond.

CALLING PHASES AND EMOTIONAL STATES: CONCLUSION

Although this chapter has been primarily concerned with the discussions of emotional states, some of the most important aspects of such states have been largely ignored. These will now be considered briefly to clarify their possible relation to the system, whose properties have been considered in preceding sections.

(a) Relation to reward and punishment, and to pleasure and its opposite

Pleasure and negative affect are usually central to discussions of human emotions. In animals introspective judgements are not available, and alternative approaches have been sought through studies of the reinforcement (*i.e.* the effects on future performance) associated with the stimulus evoking the state under investigation, or studies of EDs which appear to occur in situations comparable with those evoking the human smile and chuckle or sob (*i.e.* responses which are usually held to be associated with positive or negative affect).

In the chick, short calls are given on finding a sought-after stimulus, such as food, and on encountering, and seeking contact with a social partner. The possibility that short calls were caused by activation of the 'reward system' was tested by studying the patterns of calling which accompany intracranial self-stimulation (ICS) at sites in the hypothalamus (Andrew, 1967 and 1973). Short calls did not accompany stimulation at positive sites. They sometimes followed as an after-response but it was equally usual for peeps to be given instead. All positive sites in the lateral hypothalamus shared one feature: calling ceased or fell greatly in intensity during stimulation, suggesting a reduction in activation within the ICo—CMG. If the stimulation induces a state equivalent to the sustained and unimpeded performance of a consummatory response, then abolition or depression of a calling phase during stimulation with its reassertion at the end of stimulation seems reasonable.

The association of short calls with reinforcing situations in normal behaviour can be explained by the hypothesis that reinforcing stimuli such as food or social partners commonly evoke hasty, excited attempts to initiate or maintain response. However, positive reinforcement is also likely to be associated with feeding (for example) carried out silently and without haste. Punishment by its nature is likely almost always to evoke vigorous attempts to abolish stimulation and so is likely to be accompanied by intense ED. In the chick, peeps occur at such times. However, it is possible that this need not always be true: one can imagine in man at least stoically endured pain with little effect on ongoing behaviour which would nevertheless be avoided in the future. A final important dissociation of pleasure (and positive reinforcement) and the EDs usually thought to be characteristic of pleasure is the occurrence of weeping and other intense EDs in man when a greatly desired event has come about. The present theory would thus admit an indirect and sometimes loose association between affect state or type of reinforcement and a particular ED or level of activation of the ICo—CMG; the association actually found appears to be of this type.

In man the marked change to positive affect which accompanies the development of laughter in a previously hostile or embarassed social encounter is so clearcut as to suggest some special effect of sharing in this display. Other special instances may exist, but it would be argued here that they involve special mechanisms and need not therefore affect the main arguments which have been advanced.

(b) Does the present hypothesis explain the wide range of patterns of EDs?

It will be convenient to take chick vocalisations as an example since it is with them that I have been chiefly concerned here. The hypothesis which has been advanced is that these reflect the state of a single functional system. A considerable variety of calls can be generated if intensity and duration of activation are taken into account. Here it has been assumed that brief moderate activation will yield a short call, a brief intense activation a short trill, and a sustained activation peeps. However, at least three other types of effect must be taken into account when attempting to explain the full complexity of chick vocalisations.

Firstly, the lower level mechanisms responsible for the actual performance of calls have their own complicating properties: in particular, there is a tendency to repeat in considerable detail the pattern of calling which has just been performed (Andrew, 1969), and this appears to hold even if other factors affecting calling change.

Secondly, crowing still appears following the administration of testosterone, despite ICo lesions (Andrew and De Lanerolle, 1975); other routes independent of the ICo area may exist which are capable of initiating some patterns of calling.

Thirdly, the fact that most calls are abolished by ICo lesions does not mean that specific external stimuli may not be important in the causation of some of them in intact animals. Thus, to take one example, long alarm trills are readily evoked by long low sounds, although they also occur sometimes after purely visual startling stimuli. It may well be that activation of the CMG-ICo and an appropriate acoustic input can interact to make performance of this particular call likely.

Much the same arguments may be applied to the expressive facial movements which are lost after CMG lesions. The classification and causation of these movements have been discussed elsewhere recently (Andrew, 1972b), and will not be considered here. The important point to stress is that although there is good evidence that EDs such as vocalisation depend on the integrity of a single system, and reflect its state of activation in a relatively simple way, this does not mean that the full complexity of these responses has to be explained solely in terms of the activation of that system.

It will also be obvious that a number of categories of responses which are likely to be associated with activation of the CMG—ICo system have been almost or completely omitted from consideration here. The most important of these are the various cardiovascular, respiratory and other reflex preparations for exertion which are known to be initiated by the CMG and systems which converge on it (Abrahams *et al.*, 1964). Not only will these categories also contribute to the EDs present at various states of activation of the CMG—ICo system, but they will interact with the categories which have been discussed at greater length. Thus respiratory reflexes affect both vocalisation and facial expression (*e.g.* lip-rounding appears to be a reflex movement associated with vigorous ventilation), whilst mouth posture in the mammal and neck position in the bird affect the quality of vocalisation (Andrew, 1963b).

REFERENCES

Abrahams, V.C., Hilton, S.M. and Zbrozyna, A. (1964) The role of active muscle vasodilatation in the alerting stage of the defence reaction. *J. Physiol (Lond.)*, 171, 189—202.
Adametz, J. and O'Leary, J.L. (1959) Experimental mutism resulting from periaqueductal lesions in cats. *Neurology (Minneap.)*, 9, 636—642.
Allan, N.B. (1971) *The Central Control of Vocalization.* D. Phil Thesis, University of Sussex.
Altmann, J. and Carpenter, M.B. (1961) Fibre projections of the superior colliculus in the cat. *J. comp. Neurol.*, 116, 157—178.
Andrew, R.J. (1962) The situations that evoke vocalization in primates. *Ann. N.Y. Acad. Sci.*, 102, 296—315.
Andrew, R.J. (1963a) The effect of testosterone on the behaviour of the domestic chick. *J. comp. physiol. Psychol.*, 56, 933—940.
Andrew, R.J. (1963b) The origin and evolution of the calls and facial expressions of the primates. *Behaviour*, 20, 1—109.
Andrew, R.J. (1964) Vocalization in chicks and the concept of stimulus content. *Anim. Behav.*, 12, 64—76.
Andrew, R.J. (1967) Intracranial self-stimulation in the chick. *Nature (Lond.)*, 213, 847—848.
Andrew, R.J. (1969) The effect of testosterone on avian vocalizations. In R.A. Hinde (Ed.), *Bird Vocalizations*, Cambridge University Press, London.
Andrew, R.J. (1972) The information potentially available in mammal displays. In R.A. Hinde (Ed.), *Non-Verbal Communication.* Cambridge University Press, London, pp. 179—206.
Andrew, R.J. (1973) The evocation of calls by diencephalic stimulation in the conscious chick. *Brain Behav. Evol.*, 7, 424—446.
Andrew, R.J. (1975) Changes in visual responsiveness following intercollicular lesions and their effects on avoidance and attack. *Brain Behav. Evol.*, in press.
Andrew, R.J. and De Lanerolle, N. (1975) The effects of muting lesions on emotional behaviour and behaviour normally associated with calling. *Brain Behav. Evol.*, in press.
Archer, J. (1973) Tests for emotionality in rats and mice: a review. *Anim. Behav.*, 21, 205—235.
Bennett, T.L. (1971) Hippocampal theta activity and behavior — a review. *Commun. behav. Biol.*, 6, 37—48.

302

Berntson, G.C. (1972) Blockade and release of hypothalamically and naturally elicited aggressive behaviour in the cat following midbrain lesions. *J. comp. physiol. Psychol.*, 81, 541—554.

Black, A.H. and Young, G.A. (1972) Electrical activity of the hippocampus and cortex in dogs operantly trained to move and hold still. *J. comp. physiol. Psychol.*, 79, 128—141.

Blanchard, D.C. and Blanchard, R.J. (1972) Innate and conditioned reactions to threat in rats with amygdaloid lesions. *J. comp. physiol. Psychol.*, 81, 281—290.

Bland, B.H. and Vanderwolf, G.H. (1972), Diencephalic and hippocampal mechanisms of motor activity in the rat: effects of posterior hypothalamic stimulation on behavior and hippocampal slow wave activity. *Brain Res.*, 43, 67—88.

Brown, J.L. (1965a) Loss of vocalization caused by lesions in the nucleus mesencephalicus lateralis of the redwinged blackbird. *Amer. Zoologist*, 5, 693.

Brown, J.L. (1965b) Vocalizations evoked from the optic lobe of a song bird. *Science*, 149, 1002—1003.

Cairns, H., Oldfield, R.C., Pennybacker, J.B. and Whitteridge, D. (1941) A kinetic mutism with an epidermoid cyst of the third ventricle. *Brain*, 64, 273—290.

Candland, D.K. and Nagy, Z.M. (1969) The open field: some comparative data. *Ann. N. Y. Acad. Sci.*, 159, 831—851.

Casagrande, V.A., Harting, J.K., Hall, W.C. and Diamond, I.T. (1972) Superior colliculus of the tree-shrew: a structural and functional subdivision into superficial and deep layers. *Science*, 177, 444—447.

Chaurand, J.P., Vergnes, M. et Karli, P. (1972) Substance grise centrale du mésencéphale et comportement d'agression interspécifique du Rat. *Physiol. Behav.*, 9, 475—481.

Collias, N.E. and Joos, M. (1953) The spectrographic analysis of sound signals of the domestic fowl. *Behaviour*, 5, 175—188.

Crosby, E.C. and Woodburne, R.T. (1951) The mammalian midbrain and isthmus regions. II. The fiber connections. C. The hypothalamotegmental pathways. *J. comp. Neurol.*, 94, 1—32.

De Lanerolle, N. (1972) *Brain Mechanisms and the Causation of Calls*. D. Phil Thesis, University of Sussex.

De Lanerolle, N. and Andrew, R.J. (1975) Midbrain structures controlling vocalization in the domestic chick. *Brain Behav. Evol.* in press.

Douglas, R.J. (1967) The hippocampus and behavior. *Psychol. Bull.*, 67, 416—442.

Duncan, I.J.H. and Wood-Gush, D.G.M. (1972) Thwarting of feeding behaviour in the domestic fowl. *Anim. Behav.*, 20, 444—451.

Ewert, J.P. (1970) Neural mechanisms of prey-catching and avoidance behavior in the toad (*Bufo bufo* L.). *Brain Behav. Evol.*, 3, 36—56.

Fernandez de Molina, A. and Hunsperger, R.W. (1962) Organisation of the subcortical system governing defence and flight reactions in the cat. *J. Physiol. (Lond.)*, 160, 200—213.

Field, P.M. (1972) A quantitative ultrastructural analysis of the distribution of amygdaloid fibres in the preoptic area and the ventromedial hypothalamic nucleus. *Exp. Brain Res.*, 14, 527—538.

Fonberg, E. (1972) Control of emotional behaviour through the hypothalamus and amygdaloid complex. In *Physiology, Emotion and Psychosomatic Illness, Ciba Symposium No. 8*. Elsevier, Amsterdam, pp. 131—161.

Grastyan, E., Lissale, K., Madarosz, I. and Donhoffer, H. (1959) Hippocampal electrical activity during the development of conditioned reflexes. *Electroenceph. clin. Neurophysiol.*, 11, 409—430.

Grastyan, E., Karmos, G., Vereczkey, L. and Kellehyi, L. (1966) The hippocampal electrical correlates of the homeostatic regulation of motivation. *Electroenceph. clin. Neurophysiol.*, 21, 34—53.

Gray, J.A. (1970) Sodium amobarbital, the hippocampal theta rhythm and the partial reinforcement extinction effect. *Psychol. Rev.*, 77, 465—480.

Gray, J.A. and Ball, G.G. (1970) Frequency-specific relation between hippocampal theta rhythm, behaviour and amobarbital action. *Science*, 168, 1246—1248.

Huber, G.C., Crosby, E.C., Woodburne, R.T., Gillilan, L.A., Brown, J.O. and Tamthai, B. (1943) The mammalian midbrain and isthmus regions. Part I. The nuclear pattern. *J. comp. Neurol.*, 78, 129—536.

Holst, D. Von. (1969) Sozialer Stress bei Tupajas (*Tupaia belangeri*). *Z. vergl. Physiol.*, 63, 1—58.

Hunsperger, R.W. (1963) Comportements affectifs provoqués par la stimulation électrique du tronc cérébral et du cerveau antérieur. *J. Physiol. (Paris)*, 55, 45—98.

Ingle, D. (1973) Disinhibition of tectal neurons by pretectal lesions in the frog. *Science*, 180, 422—424.

Jarrard, L.E. (1973) The hippocampus and motivation. *Psychol. Bull.*, 79, 1—12.

Kanai, T. and Wang, S.C. (1962) Localization of the central vocalization mechanism in the brain stem of the cat. *Exp. Neurol.*, 6, 426—434.

Karten, H.J. (1966) Projections of the optic tectum of the pigeon (*Columba livia*). *Anat. Rec.*, 151, 369.

Karten, H.J. (1967) The organisation of the ascending auditory pathways (*Columba livia*). I. Diencephalic projections of the inferior colliculus (nucleus mesencephalicus lateralis pars dorsalis). *Brain Res.*, 6, 409—427.

Kelly, A.H., Magoun, H.W. and Ranson, S.W. (1946) A midbrain mechanism for facio-vocal activity. *J. Neurophysiol.*, 9, 181—189.

Kiley, M. (1972) The vocalizations of ungulates, their causation and function. *Z. Tierpsychol.*, 31, 171—222.

Klemm, W.R. (1970) Correlation of hippocampal theta rhythm, muscle activity and brain stem reticular formation activity. *Commun. behav. Biol.*, A 5, 147—151.

Klemm, W.R. (1972) Effects of electric stimulation of brain stem reticular formation on hippocampal theta rhythm and muscle activity in unanaesthetized, cervical and midbrain-transected rats. *Brain Res.*, 41, 331—344.

Kluver, H. and Bucy, P.C. (1938) An analysis of certain effects of bilateral temporal lobectomy in the rhesus monkey, with special reference to 'psychic blindness'. *J. Psychol.*, 5, 33—54.

Lynch, G. and Crain, B. (1972) Increased general activity following lesions of caudal reticular formation. *Physiol. Behav.*, 8, 747—750.

Marshall, J.F., Turner, B.H. and Teitelbaum, P. (1971) Sensory neglect produced by lateral hypothalamic damage. *Science*, 174, 523—525.

Olds, M.E. and Olds, J. (1963) Approach-avoidance analysis of rat diencephalon. *J. comp. Neurol.*, 120, 259—295.

Peek, F.W. and Phillips, R.E. (1971) Repetitive vocalization evoked by local electrical stimulation of avian brains. II. Anaesthetised chickens (*Gallus gallus*). *Brain Behav. Evol.*, 4, 417—438.

Phillips, R.E. (1964) Wildness in the mallard duck. Effects of brain lesions and stimulation on escape behaviour and reproduction. *J. comp. Neurol.*, 122, 129—155.

Phillips, R.E. (1968) Approach-withdrawal behavior of peach-faced lovebirds, *Agapornis roseicolis*, and its modification by brain lesions. *Behaviour*, 31, 163—184.

Phillips, R.E. and Youngren, O.M. (1974) A brain pathway for thalamically evoked calls in birds. *Brain Behav. Evol.*, 9, 1—6.

Rossi, G.F. and Brodal, A. (1956) Corticofugal fibers to the brain stem reticular formation. An experimental study in the cat. *J. Anat. (Lond.)*, 90, 40—62.

Schneider, G.E. (1969) Two visual systems. *Science*, 163, 895—902.

Schreiner, L. and Kling, A. (1953) Behavioral changes following rhinencephalic injury in the cat. *J. Neurophysiol.*, 16, 643—659.

Sherman, S.M. (1974) Visual fields of cats with cortical and tectal lesions *Science,* 185, 355—357.

Skultety, F.M. (1958) The behavioral effects of destructive lesions of the periaqueductal grey matter in adult cats. *J. comp. Neurol.,* 110, 337—366.

Sprague, J.M. and Meikle, T.H. (1965) The role of the superior colliculus in visually guided behavior. *Exp. Neurol.,* 11, 115—146.

Stein, W. (1964) Reciprocal action of reward and punishment mechanisms: In R. Heath (Ed.), *The Role of Pleasure in Behavior.* Harper and Row, New York, pp. 113—137.

Szentágothai, J., Flerko, B., Mess, B. and Halász, B. (1968) *Hypothalamic Control of the Anterior Pituitary.* Akademiai Kiado, Budapest.

Ursin, H. (1965) Effect of amygdaloid lesions on avoidance behavior and visual discrimination in cats. *Exp. Neurol.,* 11, 298—317.

Vanderwolf, C.H. (1971) Limbic—diencephalic mechanisms of voluntary movement. *Psychol. Rev.,* 78, 83—113.

Voneida, T.J. (1970) Behavioral changes following midline section of the mesencephalic tegmentum. *Brain Behav. Evol.,* 3, 241—260.

Whishaw, I.Q. and Vanderwolf, C.H. (1973) Hippocampal EEG and behavior: changes in amplitude and frequency of RSA (theta rhythm) associated with spontaneous and learned movement patterns in rats and cats. *Behav. Biol.,* 8, 461—484.

Whishaw, I.Q., Bland, B.H. and Vanderwolf, C.H. (1972) Hippocampal activity, behavior, self-stimulation and heart rate during electrical stimulation of the lateral hypothalamus. *J. comp. physiol. Psychol.,* 79, 115—127.

Wickelgren, B.G. and Sterling, P. (1969) Influence of visual cortex on receptive fields in the superior colliculus of the cat. *J. Neurophysiol.,* 32, 16—23.

14

GUSTATORY BEHAVIOUR OF THE CHICKEN AND OTHER BIRDS

MICHAEL J. GENTLE

Poultry Research Centre, Edinburgh (Great Britain)

It has been stressed by Kare and Ficken (1963) that there are great differences in preference behaviour among species of mammals. If a comparison is made between species of birds, similar differences in preference behaviour are found. Much of the early work on the gustatory system of birds has been reviewed by Kare (1965) and Wood-Gush (1971). I will present more recent data on four aspects of gustatory behaviour of birds, namely its importance in the life of the animal, preferences, sensitivity, and lastly the central nervous projections of the gustatory system.

One of the characteristic differences between the gustatory system of the bird and that of the mammal is the small number of taste buds present in the bird. This is clearly shown in Table I where the mean number of taste buds for a number of avian and mammalian species is presented. The distributions of the taste buds in many of the avian species studied show some interesting features. In the chicken (Lindenmaier and Kare, 1959; Gentle, 1971a), blue tit and Barbary dove (Gentle, unpublished observations), and to some extent the pigeon (Moore and Elliott, 1946) the taste buds are mainly confined to

TABLE I

Mean number of taste buds described for various species of birds and mammals

Chicken	24	Lindenmaier and Kare (1959)
Pigeon	37	Moore and Elliott (1946)
Bullfinch	46	Duncan (1960)
Starling	200	Bath (1906)
Duck	200	Bath (1906)
Parrot	350	Bath (1906)
Barbary dove	54	Gentle (unpublished observations)
Blue tit	24	Gentle (unpublished observations)
Human	9000	Cole (1941)
Rabbit	17,000	Moncrieff (1967)

the posterior part of the tongue, caudad to the tongue fold. This posterior part of the tongue is also where both the food and the water collect before the bird swallows, and this leads to the possibility, at least in the chicken, that the bird tastes the food only just prior to swallowing. The structure of the taste buds also differs in different species. In the pigeon they are essentially flask-shaped structures, the gustatory cells being long and narrow with their longitudinal axes the same as the longitudinal axis of the taste bud (Moore and Elliott, 1946). In the chicken, however, the cells are roughly columnar in shape (Gentle, 1971a). The taste buds of the chicken have an innervation similar to that found in other vertebrates with a nerve plexus at the base of the bud giving rise to fibres which enter the bud. The cells of the taste bud also show acetylcholinesterase activity and degenerate following denervation.

The fact that birds have relatively few taste buds compared with mammals would suggest that taste is a relatively unimportant sense in birds. This, however, is not the case and many experiments have shown that birds will rapidly learn to avoid distasteful objects. The chicken for example will reject quinine hydrochloride solutions at concentrations as low as 0.001 M, which suggests a mechanism to prevent the excessive consumption of toxic plant alkaloids. Work by Hughes and Wood-Gush (1971) showed that chickens fed a calcium deficient diet developed a specific appetite for calcium. This could occur using either visual or gustatory cues. Recent work by Hogan (1971) would suggest that taste is important in the development of the adult hunger—feeding system. In a series of experiments using newly hatched chicks, Hogan found that there was an initial increase in the incentive value of the feed during the first four or five days after hatching, which was most likely due to some kind of taste mechanism that may be related to the nutritional state of the animal.

Comparing the sense of taste of birds with mammals the classification of taste into the four qualities used in human work, i.e. salt, sour, bitter and sweet, serves very little purpose. Work on the preference behaviour from a number of species supports the concept of a separate taste world for each species. Saccharin serves as a good example of the danger of ascribing the same taste sensation from species to species. In man saccharin tastes sweet with a bitter aftertaste, but both pigs and rats prefer saccharin to water over a wide range of concentrations, whereas most chickens and dogs find it offensive (Kare and Ficken, 1963). It is also difficult to generalise within a group of chemical compounds. Kare and Medway (1959) reported that chicks behaved indifferently to sucrose and glucose solutions compared with water but xylose solutions were increasingly rejected from concentrations of 5 to 25%. They concluded that there was no clear-cut evidence to explain the rejections or indifference to sugars and that there was a suggestion that the discrimination was based upon the absolute specificity for the sugar involved.

It has been possible to show that in the chicken taste plays a role in the control of food and water intake as well as in the acceptance of novel foods. In a series of experiments a number of chickens were deprived of their sense of taste by sectioning the lingual nerves (Gentle, 1971b). Following surgery the chickens exhibited various degrees of aphagia which lasted from 1 to 14 days. When they began to eat spontaneously they showed a reduced intake compared with the control birds. These recovered aphagics showed no motor defects following nerve section, and the reduced intake was due to a reduction in pecking. This experiment suggests that taste plays a role in the motivation of feeding.

Although birds, unlike mammals, do not break down their food orally before swallowing and the food is only momentarily in contact with the tongue, it seems to be biologically advantageous for the bird to test new food sources orally. To test this hypothesis birds which had undergone lingual nerve section were presented with a novel diet, made from their normal mash diet by the addition of tasteless food dyes. They were presented with the diet on a single occasion for 8 h. Fig. 1 shows the results of this experiment. In the sham-operated control birds there was no significant difference in the amount or pattern of food intake when they were fed on the normal diet or on a coloured diet. The lingually denervated birds rejected the green diet for the first hour but increased their food intake in the following 3 h. The red diet was rejected for the first 3 h of presentation by the denervated birds. The green diet differed only slightly in colour from their normal diet whereas the red differed to a much greater extent, so it appears that the amount of food consumed on the first presentation of a novel diet depended on the degree of novelty. In a second experiment the same birds were presented with a pelleted green diet, which therefore differed

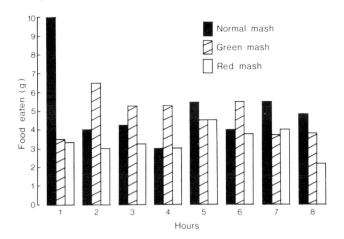

Fig. 1. The hourly mean values of the weight of food eaten by six birds with lingual nerve section on normal mash and when first presented with green and red coloured mash.

in both colour and texture from their normal diet. The denervated birds totally rejected the novel diet for several days while the control birds continued to eat normally. This suggests that when deprived of gustatory feedback the birds were reluctant to accept a novel diet and that taste is probably important in the initial screening of the food.

It has been shown by Kitchell *et al.* (1959) that in the chicken, as in many mammals, water has a specific taste. This water taste, together with other oropharyngeal factors, plays a role in the control of water intake. In a recent experiment Gentle (1974) found, by measuring the length of EEG arousal following repeated injection of 0.5 ml of water into the mouth of the chicken every 3 min, that the bird rapidly habituated to the stimulus. When the birds were deprived of water before testing, habituation took longer to occur and the length of arousal remained significantly longer from the seventh to the twelfth stimulus. This suggests that the water stimulus was rewarding for a longer period of time following deprivation, and this would reinforce drinking behaviour, enabling the animal to make good any deficit. Even after deprivation the birds habituated by the fifteenth stimulus, which suggests that there is a mechanism preventing excessive water consumption. The birds showed a different pattern of habituation following crop loading (50 ml of water at body temperature injected into the crop 15 min before testing). From the seventh to the fifteenth stimulus there was a significant increase in the length of arousal compared with that when they had free access to food and water before the test. The birds also behaved normally for the first six stimuli but from the seventh onwards they vocalised at the onset of the stimulus and attempted to escape from the box. In this experiment the water was acting as an aversive stimulus following crop loading. These two experiments suggest that the feedback from the oropharyngeal receptors, which is positive following deprivation and negative after crop loading, is modified by the state of hydration of the animal.

PREFERENCE BEHAVIOUR

In looking at taste preferences and aversions two aspects are of major importance: individual variability and postingestional feedback. Individual variability has been studied in the chicken (Ficken and Kare, 1961; Gentle, 1972) and in the pigeon (Duncan, 1960) where individuals had markedly different rejection thresholds for chemicals in solution and there were large differences in preference behaviour between individuals at low concentrations. Postingestional feedback can be very extensive and involves not only chemicals in the bloodstream but also very rapid direct sensory feedback from the crop and duodenum (Gentle and Richardson, 1972; Richardson and Gentle, 1972). Postingestional feedback is present in all trials of taste preference lasting for more than a few minutes.

The importance of postingestional feedback was demonstrated experimentally by Capretta (1961); he found that preference for different coloured foods could be altered by noxious stimulation of the crop. When the chicks were given a choice between red and blue mash some preferred one colour and some the other. He than paired crop loading with 10% sodium chloride, milk or water to their preferred and non-preferred colour. Neither the milk nor water affected the preference but the noxious salt load reversed the colour preference.

Sugars

Many mammals, though by no means all, show preferences for some sugar solutions compared with water in preference trials, and this is also wide-spread in birds. Preferences for some sugars are shown by the bob-white quail (Brindley and Prior, 1968; Hamrum, 1953), great tit (Warren and Vince, 1963), pigeon (Duncan, 1960), chicken (Gentle, 1972; Gunther and Wagner, 1971) and in parrots, budgerigars and broad-tailed hummingbirds (Kare, 1965). Indifference to sucrose has been reported in many species, including the herring gull, starling, goose, siskin and laughing gull (Kare, 1965). The response to sugars in the chicken is of interest as there have been a number of conflicting reports in the literature. Jacobs and Scott (1957) reported a marked preference for sugars by chicks and both preference and rejection was reported to sucrose and glucose solutions by Gentle (1972), whereas Kare and Medway (1959) found that the chick was totally indifferent to both sucrose and glucose solutions. Kare and Medway (1959) tested the reaction of the fowl to a series of sugars in a two-choice situation where distilled water was the alternative. The birds were tested in large groups and no account of any possible social influences were considered. The testing periods were also many hours long in which case postingestional factors would play a major role. They found indifference to glucose and sucrose at all concentrations up to 25%, whereas xylose was increasingly rejected from 5 to 25%. Use of a different testing procedure in which individual birds were given a choice between water and a sugar solution in identical waterers with a 3-min testing period gave very different results (Gentle, 1972). The mean preference values for ten birds to sucrose and glucose is shown in Fig. 2. A value of 50% means that the birds were indifferent. It can be seen that the response depends on the sugar tested and that glucose was rejected at all concentrations above 5%, although some birds showed a preference for a 1% solution. Sucrose was rejected at 2.5%, preferred at 5%, selected indifferently at 10 and 20% and strongly rejected above 20%. These data suggest that short-term trials measure taste preference whereas in long-term trials post-ingestional feedback masks the initial preferences.

The preference behaviour of the bird can be modified by the nutritional needs of the animal. Changes in sucrose preference have been reported in the chicken following caloric dilution of the diet (Kare and Maller, 1967).

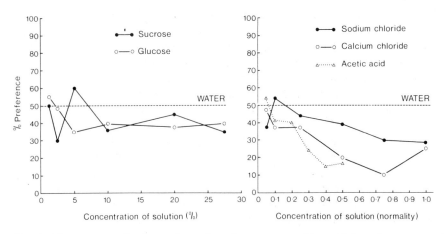

Fig. 2. The mean preference values given by a group of ten chickens to sucrose, glucose, acetic acid, sodium and calcium chlorides.

The birds were indifferent to a 10% sucrose solution compared with water in a choice situation, but following diet dilution they showed a significant preference for the sucrose solution.

Salts

A pattern of indifference—aversion rather than preference—aversion to salt solutions is common in many species of birds (Bartholomew and Cade, 1958; Hamrum, 1953; Kare and Ficken, 1963; Warren and Vince, 1963), and also in the herring gull, which has been observed to drink seawater (Harriman and Kare, 1966). In a 3-min preference test with a salt solution and distilled water the chicken also showed an indifference—aversion curve. The mean values for calcium and sodium chlorides are shown in Fig. 2. Although the results for potassium chloride are not shown in this figure the curve obtained was similar in shape to that found for sodium chloride but there was increased rejection of the potassium chloride at the higher concentrations. The birds selected calcium chloride indifferently at 0.05 N but rejected it in more concentrated solutions. They rejected sodium chloride at 0.05 N, were indifferent to it at 0.1 and 0.25 N, and rejected it at higher concentrations. Some individual chickens were observed to show a preference for a 0.1 N sodium chloride solution and marked preferences for it were seen in the pigeon and bantams (Duncan, 1962). Rats also show a preference for diluted sodium chloride solutions (Weiner and Stellar, 1951). It is interesting to note that Hughes and Wood-Gush (1970) were unable to demonstrate a specific appetite for sodium in the chicken and during long-term trials salt-deprived birds rejected a 0.7% sodium chloride solution when given in a choice situation with water.

Acids

Some birds have a very wide range of tolerance for acidity in their drinking water. Fuerst and Kare (1962) reported that over an 18-day period chicks would tolerate strong mineral acid solutions, for example pH 2. The preference behaviour shown by adult chickens towards acids is similar to that shown to salt solutions in that they tend to be indifferent to dilute solutions and reject the more concentrated solutions. Fig. 2 shows the results of testing ten birds' responses to a variety of acetic acid solutions in a 3-min period. Hydrochloric acid at similar concentrations elicits similar responses but there is increased rejection of the acid at all values from 0.2 to 0.5 N. In this respect the chicken is not so sensitive to acids as is the pigeon which shows a significant reduction in intake of hydrochloric acid at 0.0075 N and of acetic acid at 0.01 N (Duncan, 1964). This value for acetic acid is higher than that reported previously, of 0.004 N (Duncan, 1960), but is comparable with the value of 0.005 N which only just tastes sour to man (Moncrieff, 1967).

Quinine

The chicken is very sensitive to quinine hydrochloride and will reject a 0.001 M solution whereas the pigeon failed to show any response from recordings in the lingual nerves when the tongue was stimulated with a 0.02 M solution (Kitchell *et al.*, 1959).

Chemical mixtures

Although in work on mammalian taste it is commonly found that mixing sugars with acid solutions renders the acids more acceptable, Duncan (1964) reported that this did not occur in the pigeon. In a recent experiment Gentle (1972) found that in the chicken a slight rejection was shown to a 10% glucose solution (39.4% preference) and an even greater rejection of a 0.3 N acetic acid solution (24%). However, a mixture of both solutions gave an intermediate preference value (35.6%). A similar result is seen with more concentrated solutions of the two substances. A mixture of salt and acid solutions yielded very different results. Solutions of 0.5 N sodium chloride and 0.2 N acetic acid were each slightly rejected when presented separately (39.0% and 41.9% respectively), but a mixture of the two was markedly rejected (27.7%). These results in the chicken are comparable in many respects to the psychophysical studies on the human response to chemical mixtures. In general, heterogenous mixtures give a depression of the judged magnitude of each component (Halpern, 1967). This means that the two components tasted quite differently but that the judged intensity of each of the components of the mixture was smaller than the intensity of the separate

312

component. The response of adding glucose to the highly rejected acetic acid solution in the chicken would reduce the magnitude of the intensity of the acid and result in the increase in consumption of the more acceptable mixture. The addition of sodium chloride to acetic acid in the chicken gave increased rejection of the mixture and the bird was responding in a manner similar to the human response to homogenous mixtures. There are, however, a few instances of an increase in the judged intensity of heterogenous mixtures in humans (Halpern, 1967). Work on the redwinged blackbird by Rogers and Maller (1973) has suggested that the addition of sodium chloride makes a sucrose stimulus more discriminable.

SENSITIVITY

Responses from the lingual nerves following stimulation of the tongue have been recorded in the chicken and the pigeon (Kitchell *et al.*, 1959; Halpern, 1962; Landolt, 1970). Although in general all these studies show a close correlation between electrophysiological events and the behaviour of the animal, there are some important exceptions. Neither Landolt (1970) nor Kitchell *et al.* (1959) found any response in the lingual nerve of the pigeon to quinine hydrochloride, although there is marked behavioural rejection of this compound (Duncan, 1960).

Another approach has been to look at arousal changes in the EEG as a measure of taste sensitivity. From preference trials it can be seen that there are many solutions to which the chicken shows indifference even though it may reject solutions which are more, or less, concentrated. This leads to two possible hypotheses, namely (1) that they do not taste the chemical, and

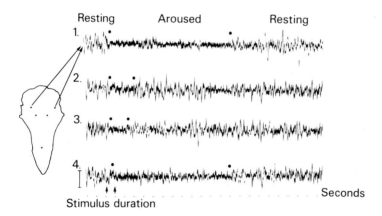

Fig. 3. The positions of the recording electrodes and the EEG pattern following stimulation with distilled water and quinine hydrochloride. 1, 2 and 3 are the responses to the first, third and sixth stimuli with distilled water and 4 is the response to the first stimulus with quinine hydrochloride solution. Calibration represents 100 μV.

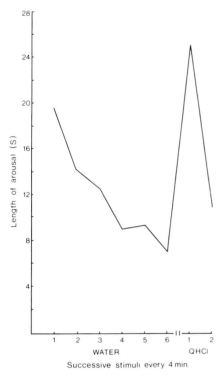

Fig. 4. The length of arousal to six successive stimuli with distilled water followed by two stimuli with 0.005 M quinine hydrochloride.

(2) that they taste the chemical but for some reason show no preference for or aversion to it. To test which of these two was correct a technique using arousal changes in the EEG was devised (Gentle, 1972). The birds were presented with a series of water stimuli at 4-min intervals and the length of arousal to each was measured. The length of arousal to each successive stimulus was reduced as the bird habituated. After the sixth water stimulus the bird was presented with a chemical solution which, if perceived by the bird, caused an increase in the length of arousal. Fig. 3 shows the EEG pattern recorded and Fig. 4 shows the habituation to distilled water and the subsequent increase in arousal following stimulation with a chemical solution. All the chemical concentrations which were selected indifferently in pre-ference trials (Fig. 2) were tested using the EEG technique. It appears from the results that both hypotheses were correct. Neither 0.05 nor 0.1 N acetic and hydrochloric acids produced any clear behavioural response, but both gave large increases in the length of arousal, whereas 10% sucrose, for example, gave neither an increase in the length of arousal nor any behavioural response.

GUSTATORY PROJECTIONS IN THE CENTRAL NERVOUS SYSTEM

Kitchell *et al.* (1959) were able to show that the impulses from the taste buds of the chicken and pigeon are conducted only in the lingual branches of the glossopharyngeal nerve. In the chicken, as in mammals, the sensory fibres from the glossopharyngeal nerve project onto the nucleus of the solitary tract (Ariëns Kappers *et al.*, 1960), and from there probably to a secondary gustatory nucleus. The site of the secondary area in mammals is in the pons (Norgren and Leonard, 1973) and in bony fish is in the isthmus region. In the lower vertebrates (reptiles, amphibians and fish), fibres run from there to the hypothalamus whereas in mammals most of the fibres run to the ventrobasal complex of the thalamus. From his work on birds, Erulkar (1955) concluded that none of the thalamic nuclei is either anatomically or functionally equivalent to the mammalian ventral thalamic group.

In the chicken the oral aversion response to distasteful stimulation is characterised by exaggerated tongue and beak movements, often followed by beak-wiping and head-shaking. Studies using electrical stimulation of the brain have produced reports of beak-wiping and beak movements from sites in the tractus quintofrontalis (Phillips and Youngren, 1971) and from the archistriatum, tractus frontoarchistriatus and tractus occipitomesencephalicus (Putkonen, 1967). The quintofrontal structures have been extensively studied by Zeigler and Karten (1973a and b) and are important in the neurosensory control of feeding but do not contain representation from the tongue. Similarly, the archistriatum has been shown to be important in feeding (see Zeier, 1971 for a further discussion on the archistriatum and for feeding see Wright, Chapter 15 in this volume) and probably involves sensory trigeminal information (Zeigler *et al.*, 1969). Work by Gentle (1973) revealed a number of sites in the diencephalon which gave rise to tongue and beak movements following electrical stimulation. A number of sites were in the lateral forebrain bundle but there was also a region in the posterioventral hypothalamus, mainly in the stratum cellulare externum, which gave pronounced tongue and beak movements during electrical stimulation.

The possibility that the archistriatum may be required for the chick to make learned gustatory associations comes from the work of Benowitz (1972). The learning task in this experiment was the suppression of an innate pecking response to a small metallic bead. The bead was coated with methyl anthranilate, a chemical which the chick finds very distasteful. By a series of ablations he found that the removal of the posteriolateral part of the telencephalon (damage mainly confined to the caudal neostriatum and much of the archistriatum) prevented the chicks from showing any retention or relearning of this simple task. The birds, however, still continued to show the head-shaking response after tasting the methyl anthranilate, although this does not necessarily indicate telencephalic recognition of taste.

More recent work on the chicken (Gentle, 1975) has shown that lesions in

EXPERIMENTAL LESIONS

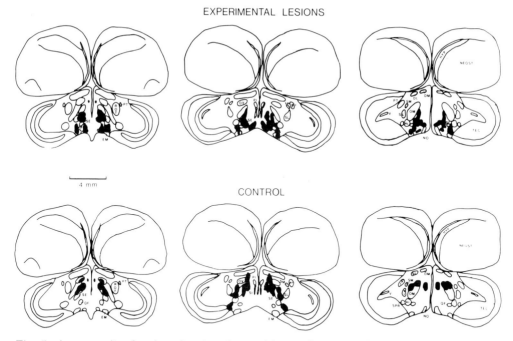

CONTROL

Fig. 5. A composite drawing showing the position and extent of all experimental and control lesions. Three different sectional planes are shown at intervals of 0.5 mm, the most anterior being on the left of the figure. The lesions are shaded in black and the experimental lesions are presented in the upper sections and the controls in the lower sections. Abbreviations used: AL, ansa lenticularis; DM, nucleus dorsomedialis; DL, nucleus dorsolateralis; EM, nucleus ectomammillaris; HYP, hyperstriatum; NEOST, neostriatum; NO, nervus oculomotorius; OM, tractus occipitomesencephalicus; OV, nucleus ovoidalis; PI, nucleus paramedianus intermedius; PT, nucleus praetectalis; QF, tractus quintofrontalis; ROT, nucleus rotundus; SE, stratum cellulare externum; SI, stratum cellulare internum; SL, nucleus spiriformes lateralis; SM, nucleus spiriformes medialis; SPR, nucleus subpraetectalis; TEC, optic tectum and V, ventricle.

TABLE II

Mean percentage preference for a 0.005 M quinine hydrochloride solution before and after lesion

Bird	Before	After
Experimental lesions		
Stratum cellulare externum	10.4	133.4
Posterioventral hypothalamus	16.4	45.6
Control lesions	39.6	25.1

the stratum cellulare externum result in gustatory deficits. A group of 43 adult chickens were tested using a single stimulus technique similar to that used by Duncan (1962). The birds were tested with both water and a solution of 0.005 M quinine hydrochloride before and after brain lesion. The amount of quinine hydrochloride was expressed as a percentage, arrived at by dividing the amount of quinine hydrochloride ingested by the amount of water ingested multiplied by 100. Eighteen experimental birds had bilateral lesions in the posterioventral hypothalamus and 25 birds acted as controls with bilateral lesions placed in a variety of other regions of the diencephalon. The position of the lesions is shown in a composite diagram in Fig. 5 and the results in Table II. Before the lesions were made the birds rejected the quinine hydrochloride solution $(P < 0.01)$. Lesions in the posterioventral hypothalamus resulted in reduced sensitivity to quinine and birds with lesions in the stratum cellulare externum showed no rejection of the quinine hydrochloride. Before the lesions all birds showed the oral aversion response. After the lesions the control birds continued to show this response whereas those experimental birds with lesions in the stratum cellulare externum failed to show this response. This suggests that in the chicken the stratum cellulare externum has a specific gustatory role. If this proves to be the case then it will mean that in the bird the pattern followed is similar to that seen in the lower vertebrates, where the diencephalic gustatory projection is hypothalamic.

REFERENCES

Ariëns Kappers, C.V., Huber, G.C. and Crosby, E.C. (1960) *The Comparative Anatomy of the Nervous System of Vertebrates Including Man.* Hafner, New York.
Bath, W. (1906) Die Geschmacksorgane der Vögel und Krokodile. *Arch. Biontol. (Berl.)*, 1, 5—47.
Bartholomew, G.A. and Cade, T.J. (1958) Effects of sodium chloride on the water consumption of house finches. *Physiol. Zool.*, 16, 304—307.
Benowitz, L. (1972) Effects of forebrain ablations on avoidance learning in chicks. *Physiol. Behav.*, 9, 601—608.
Brindley, L.D. and Prior, S. (1968) Effects of age on taste discrimination in the bobwhite quail. *Anim. Behav.*, 16, 304—307.
Capretta, P.J. (1961) An experimental modification of food preference in chickens. *J. comp. physiol. Psychol.*, 54, 238—242.
Cole, E.C. (1941) *Comparative Histology.* Blakiston, Philadelphia, Pa.
Duncan, C.J. (1960) Preference tests and the sense of taste in the feral pigeon. *Anim. Behav.*, 8, 54—60.
Duncan, C.J. (1962) Salt preferences of birds and mammals. *Physiol. Zool.*, 35, 120—135.
Duncan, C.J. (1964) The sense of taste in the feral pigeon. The response to acids. *Anim. Behav.*, 12, 77—83.
Erulkar, S.D. (1955) Tactile and auditory areas in the brain of the pigeon. *J. comp. Neurol.*, 118, 43—48.

Ficken, M.S. and Kare, M.R. (1961) Individual variation in the ability to taste. *Poultry Sci.*, 40, 1402.

Fuerst, W.F., Jr. and Kare, M.R. (1962) The influence of pH on fluid tolerance and preference. *Poultry Sci.*, 41, 71—77.

Gentle, M.J. (1971a) The lingual taste buds of *Gallus domesticus* L. *Brit. Poultry Sci.*, 12, 245—248.

Gentle, M.J. (1971b) Taste and its importance to the domestic chicken. *Brit. Poultry Sci.*, 12, 77—86.

Gentle, M.J. (1972) Taste preference in the chicken (*Gallus domesticus* L.). *Brit. Poultry Sci.*, 13, 141—155.

Gentle, M.J. (1973) Diencephalic stimulation and mouth movements in the chicken. *Brit. Poultry Sci.*, 14, 167—171.

Gentle, M.J. (1974) Changes in habituation of the EEG to water deprivation and crop loading in *Gallus domesticus*. *Physiol. Behav.*, 13, 15—19.

Gentle, M.J. (1975) Gustatory hyposensitivity to quinine hydrochloride following diencephalic lesions in *Gallus domesticus*. *Physiol Behav.*, in press.

Gentle, M.J. and Richardson, A. (1972) Changes in the electroencephalogram of the chicken produced by stimulation of the crop. *Brit. Poultry Sci.*, 13, 163—170.

Gunther, W.C. and Wagner, M.W. (1971) Preferences for natural and artificial sweetners in heat-stressed chicks of different ages. *Proc. Ind. Acad. Sci.*, 81, 401—409.

Halpern, B.P. (1962) Gustatory nerve responses in the chicken. *Amer. J. Physiol.*, 203, 541—544.

Halpern, B.P. (1967) Some relationships between electrophysiology and behaviour in taste. In M.R. Kare and O. Maller (Eds.), *The Chemical Senses and Nutrition*. Johns Hopkins Press, Baltimore, Md., pp. 213—242.

Hamrum, C.L. (1953) Experiments on the senses of taste and smell in the bob-white quail (*Colinus virginianus virginianus*). *Amer. Midl. Nat.*, 49, 872—877.

Harriman, A.E. and Kare, M.R. (1966) Aversion to saline solutions in starlings, purple grackles and herring gulls. *Physiol. Zool.*, 39, 123—126.

Hogan, J.A. (1971) The development of a hunger system in young chicks. *Behaviour*, 39, 128—201.

Hughes, B.O. and Wood-Gush, D.G.M. (1970) Investigations into specific appetites for sodium and thiamine in domestic fowls. *Physiol. Behav.*, 6, 331—339.

Hughes, B.O. and Wood-Gush, D.G.M. (1971) A specific appetite for calcium in domestic chickens. *Anim. Behav.*, 19, 490—499.

Jacobs, H.L. and Scott, M.L. (1957) Factors mediating food and liquid intake in chickens. I. Studies on the preference for sucrose and saccharine solutions. *Poultry Sci.*, 36, 8—15.

Kare, M.R. (1965) The special senses. In P.D. Sturkie (Ed.), *Avian Physiology*. Baillière, Tindall and Cassell, London, pp. 433—441.

Kare, M.R. and Ficken, M.S. (1963) Comparative studies on the sense of taste. In Y. Zotterman (Ed.), *Olfaction and Taste*. Pergamon Press, London, pp. 285—298.

Kare, M.R. and Maller, O. (1967) Taste and food intake in domesticated and jungle fowl. *J. Nutr.*, 92, 191—196.

Kare, M.R. and Medway, W. (1959) Discrimination between carbohydrates by the fowl. *Poultry Sci.*, 38, 1119—1127.

Kitchell, R.L., Ström, L. and Zotterman, Y. (1959) Electrophysiological studies of thermal and taste reception in chickens and pigeons. *Acta physiol. scand.*, 46, 133—151.

Landolt, J.P. (1970) Neural properties of pigeon lingual chemoreceptors. *Physiol. Behav.*, 5, 1151—1160.

Lindenmaier, P. and Kare, M.R. (1959) The taste end-organs of the chicken. *Poultry Sci.*, 38, 545—550.

318

Moncrieff, R.W. (1967) *The Chemical Senses.* Leonard Hill, London.

Moore, C.A. and Elliott, R. (1946) Numerical and regional distribution of taste buds on the tongue of the bird. *J. comp. Neurol.*, 84, 119—131.

Norgren, R. and Leonard, C.M. (1973) Ascending central gustatory pathways. *J. comp. Neurol.*, 150, 217—238.

Phillips, R.E. and Youngren, O.M. (1971) Brain stimulations and species-typical behaviour: activities evoked by electrical stimulation of the brains of chickens (*Gallus gallus*). *Anim. Behav.*, 19, 757—779.

Putkonen, P.T.S. (1967) Electrical stimulation of the avian brain. *Ann. Acad. Sci. fenn.* A5, 130, 9—95.

Richardson, A. and Gentle, M.J. (1972) Changes in the electroencephalogram of the chicken produced by stimulation of the duodenum. *Brit. Poultry Sci.*, 13, 171—173.

Rogers, J.G., Jr. and Maller, O. (1973) Effect of salt on the response of birds to sucrose. *Physiol. Psychol.*, 1, 199—200.

Warren, R.P. and Vince, M.A. (1963) Taste discrimination in the great tit (*Parus major*). *J. comp. physiol. Psychol.*, 56, 910—913.

Weiner, I.H. and Stellar, E. (1951) Salt preference of the rat determined by a single-stimulus method. *J. comp. physiol. Psychol.*, 44, 394—401.

Wood-Gush, D.G.M. (1971) *The Behaviour of the Domestic Fowl.* Heinemann, London.

Zeier, H. (1971) Archistriatal lesions and response inhibition in the pigeon. *Brain Res.*, 31, 327—339.

Zeigler, H.P. and Karten, H.J. (1973a) Brain mechanisms and feeding behaviour in the pigeon (*Columba livia*) I. Quinto-frontal structures. *J. comp. Neurol.*, 152, 59—82.

Zeigler, H.P. and Karten, H.J. (1973b) Brain mechanisms and feeding behaviour in the pigeon (*Columba livia*) II. Analysis of feeding behaviour deficits after lesions of quinto-frontal structures. *J. comp. Neurol.*, 152, 83—102.

Zeigler, H.P., Green, H.L. and Karten, H.J. (1969) Neural control of feeding behaviour in the pigeon. *Psychon. Sci.*, 15, 156—157.

15

THE NEURAL SUBSTRATE OF FEEDING BEHAVIOUR IN BIRDS

PETER WRIGHT

Department of Psychology, University of Edinburgh (Great Britain)

The two classical difficulties for all work on brain—behaviour relationships are, firstly, to precisely locate the critical anatomical structures associated with the behaviour and, secondly, in the case of brain damage, to analyse the nature of the deficits. These problems are in part due to the relative crudeness of the most commonly used techniques: electrical stimulation and electrolytic destruction of brain tissue. They apply as much to the neurology of motivation as to any other field of physiological psychology, and are well illustrated by an initial brief review of the effectiveness of these techniques in mammalian studies of hunger.

Electrical stimulation of the brain (ESB) through chronically implanted electrodes in free-moving animals has been extensively used in mapping regions of the brain involved in motivated behaviour. In Europe, following the pioneering work of Hess (1957), those who have used ESB as a tool have often followed the methods established by ethologists in describing animal behaviour. Although some workers have reported relatively precise anatomical zones related to agonistic behaviour (Brown and Hunsberger, 1963), others (von Holst and St. Paul, 1963) have stressed the variability of the behaviour elicited from the same electrode site, and interpreted stimulation as producing shifts in mood and central adaptation. In their work, von Holst and St. Paul considered it more important to understand the nature of the interaction between environmental stimulation and elicited behaviour patterns, before describing the details of the underlying neuroanatomy. Indeed, they considered that a premature emphasis on neuroanatomy would impede the functional analysis of the behaviour.

In direct contrast, the early work in the United States on electrically elicited feeding followed in the behaviourist tradition, and placed emphasis on the underlying mechanisms of motivation. For example, Neal Miller approached the problem from the position of a learning theorist especially concerned with the drive-reduction model of behaviour. Although the early work of Miller and his colleagues demonstrated that electrically elicited eating had many of the same characteristics as the hunger which follows food deprivation in the rat (Miller, 1963; Coons *et al.*, 1965), it was also apparent that closely adjacent electrodes in the very diffuse lateral hypothalamic area could yield behaviour which ranged from the purely reflexive acts of chewing and swallowing, to the full appetitive pattern of searching and locating food which was then eaten.

To establish that the elicited behaviour is of a motivational nature requires a great deal of testing, and in many of the subsequent experiments it was often assumed that electrodes in the lateral hypothalamus could be treated as a homogenous population. Recent reviews by Valenstein (1970), and Valenstein *et al.* (1970) have revealed a considerable anatomical overlap in the sites allegedly specific to different behaviour patterns. Moreover, the elicited behaviour shows considerable plasticity and the simple act of removing a preferred goal object from the rat's cage, but continuing to stimulate through the implanted electrode, will cause a switch from the previous stimulus-bound eating, to drinking, gnawing, or even to non-oral activities. Valenstein, in emphasising the need to examine all aspects of the behaviour elicited from a particular site, and in arguing the need to understand the interaction between the central stimulation and the environment before making statements on the localisation of function, has returned to effectively the same position as von Holst.

The second technique of lesioning specific sites has been even more extensively employed. In 1942, Hetherington and Ranson first described the obesity which follows destruction of the ventromedial nucleus (VMN) in the rat, and nine years later Anand and Brobeck (1951) reported aphagia and adipsia as a result of damage in the lateral hypothalamus. The persistent search in the following years for 'appetite' and 'satiety' centres and their exact hypothalamic location owed much to the dual excitation—inhibition model of Stellar (1954). However, almost from the first reports, many workers have argued that the syndrome which results from destruction of the anatomically complex lateral hypothalamic area (LHA) is due to damaging fibres of passage through this region rather than to some critical structure within the LHA (Morgane, 1961). More recently Ungerstedt (1971) induced selective degeneration in the terminal regions of dopaminergic and noradrenergic pathways using 6-hydroxydopamine, and has described essentially the same behavioural syndrome as that following electrolytic destruction of the LHA, but from damage throughout the nigrostriatal system.

The concept of discrete hypothalamic centres is now clearly inadequate, and many extrahypothalamic structures have considerable influence on food intake in the rat. Using a combination of chemical blocking and knife cuts, Albert *et al.* (1970) have described a feeding circuit which passes from the brain stem longitudinally through the lateral hypothalamus, from which separate bundles pass to the anterolateral hypothalamus, globus pallidus and thalamus. A limbic structure of especial significance is the amygdala, although the results obtained have often been contradictory and the effect varies markedly with species. Hyperphagia has been reported in cats from lesions of amygdala and the overlying pyriform cortex (Morgane and Kosman, 1957), in baboons and rhesus monkeys (Pribram and Bagshawe, 1953), as well as in the rat (Grossman and Grossman, 1963). Aphagia, anorexia, and loss in weight are also reported in rats (Kling and Schwartz, 1961), but similar lesions are without effect in monkeys, and produced only mild anorexia in cats. Many of the contradictions in these reports are due to problems in exactly demarcating the extent of the lesions in what is now recognised to be a very complex grouping of nuclei (Eleftheriou, 1972).

Even the same technique at the same locus can sometimes yield different results, and has made consideration of other variables essential. One such variable is the sex of the animal. Reynolds (1965) reported that radio-frequency destruction of the VMN was not as effective as electrolytic lesions in producing obesity in rats. He explained this discrepancy by suggesting that the electrolytic lesions (which result in both greater scarring and in the de-position of metallic ions, in comparison with radiofrequency lesions) set up irritative foci which result in continued excitation of the adjacent LHA, and hence lead to overeating. While such factors are clearly important (Rabin and Smith, 1968), another reason for these results was the use of male rats as subjects. It is easier to produce obesity from VMN lesions in female rats than it is in males, (Valenstein et al., 1969). This sex variable is not only apparent in rats but using a different means of destroying the VMN, by injections of goldthioglucose, it is apparent that female mice will become more obese than will males (Wright and Turner, 1973).

The exact nature of the deficits in feeding behaviour which follow LH lesions is only now becoming apparent, and is a result of undue emphasis on the effect of such lesions on food intake, and neglecting to examine other aspects of behaviour. Although many of the published experiments refer to lesion effects on feeding *behaviour*, it is often more appropriate to read 'food intake' in place of feeding behaviour, and the researcher outside this field could easily form the impression from reading the literature that LH-lesioned rats differ only from intact animals in their pathological aversion to laboratory chow! The position is now changing, Marshall et al. (1971) have shown that at least in the early postoperative phase LH-lesioned rats suffer pronounced disturbances in orientation to external stimuli, and exhibit some form of sensory neglect. In addition, such rats exhibit disturbances in their emotional responsiveness (Turner, 1973), and this also seems related to prob-lems of sensory awareness.

Earlier investigations of the underlying nature of the aphagic condition adopted a division of behaviour into motivational and sensory or motor im-pairment. Baillie and Morrison (1963) argued that as aphagic rats would continue to press a lever for either direct or intragastric food reward, then the deficit was predominantly a motor failure, whereas Rodgers et al. (1965) in a similar experiment concluded that the deficit reflected a loss in the desire for food, and hence was primarily a failure of motivation. However, as Lashley (1938) has indicated such an either/or distinction may not be a useful one, and all cases of motivation must reflect a partial excitation of a very specific sensorimotor mechanism.

Although the work outlined here suggests a similar neural organisation with respect to food intake in various mammals, detailed descriptions of the recovery from lateral hypothalamic aphagia are restricted entirely to the rat (Teitelbaum and Epstein, 1962). Impaired sensory responsiveness and lowered arousal appear to accompany the aphagic condition, and the road to successful recovery in the aphagic rat is to rearouse an interest in food by initially presenting only highly palatable foodstuffs. Rats are primarily animals which rely on taste and olfactory cues in the detection and selection of their food, whereas, with the possible exception of fruit-eating species, these senses are not so important as is vision in birds. A comparison of the behaviour and neural structures involved in feeding deficits in birds should help in understanding the contribution that sensory impairment makes to the defective motivation of the aphagic animal.

Zeigler et al. (1969) and Zeigler and Karten (1973a and b) have analysed feeding deficits in pigeons without assuming preeminence of the hypothalamus, and have lesioned pathways known to be involved in the mediation of what Edinger (1908) termed 'oral sense' in birds. They have reported aphagia from lesions in an extensive system projecting from the sensory trigeminal nucleus in the brain stem to the forebrain, and it is clearly of interest to ask how far this pathway can be related to the mammalian studies.

As outlined above, many extrahypothalamic structures have been implicated in mammals, but the hypothalamus still remains a key structure and in all mammals has a relatively constant anatomical organisation (MacLean, 1958). Mammals and birds are both descended from reptiles but have diverged and taken separate evolutionary paths. The diencephalon of birds is much closer to that of reptiles than to mammals, and birds show very little neocortical development but a massive expansion of the 'striatal' regions. The nomenclature associated with the avian telencephalon is confusing as it suggests homologies with mammals which are misleading. No attempt will be made here to review the organisation of the avian brain, as there are a number of excellent recent accounts (Karten, 1969; Pearson, 1972). However, two regions of the avian forebrain are thought to be directly comparable with areas of the mammalian brain which are mentioned in the previous discussion of feeding behaviour. The palaeostriatum primitivum on both histological and histochemical evidence is probably homologous with the mammalian globus pallidus (Fox et al., 1965), and the archistriatum has long been considered homologous with the mammalian amygdala (Ariëns Kappers et al. 1960), although more recent work suggests that this is only correct for its more medial and posterior aspects (Zeier and Karten, 1971). The present chapter will review some of the lesion and stimulation studies of the avian brain which are pertinent to feeding behaviour, and, to facilitate comparison with the mammalian literature, will emphasise the role of the hypothalamus and the amygdala.

BRAIN STIMULATION AND FEEDING BEHAVIOUR

Åkerman *et al.* (1960) elicited hyperphagia, polypnoeic panting and polydipsia, by electrical stimulation from implanted electrodes in the hypothalamus of pigeons (*Columba livia*). They concluded that the central regulation of body temperature, and of food and water intake is principally the same in birds and mammals. From the diagrams of Åkerman it appears that stimulation of the area ventralis anterior, the stratum cellulare externum, and the septum, caused eating, and prolonged stimulation caused over-eating. However, there are no quantitative observations in this paper, and no indication of the repeatability of the results. Indeed, since implantation of the electrodes, testing of the animals and their sacrifice, were all performed within 24 h, the work was semi-acute, and not comparable with the majority of mammalian studies.

Phillips (1964) stimulated the archistriatal regions of the forebrain in mallard ducks, and comments that the electrical stimulation sometimes resulted in "rapid movements of the bill and often of the head and neck, movements that looked like searching and gabbling (feeding movements)". Harwood and Vowles (1966) found that electrical stimulation of the forebrain in Barbary doves caused an increase in feeding both during, and for at least 10 min after, stimulation. The sites which produced effects on feeding were largely confined to the palaeostriatum and to the more ventral and posterior neostriatum. The most marked effect was to prolong the feeding bouts and to increase the rate of pecking during the bouts. Stimulation seemed to facilitate an existing tendency to feed, but did not have a motivational component in that it would not cause satiated birds to resume performance of a learned response for food reward.

Wright (1969) stimulated over 100 diencephalic sites in 37 Barbary doves in an attempt to find areas which would elicit feeding behaviour. Some 14 electrodes appeared to be associated with feeding behaviour, and were located in the ventral aspects of the lateral forebrain bundle, clustered around the point at which the bundle divides and distributes to the lower brain stem. The behaviour was typically highly variable. On one day the feeding might be quite vigorous and prolonged, and on another day the same site would yield either no feeding at all, or very weak pecking at food. The latter was most commonly the case, the pecking was slow and cautious and often broken by periods of walking around the cage and pecking at inanimate objects. Attentive postures were often seen in association with feeding, and in some animals signs of fearful behaviour, which at higher intensities of stimulation resulted in escape behaviour, were evident. In 12 of these sites, the feeding usually occurred *after* stimulation had ceased, sometimes with a latency of 2 or 3 min. In any one session it was quite impossible to predict whether feeding would occur or not.

This behaviour was very similar to that resulting from forebrain stimulation in the same species, characterised by its intermittency and variability in appearance. Harwood and Vowles (1966) suggested that the forebrain functions in cooperation with the hypothalamus as a positive feedback system to increase the vigour and duration of feeding. Work presented later in this chapter suggests that there is a close relationship between brain regions involved in feeding and in emotional behaviour, and it is possible that conditions which reduce the fearfulness of the birds, such as the presence of a feeding companion, would have resulted in more intensive feeding behaviour from these hypothalamic sites.

Similar negative results are apparent in other species. Phillips and Youngren (1971), in an extremely comprehensive survey of the forebrain and diencephalon in both chickens and ducks, tested 1500 brain loci in 87 birds and found no ESB evoked feeding. In an exploration primarily of the archistriatum in chickens, but with observations in other forebrain areas, and in the medial regions of the diencephalon, Putkonen (1967) found that rhythmic movements of the bill were frequently elicited from the archistriatum and its components, from the palaeostriatum, and from the basal neostriatum. Salivation was reported from sites in the archistriatum, the telencephalic part of the occipitomesencephalic tract, and from the palaeostriatum augmentatum. Salivation and rhythmic oral movements often occurred in conjunction, and at some archistriatal sites, pecking directed into the air or towards some object in the cage, accompanied the chewing movements. In no instance did Putkonen observe coordinated feeding behaviour from such stimulation. Delius (1967 and 1971) could not obtain reliable eating from stimulation of the diencephalon and forebrain in herring and lesser black-backed gulls (*Larus argentatus* and *Larus fuscus*), although he did observe a high frequency of foot-paddling accompanied by downward staring and small jerky head movements, from sites in the neostriatum. The identical behaviour is often observed in birds standing in small shallow pools or on the tide line, and probably stirs small animals living in the sand up to the surface where they can be seen by the gulls. Delius comments that the variability of this and other behaviour was high, and the paddling was easily inhibited by competing responses. Fear was a common disturbing factor.

A more recent study (Tweeton *et al.*, 1973) with chickens, thoroughly sampled the preoptic and supraoptic regions from which Åkerman *et al.* (1960) had produced compulsive eating in pigeons. The authors also looked for the appearance of behaviours which are normally closely correlated with feeding, such as ground-scratching and food-calling. The test procedures were carefully controlled and after initial screening, positive sites were intensively investigated. Of the 625 sites examined, 75 produced feeding behaviour in at least half of the initial screening trials. These positive sites were widely scattered throughout the region tested and only ten on further testing showed positive results. In the final testing, only four of these ten sites resulted in significant increases in eating mash from a dish, and this was always a poststimulation effect.

In mammals, those electrodes which elicit eating will also support self-stimulation, at least from sites in the lateral hypothalamus of the rat, (Hoebel and Teitelbaum, 1962). It is therefore worth noting that although avian studies of self-stimulation are available, a similar association between feeding and the rewarding effects of brain stimulation has not been detected. Goodman and Brown (1966), working with pigeons, were unable to elicit feeding or drinking from the septal and anterior palaeostriatal sites which yielded positive results in the Åkerman study; but did find that different regions of the brain had punishing or rewarding values. Macphail (1965 and 1967) was also unable to elicit alimentary responses in pigeons, but did find that stimulation in the lateral forebrain bundle (LFB) and in the palaeostriatum was rewarding, in that extinction of a food rewarded operant was slowed by replacing the food reward with intracranial stimulation. He also found that stimulation of the medial forebrain bundle and of the septal area was aversive. This is in sharp contrast to the results in mammals, where these regions are generally considered to be part of the reward system (Olds, 1962). Macphail's results are supported by the earlier observation of Phillips (1964), that stimulation of the septal area in mallard ducks yielded 'escape' responses. There is some evidence then, for a contrast between avian and mammalian organisation of the brain with respect to self-stimulation, although Andrew (1967) working with young chicks has described an anatomical organisation more similar to the mammalian pattern.

These studies provide no support for the conclusions of Åkerman et al. A single report of chemical stimulation in pigeons (Macphail, 1969) also showed that in contrast to rats and rabbits (Miller et al., 1964; Grossman, 1967), carbachol injections at various diencephalic sites in the Barbary dove do not induce eating or drinking. The stimulation studies provide no substantial evidence for a region in the lateral hypothalamus of the avian brain which could be classified as a feeding centre. The rat appears unique, even compared with other mammals, in the ease with which eating can be elicited from the lateral hypothalamus. The relative size of the electrode tip to brain volume is much larger in rats than in other mammals, and Andersson (1964) suggests that size of tip is important in this context. I have reduced the tip exposure down to under 0.2 mm in doves (Wright, 1969), but found it impossible to elicit any behaviour, and exposures above 0.5 mm would make accurate localisation in the dove hypothalamus out of the question. As Doty (1969) has pointed out, a further disadvantage of ESB is that stimulation is invariably unilateral. The two symmetrical structures on either side of the brain presumably normally act together, and the output from the un-

326

stimulated side may inhibit that from the stimulated side.

In conclusion, these experiments indicate that electrical stimulation of the brain in birds has little directive influence with respect to feeding behaviour, and where there have been positive results (Harwood and Vowles, 1966), the authors conclude that the results reflect a facilitation of activities which were already occurring. In the case of Åkerman et al. (1960) it is likely that the birds were in a state of mild food deprivation as the tests were carried out on the day of surgery, in which case electrical stimulation combined with an underlying low level drive would be sufficient to give the appearance of compulsive eating in a single series of tests. An alternative explanation from the various above observations is that stimulation both arouses food-seeking behaviour and also alerts the bird, making it more fearful and attentive to external stimuli. Depending on the underlying level of deprivation, the birds may then show intermittent feeding until they are certain that no potentially dangerous stimulus is around.

LESION STUDIES

As was the case in the experiments on avian brain stimulation the emphasis in the first studies was to look to the hypothalamus for effects of lesions on food and water intake. With the hindsight that comes from considering the more recent mammalian accounts, many of these early reports in birds were too selective in their reported observations. Feldman et al. (1957) placed bilateral electrolytic lesions throughout the diencephalon in some 350 domestic chickens, and reported 12 animals which lost the ability to eat. A number of these aphagic birds has accompanying disturbances in temperature regulation from which they subsequently died, and all of the birds showed almost no voluntary activity. They describe the aphagia as resulting from widely scattered lesions in the medial, lateral, anterior and posterior regions of the hypothalamus, and inspection of their photographs indicates that the lesions were present in the stratum cellulare externum and in the medial posterior hypothalamic nucleus. A later report in the same species (Smith, 1969) describes six hens becoming aphagic after lesions in the lateral hypothalamic nucleus, but provides no details of the postoperative behaviour, or whether the birds eventually recovered.

Lepkovsky and Yasuda (1966) found 'hyperphagia' in chickens from lesions "in what may be assumed to be the ventromedial area of the hypothalamus", but their report does not indicate a significant increase in the food intake of the lesioned animals, and the authors appear to use the term hyperphagia incorrectly. The graphs shown in this report and in a later review (Lepkovsky et al., 1967) clearly show that although such lesions will produce weight gains in both sexes, this was unaccompanied by increases in food intake, and the graph for female birds shows a decrease in food intake during the period of weight gain. In all cases the authors report that the birds appeared to be functional castrates, as assessed by changes in comb size.

A more convincing report of actual increases in food intake following hypothalamic lesions is that of Kuenzel and Helms (1967), who describe 11 cases of hyperphagia in the white-throated sparrow (*Zonotrichia albicollis*). The lesion damage was restricted to the ventromedial hypothalamus above the optic chiasma and supraoptic decussation, and adjacent to or across the ventral portion of the third ventricle at an anterior—posterior level between the supraoptic nuclei and the median eminence. The weight gains of the experimental birds were almost identical with those of photostimulated controls, suggesting that this region of the hypothalamus may be involved in the normal photo-induced weight increase of this species before its spring migration. A second report (Kuenzel and Helms, 1970) indicates that the hyperphagia is often accompanied by polydipsia, and associated with gonadal disturbances and moult. A number of other reports also describe polydipsia from this supraoptic region (Ralph, 1960; Koike and Lepkovsky, 1967; Wright and McFarland, 1969), but will not be discussed here. Kuenzel (1972) found no evidence in the white-throated sparrow of sex differences in hyperphagia similar to those reported in the rat by Valenstein *et al.* (1969), and no differences in weight gains, although females had a slight tendency to deposit more fat than the males. Aphagia is also reported from LH lesions, but as the birds survived not longer than two days, it is difficult to conclude very much from these observations.

With the possible exception of the ventromedial lesions in the white-throated sparrow, these reports provide little evidence for a hypothalamic organisation of feeding in birds similar to that of mammals. The examples of aphagia in chickens have so many other accompanying disturbances that it is quite likely that the absence of food intake may be secondary to other profound disturbances of homeostatic regulation. Such objections do not apply in the following reports of aphagia in doves and pigeons.

Wright (1968) placed small radiofrequency lesions throughout the diencephalon in Barbary doves, and reported aphagia from sites bordering on the lateral forebrain bundle at the level of the lateral hypothalamus. Daily measures of rectal temperature were taken, and although the aphagic birds had lowered body temperatures this was no greater than that reported in intact doves under conditions of food deprivation (McFarland and Wright, 1969). The aphagic doves, although unable to consume food, made repeated attempts to eat, and had difficulty in mandibulating, *i.e.* they were unable to convey seed from the tip of the beak to the back of the mouth prior to swallowing. The same mandibulation deficits were described in pigeons by Zeigler *et al.* (1969), but from lesions of the main sensory trigeminal nucleus, and of the nucleus basalis in the forebrain. That such deficits in the pigeon might primarily be a result of deafferentation of the beak is suggested by a single unit analysis of the trigeminal nucleus in which Zeigler and Witkovsky (1968) found that 72% of the cells were activated by mechanical stimulation of the beak, ranging from very light touch to pressure. Zeigler has suggested that the aphagia in pigeons is a consequence of interrupting a neural circuit extending from the sensory trigeminal nucleus in the brain stem up to the nucleus basalis in the forebrain, by way of the tractus quintofrontalis. In two reports Zeigler and Karten (1973a and b) have provided details of the histology of individual birds, together with complete accounts of food and water intake, and although they conducted operant experiments to assess the degree of motivational impairment in the aphagic pigeons there are no indications of other aspects of behaviour in the birds. As indicated at the beginning of this section, one of the lessons from the mammalian studies is the need to examine aspects of behaviour other than measures of food intake, and in the following account of aphagia and its recovery in the dove, I shall concentrate more on the nature of the behavioural disturbance rather than give a detailed consideration of the underlying anatomy.

Recovery from aphagia in the Barbary dove

In addition to daily measures of food and water intake in their home cages, all operated animals are routinely observed for 20 min each day in test chambers furnished with liberal quantities of seed and grit scattered on the cage floor. In these sessions the activity, preening, attentive, and sleeping behaviour, in addition to feeding activities, are monitored on an event recorder. As the period of aphagia is extremely variable, ranging from a few days to several weeks, and as the majority of animals are followed for several weeks or months on recovery from aphagia, it is often difficult to precisely demarcate the critical regions involved. However, using electrolytic lesions, fairly consistent feeding deficits result from destruction of an area extending from the medial borders of the LFB just rostral to the anterior commissure, through the lateral hypothalamus and into the stratum cellulare externum (Fig. 1). A full account of these experiments will appear elsewhere.

Aphagia refers to the period during which there is no measurable food intake from seed boxes in the home cage, and adipsia to the period when there is no measurable intake of water from the drinking tubes. The relative durations of aphagia and adipsia in 45 birds with lesions in the above area are plotted in Fig. 2 and for ease of comparison I have replotted in the same figure similar data from a study of Epstein and Teitelbaum (1964) in which rats received lesions in the lateral hypothalamic area. The two species fall into separate groups, as the doves show the reverse of the classical lateral hypothalamic syndrome. Although several birds were initially both aphagic and adipsic, drinking invariably recovers before the ability to eat, and in 13 cases the aphagia was unaccompanied by any disturbance in drinking. For doves, the mean duration of aphagia was 9.2 days (range 2—35), and for adipsia the mean duration was 2.1 days (range 0—9). In rats, the average length of aphagia was 4.9 (range 0—25), and for adipsia 50.8 days (range 3—590), and

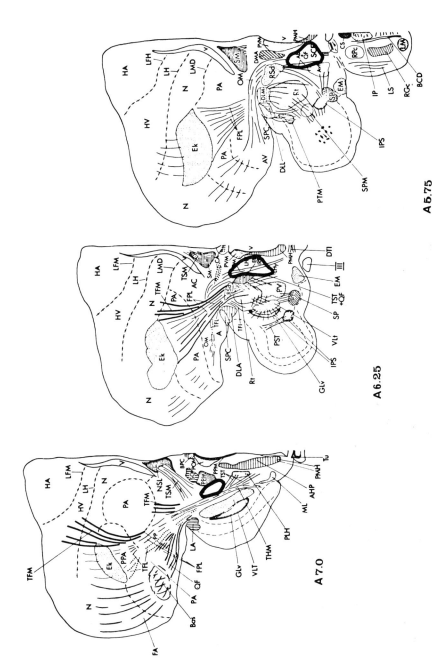

A5.75

A6.25

A7.0

Fig. 1. Frontal sections of the dove brain with anterior—posterior coordinates according to the stereotaxic atlas of the dove brain of Vowles*. Lesion areas (heavily outlined) represent a composite of damage in some 40 aphagic birds. Abbreviations: AHP, posterior hypothalamic area; AC, nucleus accumbens; AL, ansa lenticularis; Bas, nucleus basalis; CA, anterior commissure; FPL, fasciculus prosencephali lateralis (lateral forebrain bundle); LHY, lateral hypothalamus; PA, palaeostriatum augmentatum; PVM, nucleus paraventriculus magnocellularis; QF, tractus quintofrontalis; Rt, nucleus rotundus; SCE, stratum cellulare externum; SM, septal area; TSM, tractus septomesencephalicus, with additional nomenclature according to Vowles.

* See Appendix.

330

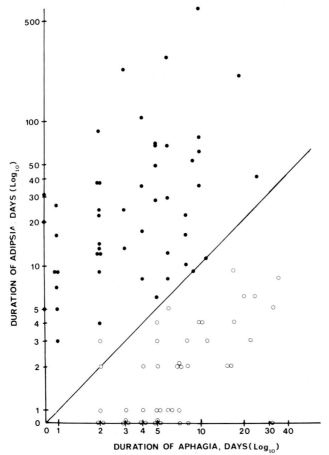

Fig. 2. Relative durations of aphagia and adipsia for individual doves, compared with the same data for rats (taken from Epstein and Teitelbaum, 1964). Doves, open circles; rats, closed circles. For all points falling above the diagonal, adipsia > aphagia; for all points falling below the diagonal, aphagia > adipsia.

in all instances, recovery from aphagia preceded recovery from adipsia.

The aphagia is considered to have ended when the rats show a return of interest in eating, as indexed by their acceptance of wet, highly palatable foodstuffs, and not the standard diet of laboratory chow. In our experience with doves, once they have resumed eating, recovery is rapid and accompanied by no more than a few days of slightly reduced food intake compared with their preoperative levels.

In the first few days following surgery, the aphagic doves spend most of their time either asleep or in a sleep-like posture with feathers puffed up, neck retracted, and eyes closed. Handling or disturbing the birds at this stage leads to brief periods of activity in which they become alert and move around the cage, eventually leading to bouts of preening followed by a resumption of the sleep-like state. After this phase, the birds demonstrate interest in food, as shown by their approach towards piles of seed, and their repeated attempts to peck and pick up seed. The characteristic feeding deficit is often one of mandibulation, but in several cases this was accompanied by inaccuracies in pecking at the seed. All birds were routinely force-fed from the third postoperative day by holding open the beak and dropping moist seed at the back of the throat, when swallowing would occur quite normally. At no stage did contact with food appear aversive in the dove, as seems to be the case with rats during Stage 1 of recovery (Teitelbaum and Epstein, 1962), although vomiting occurred spontaneously in their home cages and in the observation sessions of several birds during the early period of aphagia. Other activities involving the beak, such as preening, appeared quite normal. As noted above, such mandibulation deficits have been described in the domestic pigeon by Zeigler and Karten, and attributed to damage of quintofrontal structures. As the quintofrontal tract borders the edge of our lesions, and as we invariably find that our lesions encroach on the anterior aspects of the lateral forebrain bundle in which this tract travels, it is likely that the feeding deficits in the dove are in part due to disruption of this same system. The behaviour of a representative bird, S405, during daily observations over the course of the aphagic period is shown in Fig. 3. This bird was first seen to successfully pick up and swallow seed on the eleventh postoperative day, and the following day had consumed 8 g of seed in its home cage. On the third postoperative day, it began to make attempts to peck seed, and the amount of time spent in food-directed activity increased progressively over the period of aphagia, and showed a sharp increase on the day when successful eating was first observed.

The behaviour of the aphagic doves in these daily observation sessions indicates that following a period of immediate postoperative unresponsiveness to virtually all external stimuli, the doves remain actively interested in food and make repeated but unsuccessful attempts to ingest seed. In normal doves, the tendency to peck seed is clearly related to the level of food deprivation, and it would be reasonable to conclude from their *behaviour* that the aphagic birds remain motivated to eat, but are unable to do so because of some deficit in pecking accuracy or in mandibulation. Following Teitelbaum's analysis of motivation (1966), we would, therefore, expect that performance of an operant response for food reward in these aphagic birds would still be present.

Barbary doves were housed continuously in a Skinner-box and trained to peck separate keys for access to small quantities of food and water. Daily measures of actual water and food consumed were made in addition to recording the pattern of pecks at the food and water keys. The birds were then lesioned and returned to the Skinner-box and 24-h records of food and water pecks again recorded, each day records were also made of all behaviour

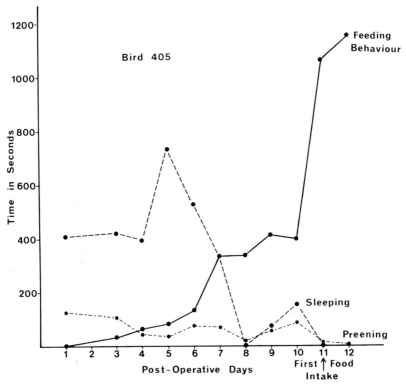

Fig. 3. Recovery from aphagia as seen in daily 20-min observation sessions.

shown in 20-min observation sessions as described previously. In all cases key-pecking for water was resumed before key-pecking for food; key-pecking for food was either absent or present at a very low level during the period of aphagia. Key-pecking for food increased dramatically following the recovery of the ability to eat as observed in the daily observation sessions. The resumption of key-pecking for water usually coincided with the first appearance of food-directed activities in the observation sessions. The time course during recovery for a representative bird, S269, is shown in Fig. 4. Although S269 did not resume water pecks until the sixth postoperative day, it was unnecessary to artificially hydrate the bird, as the simple procedure of holding the bird and slowly lowering the beak into a water trough initiated convulsive drinking from the third postoperative day. Preoperatively S269 averaged 32 pecks daily at the water key, and although this was considerably increased during the period of aphagia, the actual amount drunk remained slightly below the preoperative level. This pattern of low levels of operant response for food shown here is fairly typical; when pecking the birds are slow and hesitant, and do not attempt to put their heads down into the food hopper

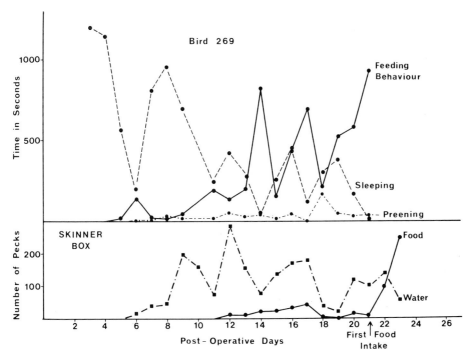

Fig. 4. Recovery from aphagia as seen in direct observation sessions (above) and operant response to food and water (below).

to peck seed. This same error is often seen in the early stages of training doves to peck a food key. The overpecking of the water key is possibly related to the behavioural contrast phenomenon reported in discrimination experiments (see Chapter 8 by Macphail in this volume), and may reflect frustration from unrewarded pecks at the food key. We have on occasion observed aphagic birds in observation cages also pecking at the surface of a water dish, and similar behaviour has been described in adipsic chickens by Lepkovsky and Yasuda (1967).

In a similar experiment with pigeons, Zeigler and Karten (1973b) also found that key-pecking for food was absent following lesions in either the sensory trigeminal nucleus or in the nucleus basalis. They report that during the period of no key-pecking (their phase 1), pigeons would not attempt to obtain 'free' reinforcement made available in their daily 30-min sessions in the Skinner-box. However, these same birds would eat a supplementary food ration in either the Skinner-box or the home cage, so that during this phase the birds were not actually aphagic at all. Key-pecking for food resumed when the pigeons responded to reinforcement. Three pigeons with basalis lesions trained to peck for water only, showed absence of key pecks for only 2—4 days, and when this resumed, pecking was slightly below the preoperative levels. In our own experiments, the continuous 24-h housing in the Skinner-box, and simultaneous training on both food and water keys, together with the difference in lesion site, makes direct comparison with the pigeon results difficult. Furthermore, as the pigeons were able to consume a food supplement they were not aphagic as were the doves, and the lack of operant response in phase 1 could have been due to the practice of providing food either in the Skinner-box or in the home cages, at the end of the daily operant sessions.

In both experiments, the choice of operant may be a poor one because of the similarity between the operant and consummatory response of pecking food, and the possible aversive consequences of the act of pecking in aphagic birds. Such an explanation is less likely in view of the overpecking on the water key observed in the aphagic doves. However, in both doves and pigeons, key-pecking for food reward is slower to return than key-pecking for water, and parallels the early recovery of adipsia compared with aphagia, when the birds are housed in their home cages. An identical observation was in fact made by Shaklee (1921) in three decerebrate pigeons, all of which after a variable period of time resumed spontaneous drinking of water, and two showed return of some feeding behaviour. Shaklee concluded,

> "It appears, therefore, that the decerebrate pigeon is practically incapable of seizing a grain with its beak. The chain of reactions which together constitute eating, appears to be broken at this point; the reaction for seizing grain appears to be missing. Moreover when the grains were placed in the anterior portion of the beak, this bird uniformly threw them out, although grains dropped well back into the throat were as uniformly swallowed. Here then there appeared to be another link missing, *viz.*, the reaction for moving objects from the anterior portion of the beak to the posterior portion."

It would seem that irrespective of the locus or extent of brain damage, there is a certain commonality in both the nature of the feeding deficit, and in the dissociation of such effects from drinking.

Changes in temperament

Doves which have recovered from aphagia continue to show pronounced changes in emotionality and in their reaction to external stimuli. They are extremely placid, and show only mild avoidance of frightening objects. As all these birds are of necessity force-fed from the third postoperative day, it is possible that the tameness on recovery is related to the experience of handling and being fed by hand. We have measured the changes in temperament by comparing the reaction of intact, but previously force-fed doves, and recovered aphagics, in their response to a disturbing stimulus. The latter consisted of a large rubber spider mounted at the end of a perspex rod, and has previously been shown by Harwood and Vowles (1967) to be extremely effective in eliciting fearful and defensive behaviour in doves.

The doves were tested in a large cage with walls of transparent perspex.

The base on two opposite sides of the cage was fitted with a loose opaque rubber curtain, arranged so that the spider could be introduced on either side of the cage through the curtains without the dove being able to see the intruder beforehand. Testing sessions consisted of ten consecutive presentations of the spider towards the bird. Each presentation lasted 5 sec, and during this time the spider was thrust vigorously at the dove. The spider was presented at 1-min intervals, and tests were conducted over four consecutive days, giving 40 presentations in all. The recovered aphagic doves were tested within 6—12 weeks of recovering their preoperative body weight, as were a control group of operated birds which did not show any aphagia postoperatively. The four intact birds were deprived of food for three days, and then force-fed for a 10-day period, they were tested 2 weeks after the end of the force-feeding. Results of this experiment are shown in Table I.

Both control groups show strong avoidance of, and react fearfully to, the presence of the spider, by wing-raising and flying upwards. In contrast, the recovered aphagic birds show a very low level of response, and are largely indifferent to its presence. They continue to show feeding in the presence of the spider, but this was never observed in either control group. On a number of occasions, the previously aphagic birds were seen to peck at the small yellow spots on the abdomen of the spider treating them as if they were food objects. Such behaviour was never observed in either of the two control groups. Six of the recovered aphagic birds gave an additional reaction to the spider, consisting of vigorous vomiting motions, although in no case was food regurgitated. This response was not consistently emitted by the birds, and when it occurred was often preceded by an intention peck towards the spider. Similar periods of vomiting have been observed during the period of aphagia, and may indicate some residual aversive component to the consummatory act of pecking in the recovered birds.

In a second experiment we trained intact and recovered aphagic doves to run down a straight alley for brief access to a dish of seed, and recorded the time taken to reach the goal box. Following several days of training, with ten

TABLE I

The mean occurrence of behaviour in response to spider presentation over 40 trials

	Recovered aphagics (n = 16)	Operated controls (n = 8)	Force-fed controls (n = 4)
Fly	1.31	18.13	27.0
Wing-raise	3.25	20.50	22.0
Feather sleek/long neck	3.75	21.75	19.8
Feeding	6.00	0	0

TABLE II

The latencies to reach a food reward goal box with approach impeded by a novel disturbing stimulus

Column headed by * shows mean latencies to reach goal box on the preceding day.

	Bird	*	Day 1			Day 2		
			Trial no.			Trial no.		
			5	7	9	5	7	9
Recovered	S469	6.0	47.0	8.5	9.5	18.0	7.5	5.5
aphagics	269	9.5	55.0	47.0	22.0	13	7	20
	464	8.5	31	9	11	15	10	9.5
	206	9.5	33	16	16	13	11	10
Intact	543	16	—	—	—	—	—	—
birds	438	21	—	—	—	58	13	10
	440	20	—	—	—	—	—	—
	475	22	—	—	—	—	20	14

trials each day, spaced 1 min apart, the birds were confronted with a mechanical toy placed halfway down the alley. The toy moved backwards and forwards emitting a loud whirring noise for about 30 sec, and was placed in the alley on trials 5, 7 and 9, over two consecutive days. The results of this experiment are given in Table II. The data are shown for individual birds and give the time in seconds to reach the goal box.

On the first day when the toy was presented, none of the intact doves left the start box (a single trial was terminated if the bird did not leave the start box within 60 sec of initiating the trial), and only one bird eventually negotiated past the toy on all three trials of the second day. In marked contrast, although the recovered aphagic birds were slowed down by the presence of the toy, all of them continued to run the alley and reach the goal box on Day 1. On Day 2, their times to reach the goal box were close to the mean of the last day of training when no toy was present.

In both these experiments, the recovered aphagic doves are characterised by an apparent lack of fearfulness when confronted by novel, disturbing stimuli. This is apparent both in the non food-deprived condition of the first experiment, and when tested under conditions of food deprivation. The next section examines the significance of these changes in terms of their effect on social influences in feeding behaviour.

Social influences on feeding

Early observations on aphagic doves (Wright, 1969) suggested that if the birds were sufficiently aroused by scattering or throwing seeds close to them, this would lead to bouts of pecking at the seed. It is well known that chickens will show substantial increments in eating when given food in the presence of others (Bayer, 1929), and this social facilitation of feeding has been extensively investigated in young chicks by Tolman (1964, 1965 and 1967). Such influences may be of particular importance in flock-feeding birds such as the pigeon. The following procedure has been used to investigate social influences on feeding in recovered aphagic doves.

The test bird is observed for 20 minutes in an unfamiliar perspex walled cage, the floor of which is covered with a mixture of seed and grit. In the first and last 5-min period of each session the test bird is alone; in the middle two 5-min periods a companion bird is introduced into an immediately adjacent cage. The companions are placed on a 48-h food-deprivation schedule, and are fully habituated to feeding in the test cage; this procedure ensures that the companion bird spends the entire 5-min period voraciously feeding, and pays no attention to the test bird in the adjacent cage. All the test birds are maintained in their home cages with continuous access to food. The test birds are observed over four consecutive daily sessions and their behaviour continuously monitored with an event recorder. All activities directed towards food — pecking and scattering seed, sweeping seed with the bill, pecking at grit, as well as eating seed and grit — were recorded together as feeding behaviour. Under these conditions intact doves reliably increase the amount of behaviour directed towards food when they are able to see another companion bird feeding in the adjacent cage (Wilcoxon test, $P < 0.005$, one tailed). Weighing the birds before and after each session indicates that very little seed is actually consumed under these conditions, and the social facilitation effect is on food-directed behaviour rather than on eating *per se*. The results of such experiments are given in Fig. 5, which shows the mean

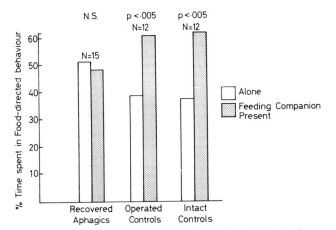

Fig. 5. Social facilitation of food-directed activities in intact and recovered aphagic doves.

time spent in food-directed behaviour over the 4-day period in the two conditions; the results are expressed as the percentage of the total time spent feeding, in the alone condition and when in the presence of the feeding companion. Doves which have recovered from aphagia were tested some 4—12 weeks postoperatively, and in marked contrast to the intact birds, they show no differences in the amount of feeding activity when alone or in the presence of a feeding companion. An operated control group which had not developed aphagia show a facilitation effect comparable with the intact group ($P < 0.005$).

There are many possible explanations for the social facilitation of feeding in birds (*e.g.* the presence of the feeding companion may act as a conditioned stimulus for eating; the feeding companion may induce a state of hunger in the satiated bird; or the feeding companion may direct the attention of the test bird to the presence of food) and Tolman (1968) has discussed these at length. In the case of wood pigeons, Murton *et al.* (1971) suggested that flocking confers a survival advantage on the individual by maximising feeding efficiency. They also point out that lone wood pigeons in the field appear very frightened and spend a good deal of their time looking around them, possibly to be alert to the approach of predators, and as a consequence their pecking rates are low. Joining the flock would have the other likely advantage of lessening the danger from a surprise attack by a predator.

If the facilitation effect is partly a disinhibition phenomenon, the combination of lack of fearfulness and the absence of social facilitation of feeding in the recovered aphagic doves may be related. The birds in our experiments are tested in an unfamiliar cage, and the intact doves when observed alone show the typical nervous posture of sleeked feathers and long neck, together with erection of the feathers over the ears, and make many quick scanning movements of the head. In the presence of the feeding companions they become more relaxed, and the feeding bouts more intensive and prolonged. It is likely, therefore, that when alone fearful behaviour predominates and may inhibit any tendency to feed; the presence of the feeding companion makes the test bird less fearful, and so disinhibits the tendency to show food-directed behaviour. Because the recovered aphagic birds are less fearful than intact birds, and if this is a sufficient explanation of the facilitation effect, then these birds would not be expected to show increments in feeding behaviour when in the presence of the feeding companion. Alternatively, the absence of a facilitation effect may be independent of any effects of the brain damage on fearfulness, and represent a permanent motivational impairment in the recovered aphagic birds.

Pronounced 'taming' effects following lesions of the ventromedial archistriatum in mallard ducks have previously been described by Phillips (1964) and in one case the lesions were of the stratum cellulare externum in the hypothalamus, a region which is included in the locus from which aphagia is obtained in doves (see Fig. 1). Feldman *et al.* (1957) also comment that lesions in this site, in addition to aphagia led to the disappearance of the flightiness common to the strain of chicken used in their experiments. Phillips concluded that all his tamed birds had in common damage to fibres of the occipitomesencephalic tract (TOM), which is the main efferent pathway from the archistriatum to the hypothalamus and midbrain. As this tract may in part be disrupted by the lesions which produce aphagia in doves, this could explain the lack of fearfulness which we have described in the re-covered aphagic doves. Partly to test the facilitation hypothesis outlined above, and also because in the mallard ducks taming was often complicated by the appearance of aphagia, we have recently investigated whether section of the TOM alone will lead to similar changes in the emotionality of doves (Wright and Spence, 1975).

In the initial studies we made electrolytic lesions at the point of entry into the diencephalon of the TOM, and this invariably led to both aphagia and adipsia in doves, and in contrast to the results from hypothalamic lesions, the adipsia was usually as pronounced as the loss of feeding. However, the histology of these birds showed that the lesions were all superficial to the TOM, with very little damage to the tract. The primary locus was in the over-lying palaeostriatum immediately lateral to the foot of the ventricles. A similar difficulty in making electrolytic lesions of the anterior commissure is reported by Zeier and Karten (1973); the dense bundle of fibres is apparently depressed by the electrode tip resulting in a lesion which circles the commis-sure but causes little direct damage. To overcome this problem, and to avoid complications in interpreting the lesion damage, we have used thin slivers of razor blade to produce a clean section through the tract. With this technique, section of the TOM is unaccompanied by aphagia, but all doves show an immediate and long-lasting change in fearfulness as measured by their response to a model predator. No such effects are seen when the knife-cuts end immediately above the TOM, or in the other necessary control group in which the anterior commissure alone is sectioned in the midline. As the changes in fearfulness were thus dissociated from aphagia, we were able to test whether such TOM-sectioned doves would show an absence of facilita-tion as predicted by the disinhibition hypothesis.

TOM-sectioned doves, and a control group in which the knife cuts ended above the tract, were tested for facilitation of feeding under identical con-ditions to those described above, and the results are shown in Fig. 6. It is clear that as a group, the TOM-sectioned doves behave in the same way as the recovered aphagic doves, and do not show any difference in the amount of feeding activity when observed alone, or when in the presence of feeding companions. The control birds show a significant increase in such activity when in the presence of the companions ($P < 0.005$). Therefore, it would seem likely that the absence of feeding facilitation which is common to both recovered aphagics and to TOM birds is due to disruption of the same neural pathways, and is a consequence of the reduced fearfulness in the two groups.

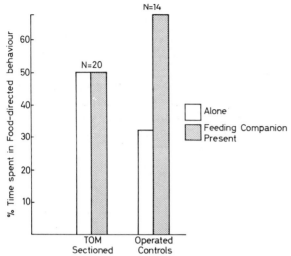

Fig. 6. Social facilitation of food-directed activity in occipitomesencephalic tract (TOM)-sectioned and operated control doves.

However, not surprisingly this explanation proves to be an oversimplification of a number of other factors which contribute to the effect, as shall be seen in the next section.

Sex differences and feeding behaviour

Section of the TOM by means of knife-cuts did not result in aphagia in doves; in fact, observations of the birds immediately following surgery provided some evidence for 'release' effects on feeding. On returning the operated birds to their home cages on the day following surgery, their behaviour was noted over the first five minutes within the home cage. Of the 22 TOM-sectioned doves, 19 were seen to put their heads into the seed boxes on the front of the cage and to eat a small amount of seed; of the 19 control-operated birds in which the TOM remained intact, only four ate within this same period, this difference in eating was significant ($\chi^2 = 15.1$; df = 1, $P < 0.001$). Records of daily food intake over the next 20 days showed that female doves exhibit a transient hyperphagia and accompanying gain in weight in the first ten days following surgery, but a similar response was not found in males (Wright, 1974). At the beginning of this chapter, a sex difference in response to ventromedial lesions in the rat was mentioned (Valenstein *et al.*, 1969), and it would appear that oestrogens exert an inhibitory effect on food intake in the rat. Preliminary experiments with doves suggest that the overeating following TOM section is also hormonally mediated, and that it is the absence of testosterone which is critical. Castrated male doves following section of the TOM also exhibited a transient hyperphagia and weight gain, whereas ovariectomised females showed the same response to TOM section as did intact females. Seventy per cent of the female doves with tract section developed a pronounced polydipsia which was longer lasting than the hyperphagia, and which did not appear in either intact or castrated males, or in any of the ovariectomised females. As Phillips (1964) also found that mallard ducks sometimes showed signs of overdrinking following lesion damage to the TOM, and that follicle size was increased postoperatively in all birds, this may well reflect increased ovarian activity following interruption of the TOM. These observations at least suggest hormonal influences on both feeding and drinking when the efferent pathways from the archistriatum are damaged.

If we now replot the results of the facilitation experiments shown in Figs. 5 and 6, but separate the birds into male and female, it is apparent that the conclusion reached in the previous section, namely that the lack of facilitation observed in both recovered aphagic and TOM-sectioned doves is due to damage to the same neural pathway, can no longer hold. In Fig. 7 this has been done for the recovered aphagic birds, and the lack of facilitation is present in both male and female doves. However, when this is applied to the TOM results, a clear sex difference emerges, and the male birds now show less food-directed activity in the presence of the companion birds than when they are alone; the females continue to show an increase in such activities when the companion birds are present (Fig. 8).

Any explanation for these sex differences must be tentative, and we think is related to the primary disturbance following TOM section, which appears to be a form of visual agnosia. Preliminary studies of the courtship behaviour

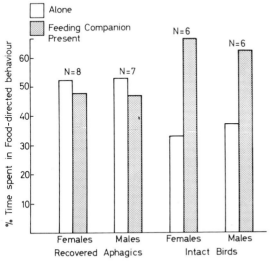

Fig. 7. Social facilitation of food-directed activity in recovered aphagic male and female doves and in operated controls.

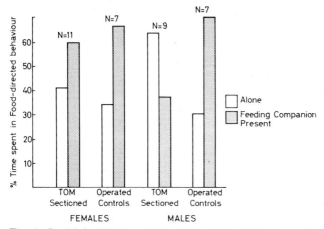

Fig. 8. Social facilitation of food directed activity in occipitomesencephalic tract (TOM)-sectioned and operated control male and female doves.

of TOM-sectioned males indicate that although they will respond to the presence of females by strutting and nest-soliciting, those components which involve a definite orientation to the female, such as hopping and charging, are either absent or very much reduced. It is possible that auditory rather than visual cues initiate much of this courtship behaviour.

Male doves with unilateral transection of the TOM will court and behave aggressively to their mirror image only through the eye ipsilateral to the cut, and when this eye is masked, they do not respond to their mirror image, but will continue to avoid obstacles and to peck accurately at seed. If TOM-sectioned doves are paired with intact mates and then followed through a complete breeding cycle, both sexes show decreased defensive behaviour to a model predator when incubating or rearing squabs (Wright and Spence, 1975). However, the early components of the defensive sequence: puffing and feather erection, can be initiated in male birds and not in females. If the predator is presented from above, or without contacting the cage floor such that the approach is now silent, the males will no longer show even this low level of defensiveness.

This difference in response to sound could explain the sex difference in social facilitation of TOM-sectioned birds; it could be argued that the male doves are more distracted by the sound of the companion pecking seed in the adjacent cage, and in consequence show less feeding behaviour in its presence.

Whatever the eventual explanation of the sex difference in the TOM-sectioned doves, these results do indicate that the absence of feeding facilitation in the recovered aphagic doves is independent of possible damage to the TOM, and represents a permanent impairment in the ability of external stimuli to arouse the tendency to feed.

THE INTERACTION BETWEEN AMYGDALA AND HYPOTHALAMUS

In order to bring these observations on the dove into some perspective, it will be necessary to first briefly summarise the involvement of the mammalian amygdala in alimentary and emotional behaviour. As described at the beginning of this chapter, the studies on food intake following lesions in the amygdala have often been contradictory, but it is now reasonably certain that a part of this influence is by modulating the activity of hypothalamic mechanisms (Kaada, 1972).

White and Fisher (1969) have shown that continuous electrical stimulation of the corticomedial—pyriform transition zone of the amygdala suppresses food intake in deprived rats, and this suppression is abolished following either destruction of the ventromedial nucleus or by severing the stria terminalis. They suggest that amygdaloid stimulation acts via the stria terminalis to increase the level of activity in the ventromedial nucleus of the hypothalamus, and that this in turn produces a suppression of activity in the lateral hypothalamus resulting in the lowered food intake. Fonberg (1972), working with dogs, has emphasised that the same behavioural syndrome which follows destruction of the ventromedial and lateral hypothalamic area, also follows from damage to the lateral amygdala and the dorsomedial amygdala. She has found that although the classical effects on food and water intake follow from such lesions, the most striking changes were in the general behaviour and motivational state of the dogs. From both lateral hypothalamic and dorsomedial amygdala lesions there follows a variable period of aphagia and adipsia. Immediately following surgery, these dogs spend most of their time half asleep; they sometimes walk round aimlessly, they do not seem to recognise the experimenter, and appear sad and indifferent. This is a rather striking parallel to some of the behaviour we have described in aphagic doves. If the lesions were placed in the ventromedial hypothalamus or in the lateral amygdala, in both instances the dogs became hyperphagic, but with the amygdala lesions, overeating was not as dramatic as from the hypothalamus. It was a transient effect, most noticeable 5—10 days after surgery. The dogs became very friendly, lively and playful after the lesion in the lateral amygdala, and these changes in emotional behaviour were long-lasting in contrast with the effects on food intake. Again there is a strong resemblance to the condition following TOM section in doves, although the changes in food intake appear only in the female doves.

Zeier and Karten (1971) have recently argued on anatomical grounds that only the posterior and more medial portions of the archistriatum of birds are truly homologous with the mammalian amygdala. They have shown that the TOM is more correctly divided into two branches: a hypothalamic component (HOM) which originates in the more posterior and medial aspects of the archistriatum, and which terminates largely in the medial hypothalamus; and the major pathway which originates in the anterior two-thirds of the archistriatum and projects to the thalamus, optic tectum, tegmentum, lateral reticular formation and sensory nuclei of the brain stem. They suggest a division of the archistriatum into four regions; the anterior and intermediate regions are classified as a 'somatic sensorimotor' system, since from here issues the long descending projections of the tractus occipitomesencephalicus to the brain stem and rostral spinal cord. Only the posterior and medial regions of the archistriatum from which the HOM originates are considered to be limbic in origin.

In our experiments we have sectioned pathways from both the sensorimotor and limbic divisions of the archistriatum, and on the basis of our observations alone, it is not possible to say which component of the tract accounts for the effects on eating, and whether the same pathway will account for the changes in emotionality. However, Zeier (1971) studied localised lesions of the archistriatum in pigeons and their effect on operant response, and noted that lesions restricted to the lateral limbic division initially produced somnolence, absence of eating for a few days, and non-reactivity to tactile stimuli. Although this description is brief and no quantitative data are given, it does resemble the early behaviour of aphagic doves following hypothalamic lesions. Zeier observed no changes in food intake when the lesions were restricted to the medial-limbic archistriatum, as judged by degeneration appearing only in the hypothalamic division of the TOM, but the birds were tamer and easier to handle.

In the light of Zeier's report, and taking Fonberg's results into consideration, it seems very likely that the hyperphagia we report is due to damage of the hypothalamic component of the TOM. It follows that lesions which are restricted to the medial archistriatum in doves should also result in transient overeating effects, but as yet we have no direct evidence to support this suggestion. It is not surprising that the hyperphagia is of small proportion, as the reports of medial hypothalamic lesions resulting in overeating in birds are slight by comparison with mammalian studies (Lepkovsky and Yasuda, 1966; Kuenzel and Helms, 1967). The multiple storage organs of crop, proventriculus, and stomach which are present in birds, and the relative importance of social influences on feeding (Harlow, 1932), reflect the need to ingest food where and when it is available, and the presence of a satiety system as clearly defined as the ventromedial nucleus in rats and other monogastric mammals, may not be necessary in birds.

The extensive work of Zeigler and his collaborators has demonstrated that aphagia in pigeons can result from the transection of trigeminal afferents in the beak alone, (Zeigler, 1973) as well as similar effects from lesioning central trigeminal structures. Effects on drinking in pigeons are reported as minimal and we similarly find little evidence for pronounced disturbances of drinking from hypothalamic lesions which produce aphagia in doves. This contrasts sharply with the association between aphagia and adipsia in the rat, and may reflect the distinct consummatory patterns of eating and drinking in pigeons.

In chickens, where drinking is not accomplished by the same pumping action as in pigeons, hypothalamic lesions do seem to result in prolonged periods of adipsia which outlast the accompanying aphagia (Lepkovsky *et al.*, 1971). Unfortunately, in these reports there is no description of behaviour other than alimentary, and it is not possible to say how far the system outlined by Zeigler can also account for the behavioural changes described here in the dove.

The feeding deficits which Zeigler has described in the pigeon following disruption of sensory trigeminal pathways lead to associated motivational impairments, and there is some evidence that section of the lingual nerve in chickens also results in loss of eating for 1—4 days (see Chapter 14 by Gentle in this volume). Similar sensory involvement is now well described in the early stages of aphagia in the rat (Marshall and Teitelbaum, 1974), and is possibly due not to damage intrinsic to the lateral hypothalamus, but to interruption of pallidohypothalamic and amygdalohypothalamic pathways. Welker (1964) found that rats with section of somatic snout afferents had difficulty in locating food pellets; they made exaggerated head movements and appeared to locate the pellets by contact with the forepaws and the ventral aspects of the lower jaw. It is very likely that the early stages of the aphagic syndrome in both birds and mammals especially reflect such sensory disturbances.

It may well be that the changes in emotionality of recovered aphagic doves are also due to a continued form of sensory neglect since we have found such changes in temperament and in feeding facilitation are invariably associated with aphagia. The neural systems which are involved in feeding behaviour in birds may well reveal a picture analogous to that emerging from Fonberg's studies in dogs, namely that emotions and general arousal are closely associated with alimentary functions.

Acknowledgement

The author wishes to thank the Agricultural Research Council for supporting part of the research reported here.

REFERENCES

Åkerman, B., Anderson, B., Fabricius, E. and Svensson, L. (1960) Observations on central regulation of body temperature and of food and water intake in the pigeon (*Columba livia*). *Acta physiol. scand.*, 50, 328—336.
Albert, D.J., Storlein, L.H., Wood, D.J. and Ehman, G.K. (1970) Further evidence for a complex system controlling feeding behaviour. *Physiol. Behav.*, 5, 1075—1082.
Anand, B.K. and Brobeck, J.R. (1951) Hypothalamic control of food intake in rats and cats. *Yale J. Biol. Med.*, 24, 123—140.

Andersson, B. (1964) Comments following paper by B.W. Robinson on forebrain, alimentary responses: some organisational principles: In M.J. Wayner (Ed.), *Thirst*, Pergamon Press, London, pp. 425—427.

Andrew, R.J. (1967) Intracranial self-stimulation in the chick. *Nature (Lond.)*, 213, 847—848.

Ariëns Kappers, C.U., Huber, G.C. and Crosby, E.C. (1960) *The Comparative Anatomy of the Nervous System of Vertebrates, Including Man, Vol. I—III*. Haffner, New York.

Baillie, P. and Morrison, S.D. (1963) The nature of the suppression of food intake by lateral hypothalamic lesions in rats. *J. Physiol. (Lond.)*, 165, 227—245.

Bayer, E. (1929) Beitrage zue Zweikumponentheorie des Hungers. *Z. Psychol.*, 112, 1—54.

Brown, J.L. and Hunsperger, R.W. (1963) Neuroethology and the motivation of agonistic behaviour. *Anim. Behav.*, 11, 439—448.

Coons, E.E., Levak, M. and Miller, N.E. (1965) Lateral hypothalamus: learning of food-seeking response motivated by electrical stimulation. *Science*, 150, 1320—1321.

Delius, J.D. (1967) Displacement activities and arousal. *Nature (Lond.)*, 214, 1259—1260.

Delius, J.D. (1971) Foraging behaviour patterns of herring gulls elicited by electrical forebrain stimulation. *Experientia (Basel)*, 27, 1287—1289.

Doty, R.W. (1969) Electrical stimulation of the brain in behavioural context. *Ann. Rev. Psychol.*, 20, 289—320.

Edinger, L. (1908) The relations of comparative anatomy to comparative psychology. *J. comp. Neurol.*, 18, 437—457.

Eleftheriou, B.E. (Ed.) (1972) *The Neurobiology of the Amygdala*. Plenum Press, New York.

Epstein, A.N. and Teitelbaum, P. (1964) Severe and persistent deficits in thirst. In M.J. Wayner (Ed.), *Thirst*. Pergamon Press, London, pp. 395—406.

Feldman, S.E., Larsson, S., Dimick, M.K. and Lepkovsky, S. (1957) Aphagia in chickens. *Amer. J. Physiol.*, 191, 259—261.

Fonberg, E. (1972) Control of emotional behaviour through the hypothalamus and amydaloid complex. In *Physiology, Emotion and Psychosomatic Illness, Ciba Foundation Symposium 8*. Elsevier, Amsterdam, pp. 131—161.

Fox, C.A., Hillman, D.E., Siegesmund, K.A. and Sether, L.A. (1965) The primate globus pallidus and its feline and avian homologues: a Golgi and electron microscopic study. In Hassler and Stephan (Eds.), *Evolution of the Forebrain*. Thieme, Stuttgart, pp. 237—248.

Goodman, I.J. and Brown, J.L. (1966) Stimulation of positively and negatively reinforcing sites in the avian brain. *Life Sci.*, 5, 693—704.

Grossman, S.P. (1967) *A Textbook of Physiological Psychology*. Wiley, London.

Grossman, S.P. and Grossman, L. (1963) Food and water intake following lesions or electrical stimulation of amygdala. *Amer. J. Physiol.*, 205, 761—765.

Harlow, H.F. (1932) Social facilitation of feeding in the albino rat. *J. genet. Psychol.*, 39, 258—277.

Harwood, D. and Vowles, D.M. (1966) Forebrain stimulation and feeding behaviour in the ring dove (*Streptopelia risoria*). *J. comp. physiol. Psychol.*, 62, 388—396.

Harwood, D. and Vowles, D.M. (1967) Defensive behaviour and the after effects of brain stimulation in the ring dove (*Streptopelia risoria*). *Neuropsychologia*, 5, 345—366.

Hess, W.R. (1957) *The Functional Organisation of the Diencephalon*. Grune and Stratton, New York.

Hetherington, A.W. and Ranson, S.W. (1942) The spontaneous activity and food intake of rats with hypothalamic lesions. *Amer. J. Physiol.*, 136, 609—617.

Hoebel, B.G. and Teitelbaum, P. (1962) Hypothalamic control of feeding and self-stimulation. *Science*, 135, 375—377.

Holst, E. von and St. Paul, U. von (1963) On the functional organisation of drives. *Anim. Behav.*, 11, 1—20.

Kaada, B.R. (1972) Stimulation and regional ablation of the amygdaloid complex with reference to functional representations. In B.E. Eleftheriou (Ed.), *The Neurobiology of the Amygdala.* Plenum Press, New York, pp. 205—281.

Karten, H.J. (1969) The organisation of the avian telencephalon and some speculations on the phylogeny of the amniote telencephalon. *Ann. N.Y. Acad. Sci.*, 167, 164—179.

Kling, A. and Schwarz, N.B. (1961) Effect of amygdalectomy on feeding in infant and adult animals. *Fed. Proc.*, 20, 335.

Koike, T. and Lepkovsky, S. (1967) Hypothalamic lesions producing polyuria in chickens. *Gen. comp. Endocr.*, 8, 397—402.

Kuenzel, W.J. (1972) Dual hypothalamic feeding system in a migratory bird, *Zonotrichia albicollis. Amer. J. Physiol.*, 223, 1138—1142.

Kuenzel, W.J. and Helms, C.W. (1967) Obesity produced in a migratory bird by hypothalamic lesions. *Bio. Science*, 4, 395—396.

Kuenzel, W.J. and Helms, C.W. (1970) Hyperphagia, polydipsia and other effects of hypothalamic lesions in the white-throated sparrow (*Zonotrichia albicollis*). *Condor*, 72, 66—75.

Lashley, K.S. (1938) Experimental analysis of instinctive behaviour. *Psychol. Rev.*, 45, 445—471.

Lepkovsky, S. and Yasuda, M. (1966) Hypothalamic lesions, growth and body composition of male chickens. *Poultry Sci.*, 45, 582—588.

Lepkovsky, S. and Yasuda, M. (1967) Adipsia in chickens. *Physiol. Behav.*, 2, 45—47.

Lepkovsky, S., Feldman, S.E. and Sharon, I.M. (1967) Food and water intake of the fowl. In C.F. Code (Ed.), *Handbook of Physiology, Section 6, Vol. I, Food and Water Intake.* American Physiological Society, Washington, D.C., pp. 117—128.

Lepkovsky, S., Furuta, F., Sharon, I.M. and Snapir, N. (1971) Thirst and behaviour in adipsic chickens with hypothalamic lesions before and after intravenous injection of hypertonic NaCl solution. *Physiol. Behav.*, 6, 477—480.

McFarland, D.J. and Wright, P. (1969) Water conservation by inhibition of food intake. *Physiol. Behav.*, 4, 95—99.

MacLean, P.D. (1958) The limbic system with respect to self-preservation and the preservation of the species. *J. nerv. ment. Dis.*, 127, 1—11.

Macphail, E.M. (1965) *The Physiological Analysis of the Motivational Systems of Pigeons.* D.Phil. Thesis, Oxford University.

Macphail, E.M. (1967) Positive and negative reinforcement from intracranial stimulation in pigeons. *Nature (Lond.)*, 213, 947—948.

Macphail, E.M. (1969) Cholinergic stimulation of the dove diencephalon: a comparative study. *Physiol. Behav.*, 4, 655—657.

Marshall, J.F. and Teitelbaum, P. (1974) Further analysis of sensory inattention following lateral hypothalamic damage in rats. *J. comp. physiol. Psychol.*, 86, 375—395.

Marshall, J.F., Turner, B.H. and Teitelbaum, P. (1971) Sensory neglect produced by lateral hypothalamic damage. *Science*, 174, 523—525.

Miller, N.E. (1963) Some motivational effects of electrical and chemical stimulation of the brain. *Electroenceph. clin. Neurophysiol.*, Suppl., 24, 247—259.

Miller, N.E., Gottesman, K.S. and Emery, N. (1964) Dose response to carbachol and norepinephrine in rat hypothalamus. *Amer. J. Physiol.*, 206, 1384—1388.

Morgane, P.J. (1961) Medial forebrain bundle and 'feeding centres' of the hypothalamus. *J. comp. Neurol.*, 117, 1—25.

Morgane, P.J. and Kosman, A.J. (1957) Alterations in feline behaviour following bilateral amygdalectomy. *Nature (Lond.)*, 180, 598—600.

Murton, R.K., Isaacson, A.J. and Westwood, N.J. (1971) The significance of gregarious feeding behaviour and adrenal stress in a population of wood-pigeons, *Columba palumbus. J. Zool.*, 165, 53—84.

Olds, J. (1962) Hypothalamic substrates of reward. *Physiol. Rev.*, 42, 554—604.

348

Pearson, R. (1972) *The Avian Brain.* Academic Press, New York.

Phillips, R.E. (1964) 'Wildness' in the mallard duck; effects of brain lesions and stimulation on 'escape behaviour' and reproduction. *J. comp. Neurol.*, 122, 139—195.

Phillips, R.E. and Youngren, O.M. (1971) Brain stimulation and species-typical behaviour: activities evoked by electrical stimulation of the brains of chickens (*Gallus gallus*). *Anim. Behav.*, 19, 759—779.

Pribram, K.H. and Bagshawe, M. (1953) Further analysis of the temporal lobe syndrome utilizing fronto-temporal ablations. *J. comp. Neurol.*, 99, 347—375.

Putkonen, P.T.S. (1967) Electrical stimulation of the avian brain. *Ann. Acad. Sci. fenn.* A5, 130, 9—95.

Rabin, B.M. and Smith, C.J. (1968) Behavioural comparison of the effectiveness of irritative and non-irritative lesions in producing hypothalamic hyperphagia. *Physiol. Behav.*, 3, 417—420.

Ralph, C.L. (1960) Polydipsia in the hen following lesions in the supraoptic hypothalamus. *Amer. J. Physiol.*, 198, 528—530.

Reynolds, R.W. (1965) An irritative hypothesis concerning the hypothalamic regulation of food intake. *Psychol. Rev.*, 72, 105—116.

Rodgers, W.L., Epstein, A.N. and Teitelbaum, P. (1965) Lateral hypothalamic aphagia: motor failure or motivational deficit? *Amer. J. Physiol.*, 208, 334—342.

Shaklee, A.O. (1921) The relative height of the eating and drinking arcs in the pigeon's brain, and brain evolution. *Amer. J. Physiol.*, 55, 65—83.

Smith, C.J.V. (1969) Alterations in the food intake of chickens as a result of hypothalamic lesions. *Poultry Sci.*, 48, 475—477.

Stellar, E. (1954) The physiology of motivation. *Psychol. Rev.*, 61, 5—22.

Teitelbaum, P. (1966) The use of operant methods in the assessment and control of motivational states. In W.K. Honig (Ed.), *Operant Behaviour: Areas of Research and Application.* Appleton-Century-Crofts, New York, pp. 565—608.

Teitelbaum, P. and Epstein, A.N. (1962) The lateral hypothalamic syndrome: recovery of feeding and drinking after lateral hypothalamic lesions. *Psychol. Rev.*, 69, 74—90.

Tolman, C.W. (1964) Social facilitation of feeding behaviour in the domestic chick. *Anim. Behav.*, 12, 245—251.

Tolman, C.W. (1965) Emotional behaviour and social facilitation of feeding behaviour in domestic chicks. *Anim. Behav.*, 13, 493—496.

Tolman, C.W. (1967) The feeding behaviour of domestic chicks as a function of rate of pecking by a surrogate companion. *Behaviour*, 29, 57—62.

Tolman, C.W. (1968) The role of the companion in social facilitation of animal behaviour. In E.C. Simnel, R.A. Hoppe and G.A. Milton (Eds.), *Social Facilitation and Imitative Behavior.* Allyn and Bacon Inc., Boston, Mass., pp. 33—54.

Turner, B.H. (1973) Sensorimotor syndrome produced by lesions of the amygdala and lateral hypothalamus. *J. comp. physiol. Psychol.*, 82, 37—47.

Tweeton, J.R., Phillips, R.E. and Peek, F.W. (1973) Feeding behaviour elicited by electrical stimulation of the brain in chickens, *Gallus gallus. Poultry Sci.*, 52, 165—172.

Ungerstedt, U. (1971) Adipsia and aphagia after 6-hydroxydopamine induced degeneration of the nigro-striatal dopamine system. *Acta physiol. scand.*, 82, Suppl. 567, 95—122.

Valenstein, E.S. (1970) Stability and plasticity of motivation systems. In F.O. Schmitt (Ed.), *The Neurosciences: Second Study Program.* Rockefeller University Press, New York, pp. 207—217.

Valenstein, E.S., Cox, V.C. and Kakolewski, J.W. (1969) Sex differences in hyperphagia and bodyweight following hypothalamic damage. *Ann. N.Y. Acad. Sci.*, 157, 1030—1048.

Valenstein, E.S., Cox, V.C. and Kakolewski, J.W. (1970) Reexamination of the role of the hypothalamus in motivation. *Psychol. Rev.*, 77, 16—31.

Welker, W.I. (1964) Analysis of sniffing of the albino rat. *Behaviour*, 22, 223—244.

White, N.M. and Fisher, A.E. (1969) Relationship between amygdala and hypothalamus in the control of eating behavior. *Physiol. Behav.*, 4, 199—205.

Wright, P. (1968) Hypothalamic lesions and food/water intake in the Barbary dove. *Psychon. Sci.*, 13, 133—134.

Wright, P. (1969) *Physiology of Feeding and Drinking in the Barbary dove (Streptopelia risoria)*. Unpublished D. Phil. Thesis, Oxford University.

Wright, P. (1974) Role of the avian amygdala and hypothalamus in the feeding behaviour of pigeons. In *Proceedings of XXVI International Congress of Physiological Sciences, Physiology of Food and Fluid Intake.* p. 140.

Wright, P. and McFarland, D.J. (1969) A functional analysis of hypothalamic polydipsia in the Barbary dove (*Streptopelia risoria*). *Physiol. Behav.*, 4, 877—883.

Wright, P. and Spence, A.M. (1975) Changes in emotionality following section of the tractus occipito-mesencephalicus in the Barbary dove (*Streptopelia risoria*). Submitted to *Anim. Behav.*

Wright, P. and Turner, C. (1973) Sex differences in bodyweight following gonadectomy and goldthioglucose injections in mice. *Physiol. Behav.*, 11, 155—159.

Zeigler, H.P. (1973) Trigeminal deafferentiation and feeding in the pigeon: sensorimotor and motivational effects. *Science*, 182, 1155—1158.

Zeigler, H.P. and Karten, H.J. (1973a) Brain mechanisms and feeding behaviour in the pigeon (*Columba livia*). I Quinto-frontal structures. *J. comp. Neurol.*, 152, 59—82.

Zeigler, H.P. and Karten, H.J. (1973b) Brain mechanisms and feeding behaviour in the pigeon (*Columba livia*). II Analysis of feeding behaviour deficits after lesions of quinto-frontal structures. *J. comp. Neurol.*, 152, 83—102.

Zeigler, H.P. and Witkovsky, P. (1968) The main sensory trigeminal nucleus in the pigeon: a single unit analysis. *J. comp. Neurol.*, 134, 255—264.

Zeigler, H.P., Karten, H.J. and Green, H.L. (1969) Neural control of feeding in the pigeon. *Psychon. Sci.*, 15, 156—157.

Zeier, H. (1971) Archistriatal lesions and response inhibition in the pigeon. *Brain Res.*, 31, 327—339.

Zeier, H. and Karten, H.J. (1971) The archistriatum of the pigeon: organization of afferent and efferent connections. *Brain Res.*, 31, 313—326.

Zeier, H.J. and Karten, H.J. (1973) Connections of the anterior commissure in the pigeon (*Columba livia*). *J. comp. Neurol.*, 150, 201—216.

APPENDIX

A STEREOTAXIC ATLAS OF THE BRAIN OF THE BARBARY DOVE
(*STREPTOPELIA RISORIA*)

D.M. VOWLES, L. BEAZLEY and D.H. HARWOOD

In the early 1960s a number of workers were stimulated by the work of Lehrman on hormones and behaviour in the Barbary dove (blond ring dove) — *Streptopelia risoria*, to start investigations into the functions of the brain in this species.

At that time the only stereotaxic atlas of an avian brain was for the domestic chicken (van Tienhoven and Juhász, 1962), and this proved difficult to apply to the dove. It was therefore necessary to construct a stereotaxic atlas for the dove, and although in the intervening period the elegant atlas of the pigeon brain by Karten and Hodos (1967) was published, there is sufficient difference between the two species to justify the preparation of a separate atlas. Not only is the dove brain much smaller than the White Carneaux pigeon used by Hodos and Karten, but the brain stem is more compressed along the rostro-occipital atlas, forming a tighter S-shape. The distribution of some of the nuclei and tracts is also slightly different, even after adjustments for relative brain size.

However, the present authors have relied very heavily on the existing pigeon atlas, supplemented by the studies of Huber and Crosby (1929), Ariëns Kappers *et al.* (1936), Jungherr (1945) and Stingelin (1958). The terminology adopted has been that of Karten and Hodos, supplemented, where appropriate, by that of Huber and Crosby.

METHODS

Head position

Two stereotaxic instruments have been used in the present study: the Trent H. Wells and the Kopf for small mammals. The head holder for use with both machines was constructed by Forth Instruments Ltd. (Edinburgh) to a design by the authors. The Wells ear plugs (for rats) are used in conjunction with normal ear bars. The front part of the skull is tightly clamped by means of two symmetrical bars which are shaped to fit into a groove in the rostral aspect of the orbit (in which the lachrymal duct normally runs).

352

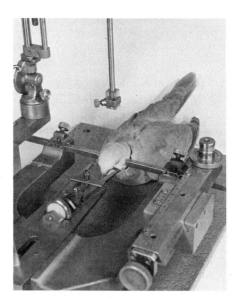

Fig. 1. The Forth Instruments' head holder in place on the Kopf stereotaxic instrument.

Fig. 2. An anaesthetized dove in position in the stereotaxic instrument.

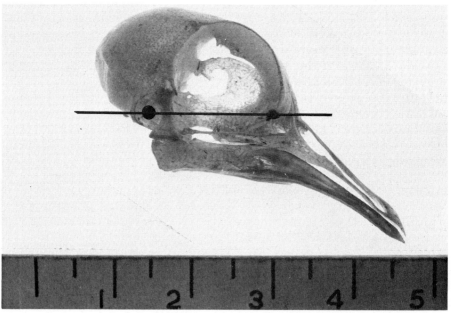

Fig. 3. A dove skull showing location of ear bar placements (external auditory meatus), and orbital fixation points (darkened areas) — the position determining the horizontal plane which is indicated by the black line. The scale is in millimetres.

When tightly clamped the head is held rigidly. The upper surface of the orbital bars is level with the centre of the ear bars, this determining the horizontal plane. The head holder and its position is illustrated in Figs. 1, 2 and 3. The head position was chosen purely for the convenience of the experimenter. With the head at this angle a simple vertical movement of an electrode can reach almost any part of the brain without encountering bone, except in the lateral tectal region. The angle of the brain to the vertical differs considerably from that which was chosen by Karten and Hodos on rational anatomical grounds: an orientation which makes comparison with mammals more direct. In Fig. 4 sagittal sections from Karten and Hodos and from the dove atlas are shown to illustrate the different angles.

Selection of anatomical reference point

In stereotaxic techniques it is necessary to define a zero reference point relative to the anterioposterior (horizontal), the lateral (transverse) and the vertical (sagittal) planes. For convenience such a reference point is most suitably determined by structures on the outside of the cranium. In the dove, skull growth was found to proceed rapidly for the three months after hatching, and skull dimensions were stabilised by nine months. All birds were therefore studied after reaching this age. However, there was considerable inter-individual variability in skull size, and even after allowing for this no external landmark could be found which produced a relatively invariant position of internal brain structures, such as the anterior or posterior commissures; a common finding was that the brain lies slightly skewed in the skull. It was therefore decided to use an internal reference point. After exploring several possibilities we selected the junction between the anterior, dorsal lobe of the cerebellum and the medial cleft between the two posterior lobes of the cerebral hemispheres: this junction makes a small triangular cleft leading down to the top of the pineal body. The position at which this cleft becomes too narrow to penetrate without damaging the brain tissue was selected as the zero-reference point. The variability of the reference point relative to the instrument zero is shown in Table I — the figures refer to height vertically above the ear bars (vertical zero), position lateral to the central meeting point of the ear bars (lateral zero) and the anterior posterior position relative to the zero point on the scale of the Wells instrument (anterior—posterior zero).

The use of this zero point entails some dangers, since a small hole has to be trephined in the bone in order to locate it: a large blood sinus runs in the spongy bone posterior to this point, and the position is best approached from a slightly anterior position. After an operation was completed the core of bone which had been extracted was normally replaced and held in position with dental cement. Because of the occasional skewness of the brain in the skull it was found desirable to check the position of the cleft between

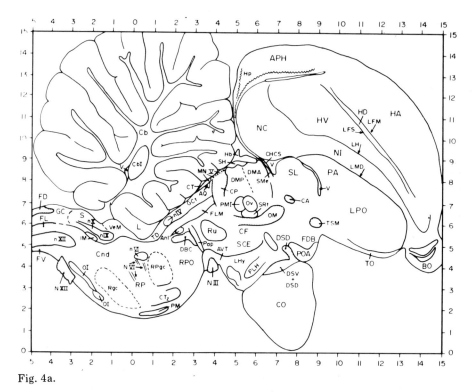

Fig. 4a.

the two cerebral hemispheres at their anterior end, and to adjust the head holder to locate this symmetrically to the posterior zero point.

Histological techniques

All brains were flushed with saline and perfused with formol—saline; they were kept in fixation for four days with a daily change of solution. In an initial series, 40 brains were sectioned in either the transverse, sagittal or horizontal planes as set by the stereotaxic instrument. Complete serial sections of 25 μm thickness were made through the whole brain following paraffin embedding. In some cases all the sections in one series were stained similarly, and in others alternate sections were stained with silver for tracts, and cresyl violet for nuclei. The most frequently used stains were Glees' silver stain (modified for wax embedding), and cresyl violet. We also used Holmes' stain, Luxol fast blue, Weigert's, Golgi, haemotoxylin and some special neurosecretory stains.

After the main structures in the brain had been identified, twelve brains were prepared to determine precise coordinates of brain structures and the amount of shrinkage caused by the method of fixation. All preparations involved the same standard times in alcohols, etc., following four days fixa-

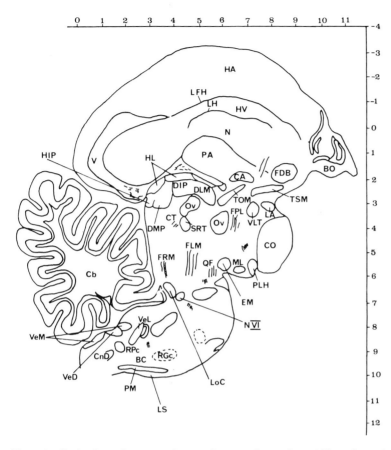

Fig. 4b. Sagittal sections at 1.0 mm distance from the midline through the brain of the pigeon (Fig. 4a), taken from Karten and Hodos (1967), and through the brain of the Barbary dove (Fig. 4b, above), to illustrate the difference in the stereotaxic planes.

TABLE I

Variation in position of anatomical reference point (junction between cerebellum and cerebral hemispheres) from zero on the stereotaxic instrument

Each figure is the mean of 50 birds.

	Mean position	Standard deviation
Anterioposterior axis	5.98 mm	2.02 mm
Left—right axis	0.04	0.39
Vertical axis	10.87	1.12

tion in formol—saline. In the fresh brain steel electrodes were used to deposit small amounts of iron by passing a DC current at points in a three dimensional grid with 2 mm separation between the individual points in the grid. Deposits were made throughout the brain tissue, each point being located with reference to the posterior anatomical reference point. The iron deposits were subsequently stained as Turnbull's blue with ferrocyanide; otherwise the brains were stained and sectioned in the usual way.

This grid of blue spots was then used in two ways. Firstly, to locate accurately actual structures with reference to the zero point, and, secondly, to determine the shrinkage caused by the methods of preparation. Shrinkage was found to be fairly consistent between brains, but varied with the parts of the brain involved and with the plane studied. Thus the greatest shrinkage, approximately 13%, was along the anterioposterior axis in the forebrain, and the least, 2—3%, lay along the vertical axis in the mid-brain. Shrinkage factors measured in this way were used to scale up appropriately those structures which are reproduced in the diagrams.

All sections were drawn using a macro-projector. After the drawings had been made similar sections were prepared using frozen sections (which shrink very little). Good agreement was found between the scaled drawings from the serial paraffin sections and the unscaled from frozen sections. As a further check some imaginary sagittal sections were reconstructed by a synthesis from the drawings of transverse sections. These were checked against real sagittal sections. Again there was good agreement.

ACCURACY OF THE ATLAS

The atlas has been indirectly tested many times by experimental use, and has proved reliable in that context. A more formal test of reliability was made by placing radiofrequency lesions at preselected points, chosen for their ease of anatomical identification, and specified by their position in the atlas. In 5 subjects lesions were placed bilaterally in the forebrain aiming at the point where the lateral forebrain bundle intersected the tractus occipitomesencephalicus, and in 9 subjects the lesions were directed at the nucleus rotundus. The actual locations of the lesions are shown in Figs. 5 and 6.

CHOICE OF SECTIONS

The most useful sections were found to be in the transverse plane. These have been drawn at 0.5 mm intervals except in areas where small anatomical nuclei are densely packed, when 0.25 mm intervals are used. Some stippling and cross-hatching is used, this has no functional significance but is to aid the easy identification of a nucleus which runs through several sections. Tracts are indicated in a partly realistic manner with diagrammatic indica-

357

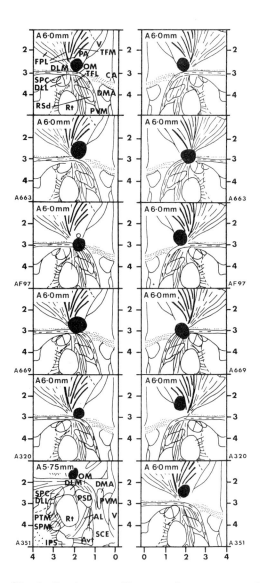

Fig. 5. Sections to illustrate the accuracy of lesion placements. Lesions were placed bilaterally in 5 subjects at the intersection of the occipitomesencephalic tract and the lateral forebrain bundle (FPL) (see text for details). The top pair of diagrams show the intended location of the lesions, and the remainder the actual lesions (in black). The numbers for each section indicate the position along the anterioposterior axis.

tions of fibre distributions. When several bundles of fibres lie together in a compound tract these separate bundles are indicated with lines of different thickness. Sections in the sagittal plane are shown at 0.5 and 1 mm intervals.

358

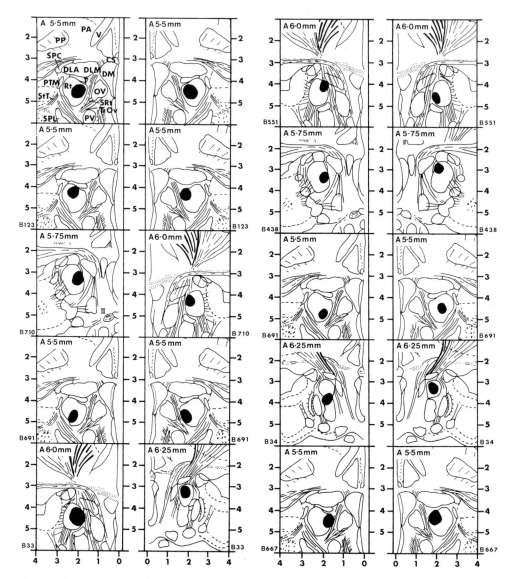

Fig. 6. Sections to illustrate the accuracy of lesion placements. Lesions were placed bilaterally in 9 subjects in the nucleus rotundus (Rt) (see text for details). The top pair of diagrams in the two left-hand columns show the intended location of the lesions, and the remainder the actual lesions (in black). The numbers for each section indicate the position along the anterioposterior axis.

Acknowledgements

The authors gratefully acknowledge the continued financial support of the Science Research Council (U.K.) throughout this project. The histological preparations were made by Messrs W. McDougal, B. Purvis, and Miss E. Forrest, whose skill and patience are much appreciated.

REFERENCES

Ariëns Kappers, J., Huber, G.C. and Crosby, E.C. (1936) *The Comparative Anatomy of the Nervous System in Vertebrates, Including Man.* Hafner, New York.
Huber, G.C. and Crosby, E.C. (1929) The nuclei and fibre paths of the avian diencephalon, with consideration of telencephalic and certain mesencephalic centers and connections. *J. comp. Neurol.,* 48, 1—25.
Jungherr, E. (1945) Certain nuclear groups of the avian mesencephalon. *J. comp. Neurol.,* 82, 55—175.
Karten, H.J. and Hodos, W. (1967) *A Stereotaxic Atlas of the Brain of the Pigeon (Columba livia).* Johns Hopkins, Baltimore, Md.
Stingelin, W. (1958) *Vergleichend-morphologische Untersuchungen am Vorderhirn der Vögel auf cytologischer und cytoarchitektonischer Grundlage.* Helbing und Lichtenhahn, Basel.
van Tienhoven, A. and Juhász, L.P. (1962) The chicken telencephalon, diencephalon and mesencephalon in stereotaxic coordinates. *J. comp. Neurol.,* 118, 185—197.

KEY TO ABBREVIATIONS IN FIGURES

Abbreviation	Structures
A	Archistriatum
AC	Nucleus accumbens
AHP	Area hypothalami posterioris
AL	Ansa lenticularis
AMA	Nucleus anterior medialis hypothalami
AP	Area pretectalis
APH	Area parahippocampalis
AV	Archistriatum, pars ventralis
Avt	Area ventralis (Tsai)
Bas	Nucleus basalis
BC	Brachium conjunctivum
BCA	Brachium conjunctivum ascendens
BCD	Brachium conjunctivum descendens
BCS	Brachium colliculi superioris
BO	Bulbus olfactorius
BPC	Bed nuclei of pallial commissure
CA	Commissura anterior
CB	Cerebellum
CbL	Nucleus cerebellaris internus
CbM	Nucleus cerebellaris intermedius
CDL	Area corticoidea dorsolateralis
Cnd	Nucleus centralis medullae oblongatae, pars dorsalis
Cnv	Nucleus centralis medullae oblongatae, pars ventralis
CO	Chiasma opticum
COS	Nucleus commissuralis septi
CP	Commissura posterior
CPA	Commissura pallii
CS	Nucleus ventralis superior (Bechterew)
CT	Commissura tectalis
DBC	Decussatio brachiorum conjunctivorum
DIP	Nucleus dorsointermedius posterior thalami
DLA	Nucleus dorsolateralis anterior thalami

Abbreviation	Structure
DLL	Nucleus dorsolateralis anterior thalami, pars lateralis
DLM	Nucleus dorsolateralis anterior thalami, pars medialis
DLP	Nucleus dorsolateralis posterior thalami
DMA	Nucleus dorsomedialis anterior thalami
DMP	Nucleus dorsomedialis posterior thalami
DTI	Decussatio tractus infundibulus
Ek	Ectostriatum
EM	Nucleus ectomamillaris
EP	Nucleus entopeduncularis
EW	Nucleus of Edinger—Westphal
FA	Tractus fronto-archistriatalis
FDB	Fasciculus diagonalis Brocae
FL	Fundiculus lateralis
FLM	Fasciculus longitudinalis medialis
FPL	Fasciculus prosencephali lateralis
FRL	Formatio reticularis lateralis mesencephali
FRM	Formatio reticularis medialis mesencephali
FUm	Fasciculus uncinatus, pars medialis
GCt	Substantia grisea centralis
Glv	Nucleus geniculatis lateralis, pars ventralis
HA	Hyperstriatum accessorium
HD	Hyperstriatum dorsale
Hb	Nucleus habenularis
HL	Nucleus habenularis lateralis
HM	Nucleus habenularis medialis
HIP	Tractus habenulo-interpeduncularis
HP	Hippocampus
HV	Hyperstriatum ventrale
IC	Nucleus intercalatus
Ico	Nucleus intercollicularis

Abbreviation	Structures
Imc	Nucleus isthmi, pars magnocellularis
IN	Tractus infundibularis
IO	Nucleus isthmo-opticus
Ip	Nucleus interpeduncularis
Ipc	Nucleus isthmi, pars parvocellularis
Ips	Nucleus interstitio-pretecto-subpretectalis
IS	Nucleus interstitialis (Cajal)
La	Nucleus laminaris
LA	Nucleus lateralis anterior thalami
LC	Nucleus linearis caudalis
LFM	Lamina frontalis suprema
LFS	Lamina frontalis superior
LFSm	Lamina frontalis superior medialis
LH	Lamina hyperstriatica
LHy	Nucleus lateralis hypothalami
LL	Lamina lateralis
LLd	Nucleus lemnisci lateralis, pars dorsalis (Groebbels)
LM	Lemniscus medialis
LMD	Lamina medullaris dorsalis
LoC	Locus coeruleus
LPO	Lobus parolfactorius
LS	Lemniscus spinalis
MC	Nucleus magnocellularis
ML	Nucleus mamillaris lateralis
MLd	Nucleus mesencephalicus lateralis, pars dorsalis
MLv	Nucleus mesencephalicus lateralis, pars ventralis
MV	Nucleus motorius nervi trigemini
N	Neostriatum
NIII	Nervus oculomotorius
nIII	Nucleus nervi oculomotorius
NIV	Nervus trochlearis
nIV	Nucleus nervi trochlearis
NV	Nervus trigeminus
NVI	Nervus abducens
nVI	Nucleus nervi abducentis
NVII	Nervus facialis
nVII	Nucleus nervi facialis
NVIII	Nervus octavus
NIX—X	Nervi glossopharyngeus et vagus

Abbreviation	Structures
nIX—X	Nucleus nervi glossopharyngei et nucleus motorius dorsalis nervi vagi
nX	Nucleus motorius dorsalis nervi vagi
NXII	Nervus hypoglossus
nXII	Nucleus nervi hypoglossi
NC	Neostriatum caudale
nDBC	Nucleus decussationis brachiorum conjunctivorum
NOM	Nuclei nervi oculomotorii
NSL	Nucleus septalis lateralis
OI	Nucleus olivaris inferior
OM	Tractus occipitomesencephalicus
OMd	Nucleus nervi oculomotorii, pars dorsalis
OMDL (ODL)	Nucleus nervi oculomotorii, pars dorsolateralis
OMv	Nucleus nervi oculomotorii, pars ventralis
Ov	Nucleus ovoidalis
PA	Palaeostriatum augmentatum
PaM	Nucleus paramedianus
PD	Nucleus pretectalis diffusus
PGL	Nucleus paragigantocellularis lateralis
PH	Plexus of Horsley
PI	Nucleus paramedianus internus thalami
PIT	Pituitary body
PL	Nucleus pontis lateralis
PLH	Nucleus lateralis hypothalami posterioris
PM	Nucleus pontis medialis
PMH	Nucleus medialis hypothalami posterioris
POA	Nucleus preopticus anterior
POM	Nucleus preopticus medialis (van Tienhoven)
PP	Palaeostriatum primitivum
PPa	Palaeostriatum primitivum augmentatum
PPC	Nucleus principalis precommissuralis
PPM	Nucleus preopticus paraventricularis magnocellularis (van Tienhoven)
PrV	Nucleus sensorius principalis nervi trigemini

362

Abbreviation	Structures
PST	Tractus pretecto-subpretectalis
PT	Nucleus pretectalis
PTM	Nucleus pretectalis medialis
PV	Nucleus posteroventralis thalami (Kuhlenbeck)
PVM	Nucleus periventricularis magnocellularis
QF	Tractus quintofrontalis
R	Nucleus raphes
RGc	Nucleus reticularis gigantocellularis
RL	Nucleus reticularis lateralis
RP	Nucleus reticularis pontis caudalis
RPc	Nucleus reticularis parvocellularis
Rpgc	Nucleus reticularis pontis caudalis, pars gigantocellularis
RPO	Nucleus reticularis pontis oralis
RSd	Nucleus reticularis superior, pars dorsalis
RSv	Nucleus reticularis superior, pars ventralis
Rt	Nucleus rotundus
Ru	Nucleus ruber
SAC	Stratum album centrale
SCa	Nucleus subcoeruleus ventralis
SCE	Stratum cellulare externum
SCI	Stratum cellulare internum
SGc	Stratum griseum centrale
SGP	Substantia grisea et fibrosa periventricularis
SH	Nucleus subhabenularis
SL	Nucleus septalis lateralis
SM	Nucleus septalis medialis
SMe	Stria medullaris
SO	Nucleus supraopticus (Ralph)
SODS	Supraoptic decussation
SP	Nucleus subpretectalis
SPC	Nucleus superficialis parvocellularis (Nucleus tractus septomesencephalici)
SPL	Nucleus spiroformis lateralis
SPM	Nucleus spiroformis medialis
SRt	Nucleus subrotundus
StT	Tractus stria-tegmentalis

Abbreviation	Structures
T	Nucleus triangularis
Ta	Nucleus tangentialis (Cajal)
TD	Nucleus tegmenti dorsalis (Gudden)
TFI	Tractus thalamo-frontalis internus
TFL	Tractus thalamo-frontalis lateralis
TFM	Tractus thalamo-frontalis medialis
THM	Tractus cortico-habenularis medialis
TIC	Tractus isthmocerebellaris
TIO	Tractus isthmo-opticus
TO	Tuberculum olfactorium
TPC	Nucleus tegmenti pedunculo-pontinus, pars compacta
TPO	Area tempero-parieto-occipitalis (Edinger, Wallenberg, and Holmes)
TrO	Tractus opticus
TrOv	Tractus nuclei ovoidalis
TSHM	Tractus strio-hypothalmicus medialis
TSM	Tractus septomesencephalicus
TST	Tractus strio-tegmentalis
TT	Tractus tecto-thalamicus
TTBd	Tractus tecto-bulbaris dorsalis
TTd	Nucleus and tractus descendens nervi trigemini
TTD	Nucleus et tractus descendens nervi trigemini
TTS	Tractus thalamostriatus
TTSp	Tractus tecto-thalamicus dorsalis
Tu	Nucleus tuberis
V	Ventriculus
VdL	Nucleus vestibularis dorsolateralis (Sanders)
VeD	Nucleus vestibularis descendens
VeL	Nucleus vestibularis lateralis
VeM	Nucleus vestibularis medialis
VLT	Nucleus ventrolateralis thalami
VLV	Nucleus ventralis lemnisci lateralis
VS	Nucleus vestibularis superior

A STEREOTAXIC ATLAS OF THE DOVE BRAIN

Figs. 1A—13B are sections in the frontal plane illustrating the left half of the brain. The first section A 0.0 mm passes through the anatomical zero and Fig. 13B, A 10.0 mm passes through a plane 10 mm anterior to the anatomical zero. Figs. 14—22 are sections in the sagittal plane. Fig. 14 Lat 1.0 is 1.0 mm from the midline and Fig. 22 Lat 6.0 is 6 mm lateral to the midline.

In both series of sections the scale is indicated in millimetres from the anatomical zero.

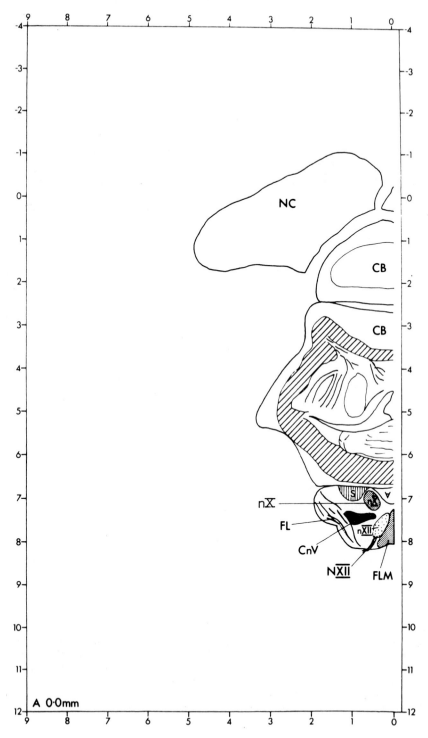

Fig. 1A.

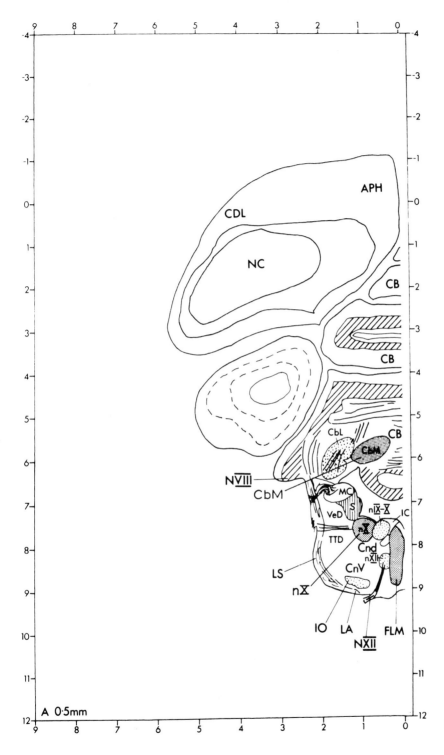

Fig. 1B.

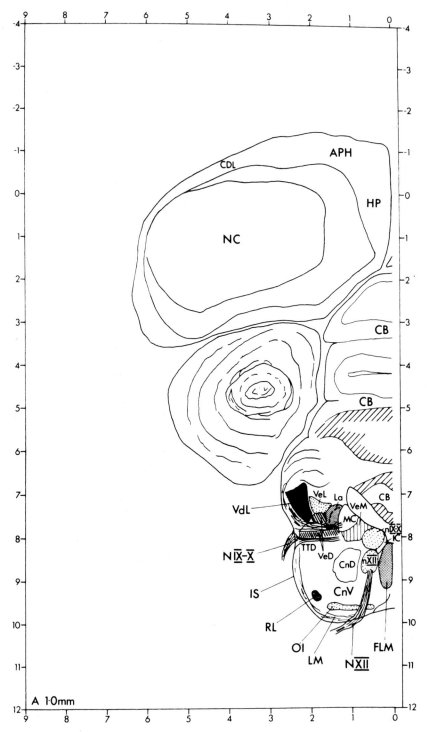

A 1·0mm

Fig. 2A.

367

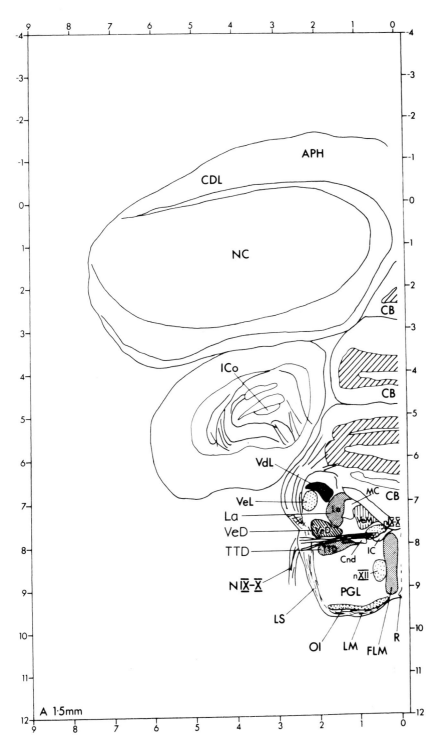

Fig. 2B.

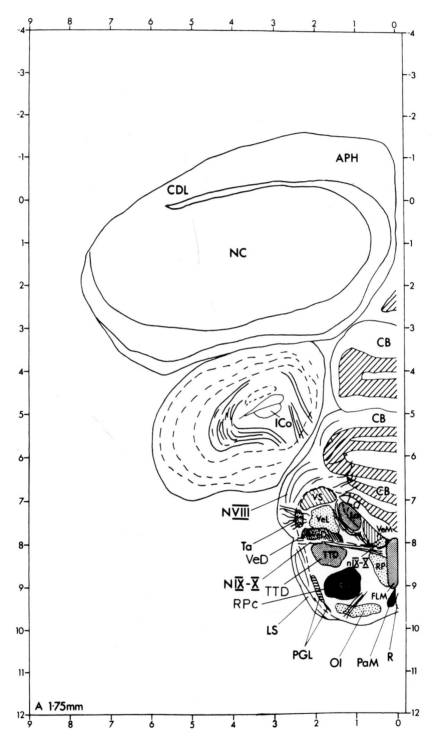

Fig. 3A.

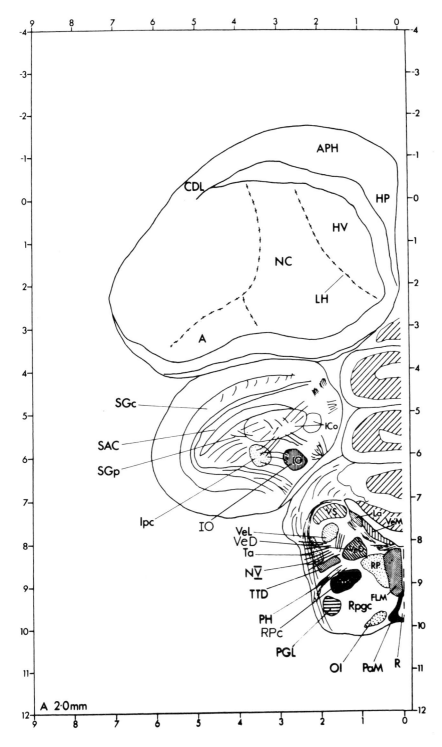

A 2·0mm

Fig. 3B.

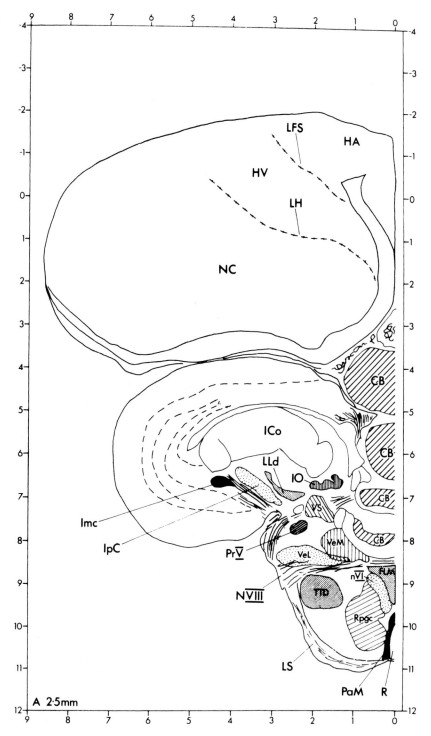

Fig. 4A.

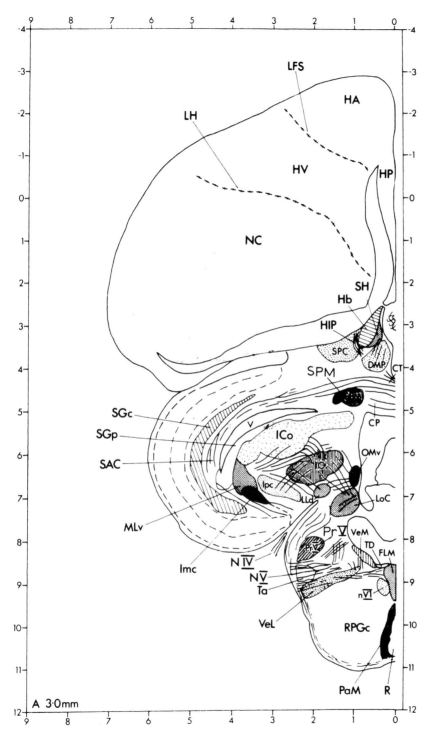

A 3·0mm

Fig. 4B.

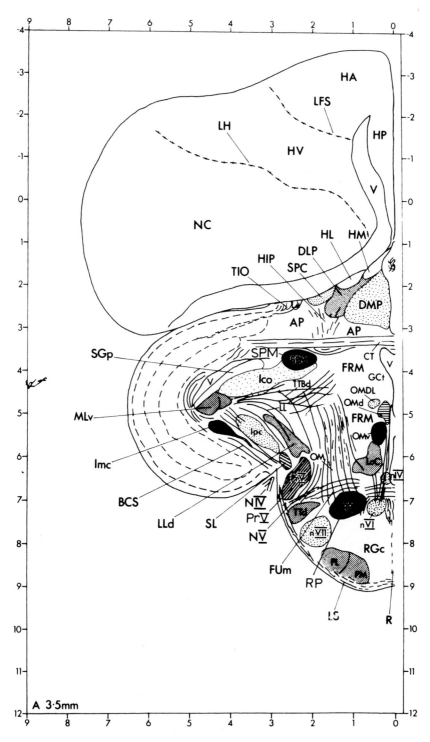

A 3·5mm

Fig. 5 A.

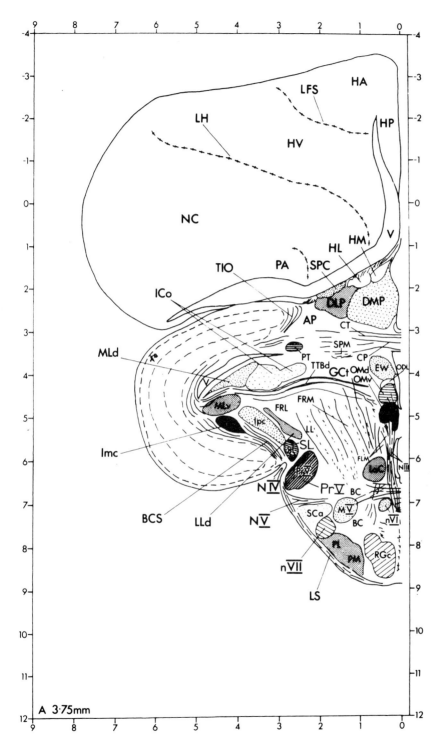

Fig. 5B.

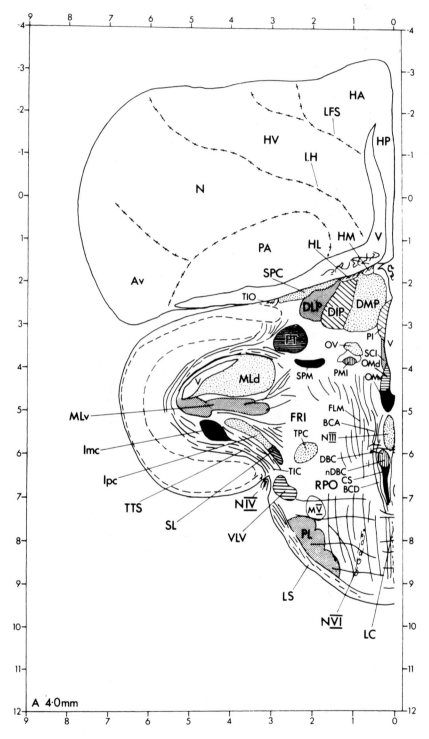

Fig. 6A.

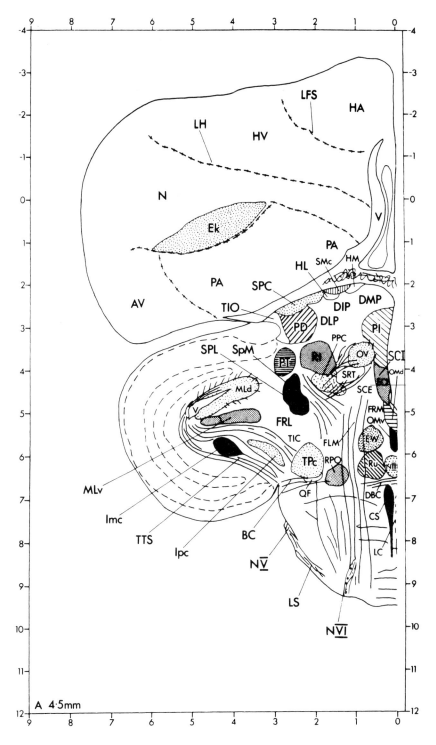

A 4·5mm

Fig. 6B.

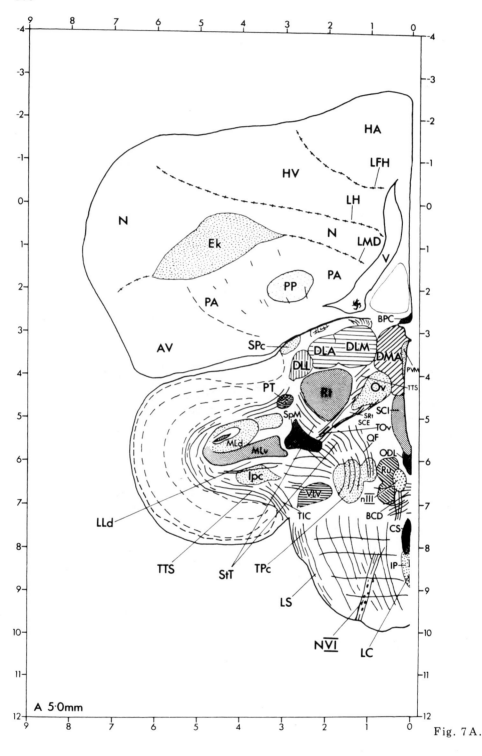

A 5·0mm

Fig. 7A.

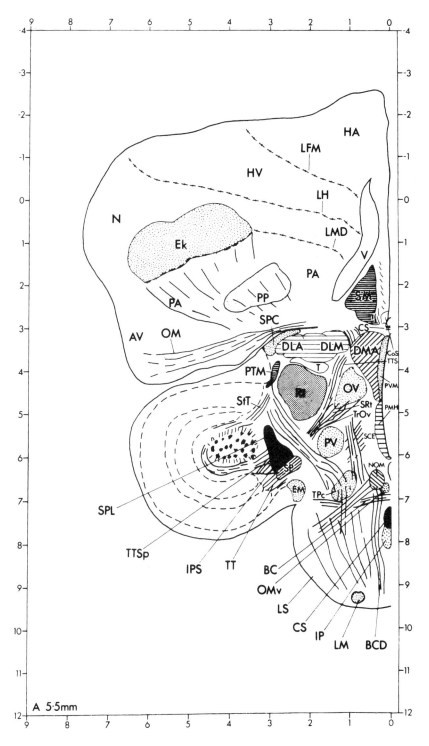

Fig. 7B.

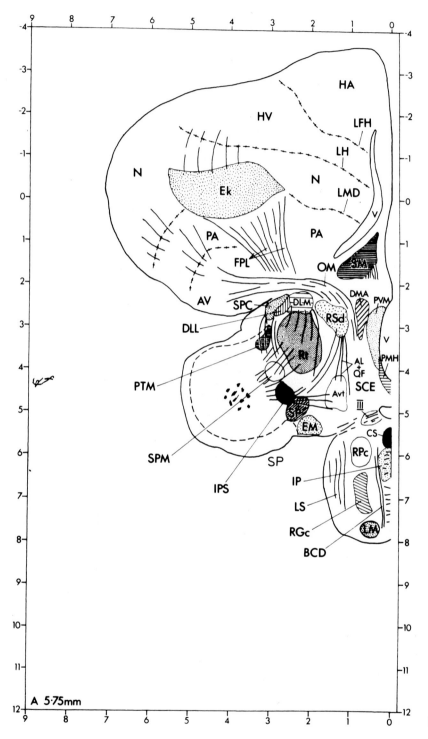

Fig. 8A.

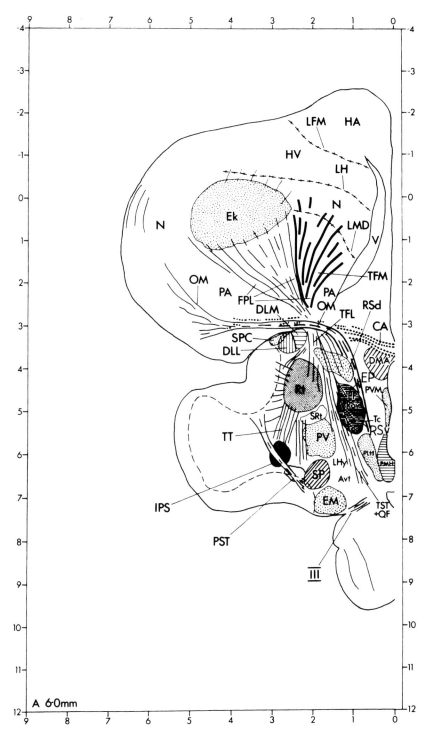

Fig. 8B.

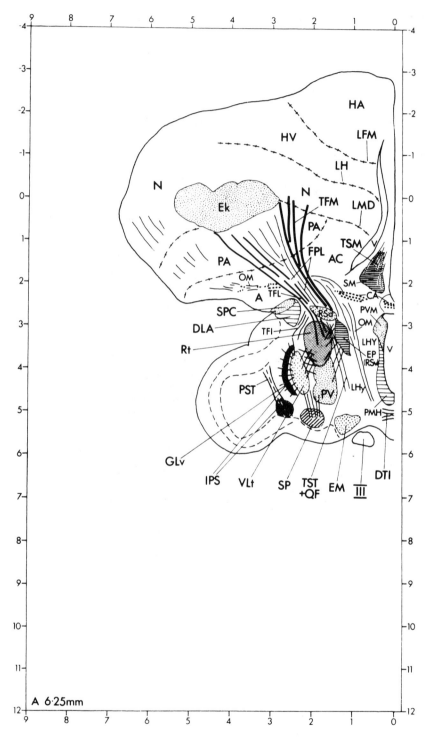

Fig. 9A.

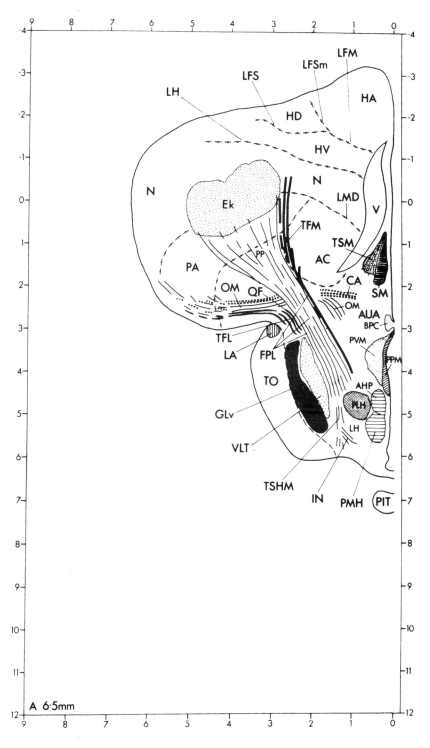

A 6·5mm

Fig. 9B.

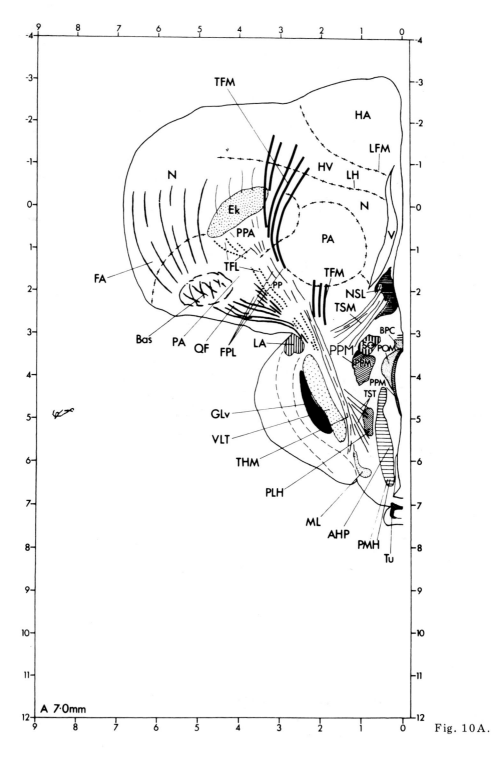

A 7·0mm

Fig. 10A.

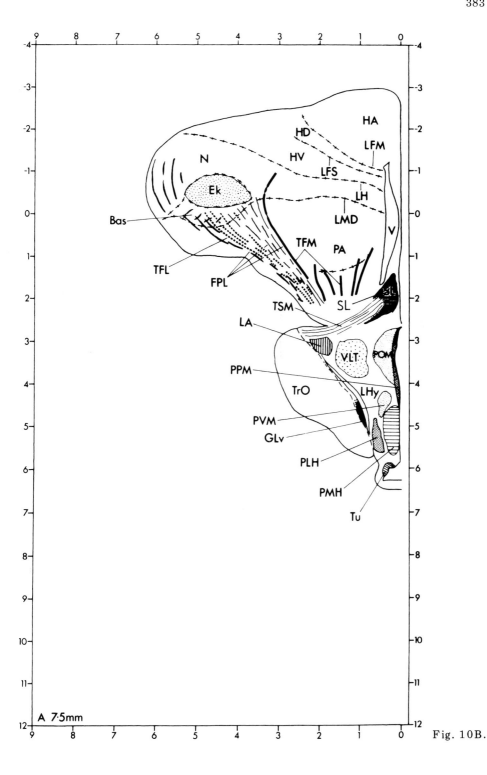

A 7·5mm

Fig. 10B.

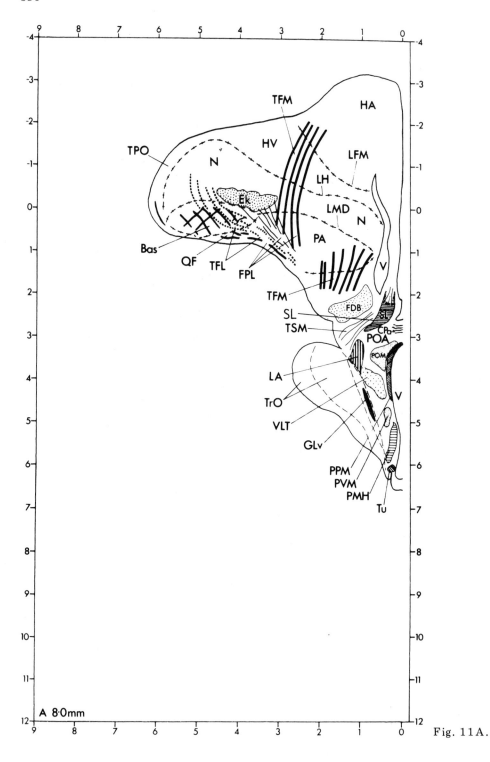

A 8·0mm

Fig. 11A.

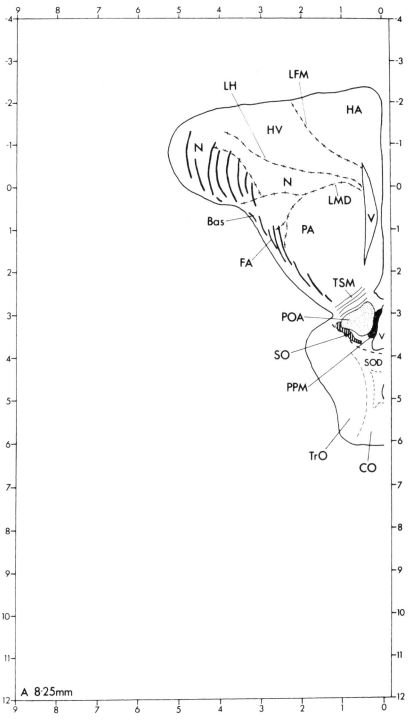

A 8·25mm

Fig. 11B.

386

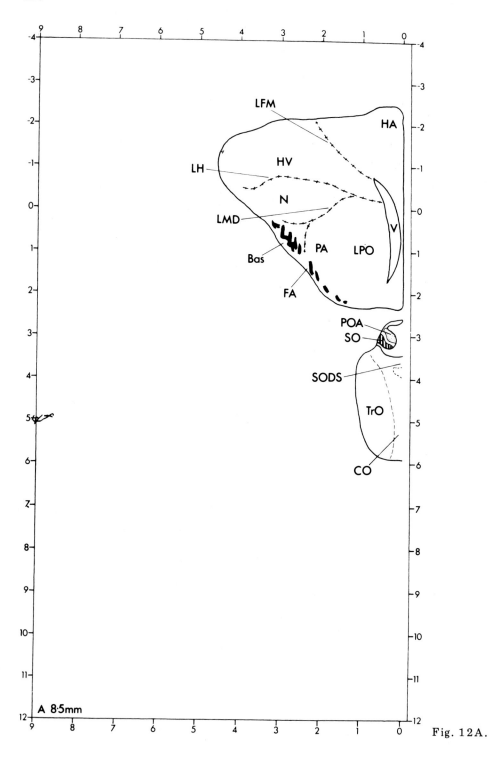

Fig. 12A.

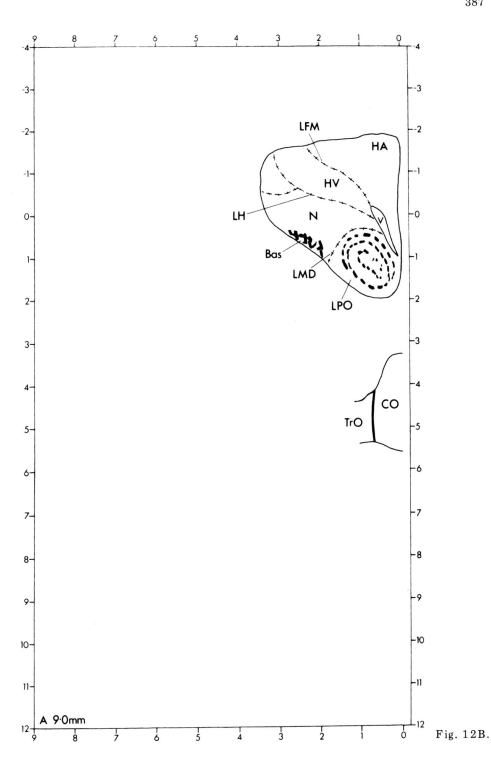

A 9·0mm

Fig. 12B.

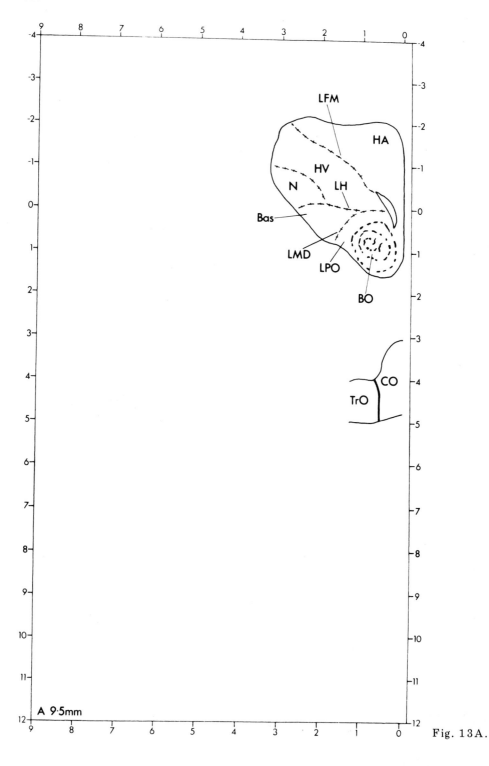

A 9·5mm

Fig. 13A.

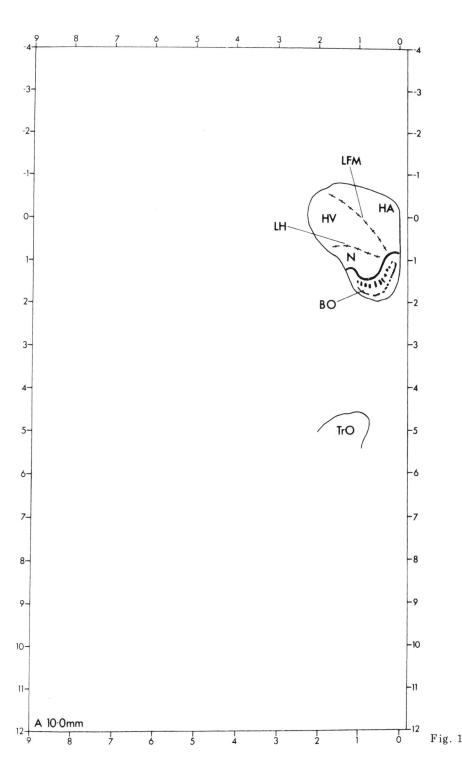

A 10·0mm

Fig. 13B.

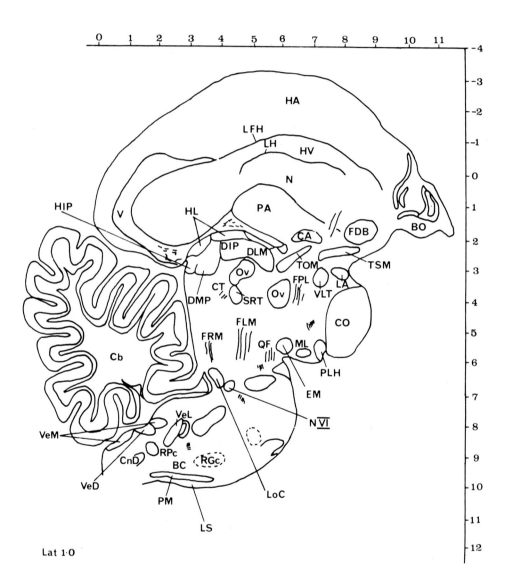

Figs. 14—22. Sagittal sections.

Fig. 14.

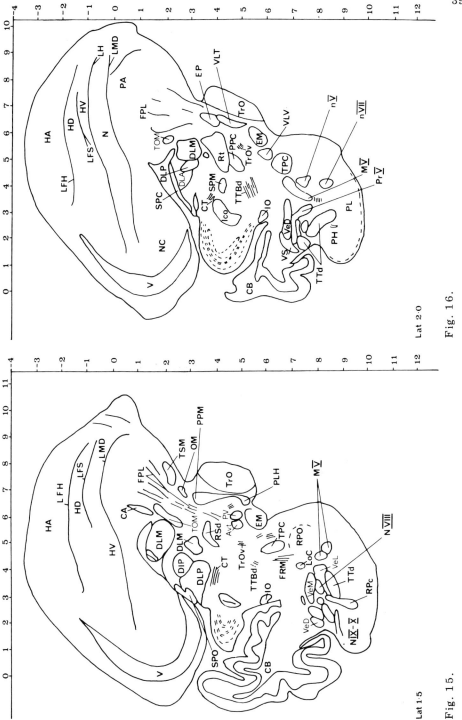

Fig. 16.

Lat 2·0

Fig. 15.

Lat 1·5

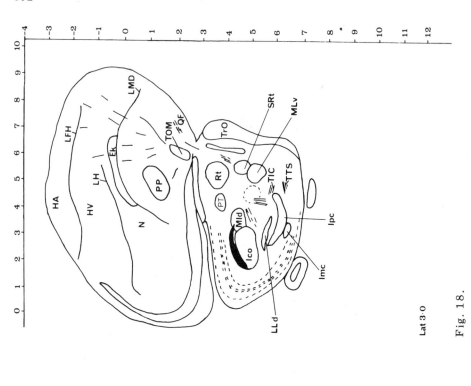

Lat 3·0

Fig. 18.

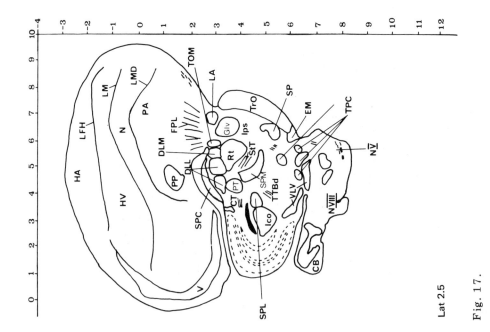

Lat 2.5

Fig. 17.

393

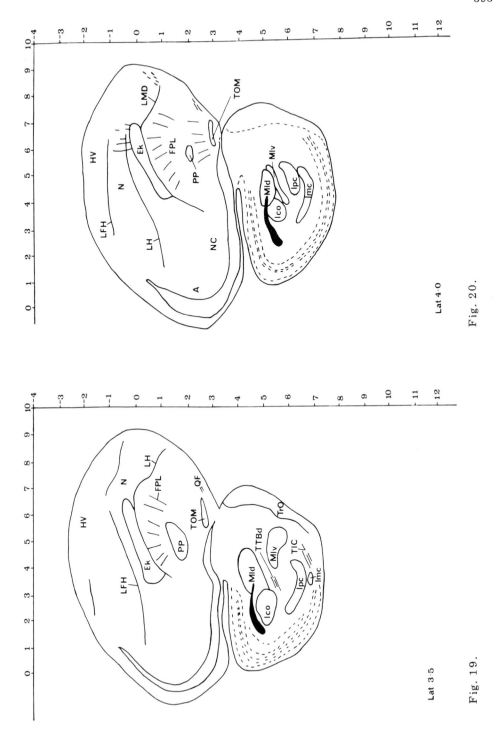

Lat 4·0   Fig. 20.

Lat 3·5   Fig. 19.

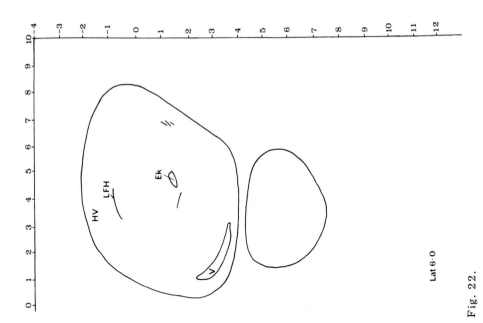

Lat 6·0

Fig. 22.

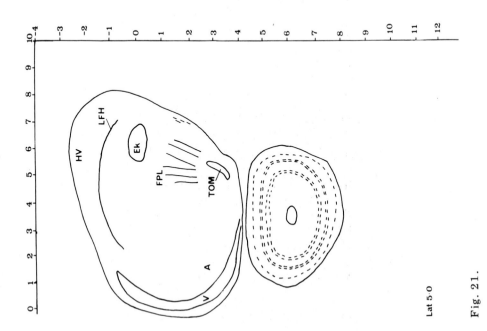

Lat 5·0

Fig. 21.

AUTHOR INDEX*

---

*Suffix *R* refers to a reference list entry.

SUBJECT INDEX